Innovation for
Energy Efficiency

Other Pergamon Titles of Interest

Pergamon Related Journals *(free specimen copy gladly sent on request)*

Innovation for Energy Efficiency

Proceedings of the European Conference
Newcastle upon Tyne, UK
15–17 September 1987

Edited by

D A REAY
David Reay & Associates, Whitley Bay, UK

and

A WRIGHT
NEI-International Research & Development Co. Ltd,
Newcastle upon Tyne, UK

PERGAMON PRESS

OXFORD · NEW YORK · BEIJING · FRANKFURT
SÃO PAULO · SYDNEY · TOKYO · TORONTO

U.K.	Pergamon Press plc, Headington Hill Hall, Oxford OX3 0BW, England
U.S.A.	Pergamon Press, Inc., Maxwell House, Fairview Park, Elmsford, New York 10523, U.S.A.
PEOPLE'S REPUBLIC OF CHINA	Pergamon Press, Room 4037, Qianmen Hotel, Beijing, People's Republic of China
FEDERAL REPUBLIC OF GERMANY	Pergamon Press GmbH, Hammerweg 6, D-6242 Kronberg, Federal Republic of Germany
BRAZIL	Pergamon Editora Ltda, Rua Eça de Queiros, 346, CEP 04011, Paraiso, São Paulo, Brazil
AUSTRALIA	Pergamon Press Australia Pty Ltd., P.O. Box 544, Potts Point, N.S.W. 2011, Australia
JAPAN	Pergamon Press, 5th Floor, Matsuoka Central Building, 1-7-1 Nishishinjuku, Shinjuku-ku, Tokyo 160, Japan
CANADA	Pergamon Press Canada Ltd., Suite No. 271, 253 College Street, Toronto, Ontario, Canada M5T 1R5

Copyright © 1988 Pergamon Press plc

First edition 1988

Library of Congress Cataloging in Publication Data
Innovation for energy efficiency.
Includes index.
1. Energy policy—Europe—Congresses.   2. Energy conservation—Europe—Congresses.   3. Energy policy—Congresses.   4. Energy conservation—Congresses.
I. Reay, D. A. (David Anthony)
HD9502.E8I55   1987   333.79′094   87-19016

British Library Cataloguing in Publication Data
Innovation for energy efficiency: proceedings of the European conference, Newcastle-upon-Tyne, UK, 15–17 September 1987.
1. Energy conservation.
I. Reay, David A.
333.79′16   TJ163.3

ISBN 0-08-034798-3

Printed in Great Britain by A. Wheaton & Co. Ltd., Exeter

Innovation for Energy Efficiency

The Conference and associated Exhibition entitled 'Innovation for Energy Efficiency' were organised within the framework of a series of European Conferences on Technology and Innovation, aimed at encouraging innovation and new approaches to energy efficiency. The Conference which was held at Newcastle Upon Tyne was supported by The Commission of the European Communities, NEI-International Research and Development Co Ltd, Newcastle upon Tyne City Council and British Gas plc Northern.

Energy efficiency is not just as a matter of engineering and managerial performance, but also is a key target for both regional industrial investment and neighbourhood self-help. As the long-term trend remains towards energy-expensive economies, then the need for decision-makers to promote comprehensive and responsible approaches to the efficient use of energy in all sectors is vital.

The Conference was an opportunity to bring together on a European and International basis the three major interests of economic regeneration, neighbourhood-based action and energy efficiency.

The main themes highlighted by the papers in these Proceedings are:

* The promotion of practical policies for energy efficiency and economic investment in areas of major structural change.

* The demonstration of the scope for the use of existing, improved and new technologies and services in the pursuit of energy efficiency.

* The scope for mobilising citizen, municipal and business investment in energy efficiency, through neighbourhood, regional, national and international programmes.

David A Reay

Alan Wright

The Organising Committee

David Green, M.B.E.

Newcastle Upon Tyne
City Council

Arthur Hoare

Regional Energy
Efficiency Office

Trevor Jones

European Commission
Luxembourg

Rick O'Farrell

Regional Energy
Efficiency Office

Linda Pickering

Energy Inform Ltd

David Reay

David Reay & Associates

Alan Wright

NEI-International
Research and Development
Co Ltd

Contents

The Plant Manufacturer in the Energy Market

P. C. Warner, M.A., C.Eng., F.I.Mech.E., F.Inst.E., F.B.I.M.

Director of Corporate Engineering, Northern Engineering Industries plc

ABSTRACT

The manufacturer of energy plant aims to be energy efficient, not only in his own manufacturing operations, but also by designing the plant to be efficient in service. Increasingly his customers are choosing to weigh up the total cost of ownership of the plant they purchase, covering both its first cost and its energy consumption (and of course its reliability); so low losses become an objective of product design. The energy industries in the United Kingdom are well used to assessing total cost of ownership and to taking initiatives in developments, in which the plant manufacturer is able to collaborate, in both traditional and new energy systems. The resulting advances in the technology are to the benefit of the general customer both at home and in the export market.

KEYWORDS

manufacturer, plant, energy, efficiency, fuel consumption, service, lifetime

INTRODUCTION

Conventional analysis of energy matters commonly envisages two parties only: the consumer of the energy, and its supplier. Almost invariably there is a third one, namely the provider of a whole variety of mechanical and electrical plant and systems for the industrial, commercial, and domestic markets, which are needed to bring the energy to the point where it is needed by the consumer, and to enable him to employ it as he wishes. The provider is, in other words, the plant manufacturer, and he is the subject of this paper. As in all businesses, the overall objective is to satisfy the market and thereby to earn enough money to service the borrowings and to renew the enterprise.

The theme of energy efficiency pervades this particular field, and there are three aspects to its impact on the manufacturer:

1

1. The energy consumed in the whole cycle of production from design and development to delivery.

2. The energy that the products actually use in service (they are designed to use as little as possible).

3. The energy losses in the operation of the plant provided to the energy supply industry and employed in the handling and conversion of energy itself (that plant too is designed to be as efficient as possible).

All three aspects are considered in this paper.

WHY ENERGY EFFICIENCY

Efficiency was not always the major theme it is today, because there seemed to be plenty of energy around. The Industrial Revolution was in a period of cheap coal, to be displaced as the dominant fuel by cheap oil in the years following the second world war; oil is now so established as a world commodity that shifts in its price are reflected by coal and gas and other fuels.

We saw two big jumps in oil prices in 1973/74 and in 1979 as the producer nations came to realise and exert their collective marketing strength. Then in 1986 they found they had overreached themselves and oil prices dropped quite sharply. The two upward moves had been useful shocks to the Western economies, who had grown complacent from the long years of apparently plentiful energy at low prices and the enormous growth in worldwide oil production. Most countries started to take a grip on their energy consumption, and to invest in ways of employing it more efficiently. The more recent fall in oil prices has eased the pressure, especially on countries in the developing world that are not themselves producers. On the other hand, some commentators and makers of policy have been treating the reduction as permanent, almost as though we could now relax our efforts to use energy efficiently. That would be a mistake. Looking ahead, we must prepare for a scarcity of fossil energy resources that is genuine, not merely caused by political factors, because by and large we are burning fossil fuels faster than we are discovering new reserves: exploration and recovery costs will rise, carrying prices with them. The shortages of the 1970s were man-made in the sense that they were the creation of the OPEC cartel. The political forces behind that may be in temporary retreat but only until the underlying shortages assert themselves. It may turn out to have been a useful rehearsal.

Obviously oil is not the only fuel. Others are more abundant, but they have their drawbacks, which count against them economically. Coal is environmentally difficult; so, it is widely supposed, is nuclear power, but it is more realistic to say that it has problems of public acceptance; the very large reserves of gas are not in convenient geographical places, and while pipelines are a solution overland, gas is not easy to transport by sea; and finally, renewables are a relatively small contribution, some of them still requiring much development.

On energy supply grounds alone, there is a strong case for using efficiently what we have. When environmental considerations are thrown in, it becomes even stronger: no source of energy is without some impact and we should keep the level as low as possible.

ENERGY IN PLANT MANUFACTURE

As a rule, plant of the type discussed in this paper is produced as one-off or in batches, with little quantity production except perhaps for small bought-in components or instruments. The processes of production are not energy intensive in the same way as, for instance, in the materials or chemical industries, and it is not an explicit objective of product design to reduce the energy actually going into its manufacture. This is not to imply that production as a whole does not consume a fair amount of energy, but it does not depend directly on the design of the product. There is scope for economy within some of the production processes; and in matters like lighting, heating, and the provision of services.

Savings designed to reduce energy consumption normally require some investment, which ought to be viewed in the same way as other investment proposals. The ultimate objective is to improve the efficiency with which the organisation's resources are being used, and to the extent that production costs will fall if the intended energy savings are achieved, then normal financial criteria (return on capital employed and availability of cash) will determine whether the investment is worthwhile. It ought not to be necessary to impose special rulings giving energy saving investment preference over other forms, and the criteria set by management in a properly run business should produce the desired results. Even so, it has to be recognised that attention to opportunities for energy saving have not until the mid 1970s been part of the normal experience of management and it was a new concept to many staff. It was therefore found valuable to appoint Energy Managers who could perform energy audits, advise the energy users throughout the offices and factory, calculate for them their energy costs and potential savings, suggest improvements that would not perhaps be immediately obvious, and perform technical evaluations. In a group of companies heavily engaged in the supply of mechanical and electrical engineering plant directed at the use and conversion of energy, there is ample technical knowledge available to ensure that all activities are duly economic thermodynamically, but it still pays to direct attention specifically to that end.

THE ENERGY INDUSTRY

We now turn to two areas concerned with the behaviour of the products in service, but distinguishing between those for the general customer who is using energy, and those for the energy specialist to help him in his task as a supplier of fuel or electricity. By way of background, it is helpful to look in outline at the structure of the energy industry.

Fig. 1 shows the three groups: consumers, suppliers, and manufacturers. The consumers include people, who live in houses, drive cars – and also obtain services and buy goods; other consumers are the industrial and commercial organisations large and small who provide those goods and services and use energy to do it. Consumers choose between electricity, or gas, or coal, and decide how much. They have an influence on the energy suppliers, essentially the electricity supply industry and the fuel companies (coal, oil, and gas). Consumers must equip themselves with devices appropriate to the particular form of energy they have selected, and they get those from plant manufacturers. As an obvious example, a railway must choose whether to operate its trains on electricity or on

4

oil, and that decides whether it will purchase electrical or diesel locomotives. It does not matter why the consumer behaves as he does: whatever his reasons, they lead to those two influences, a fuel choice influence on the energy supplier, and a plant choice influence on the manufacturer. It is not open to a consumer to switch to a cheaper fuel at short notice unless he has already installed the relevant plant. For instance, if he has gas central heating, he cannot take advantage of lower electrical prices without storage heaters, which may be expensive to hold merely for stand-by. Few general consumers do have alternative equipment.

The manufacturers of plant are a composite group made up of many different firms: they supply equipment and systems to one or more of the consumer markets, domestic, commercial and industrial; and to the energy suppliers. A manufacturer reacts to perceived market need in the classic manner by trying to adapt his products to it and to improve competitiveness for price, quality, and delivery. That would be backed up by new product investment, design and development work on the products, and by regular up-dating of the manufacturing facilities.

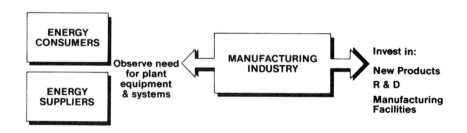

FIG. 1 THE ENERGY INDUSTRY

So much for background; we shall be returning to a number of these points in what follows.

THE CONSUMER'S PURCHASING PRACTICES

We have seen that from time to time the consumer of energy will also purchase the devices in which the fuel will be used, namely the engineering plant which are products for instance of NEI companies; and that in principle he has two sets of choices, one about the fuel, and one about the devices to select. For many purposes (machine tools, cranes, etc) there is no real choise of fuel: they have to be driven electrically, so the energy plant takes the form of electric motors, with the switchgear, transformers, and other distribution devices inseparable from its use. For other purposes associated either with the production process or with the heating or ventilation of the works and offices, a fossil fuel may be more appropriate, and the list of devices lengthens to include boilers, heaters, fuel handing and ash disposal plant, etc. Either way, the consumer of energy is also a purchaser of devices and it is important for manufacturers to understand the factors that may influence him.

Traditionally, preference was given to the cheapest plant; and purchasing rules would stipulate a minimum of three tenders with the contract awarded to the lowest except in special circumstances. Fortunately, it is now being more widely realised that it is the total cost of ownership over the effective life of the purchase that should be the determining argument, not merely the first cost.

The emphasis on plant reliability and the recognition that enforced outages and remedial work represent potentially high costs throughout its life have spurred the development of Quality Assurance with its concept of minimum quality cost, i.e. the aggregate of production, inspection, and remedial costs. At the tender evaluation stage, it is difficult to be confident numerically about remedial costs, so they are usually dealt with through an assessment of the vendor's quality performance, giving a limited list of approved tenderers. Quality requirements are imposed increasingly by customers, especially by the more technically conscious, but a surprising number are reluctant and there are obvious contrary pressures. For instance, the personal performance of a procurement executive is unlikely to be measurable by his achievement of the lowest total cost of ownership, for the obvious reason that the proof of success lies in the future - and the benefit will be credited to an operating rather than a purchasing department. It is altogether easier to measure success by relating the combined purchases to the allowed budget. The purchasing process as a whole remains biased in favour of low first cost.

ENERGY PLANT FOR THE GENERAL CONSUMER

We have seen that purchasing departments in industry and commercial organisations, and for that matter the ordinary domestic householder, have traditionally had a bias towards low first cost, but that better understanding of the total cost of ownership is beginning to favour plant with good reliability. Assessing the cost of ownership has a bearing also on energy considerations. For one thing, there are differentials in price between the various fuels, which will not remain constant throughout life (expressed per unit of energy). For another, different devices have differing efficiencies, either because of the choice of fuel (a gas fired boiler being more efficient than a coal fired one, for instance, and neither as efficient as an electrical heater), or because of features in the design of the device itself. The plant purchases are capital investments that commit their owner to a definite type of fuel and to a particular efficiency of its use for the period of exploitation. Higher efficiency usually implies higher first cost, but if it gives a lower cost of ownership it is still worthwhile.

The prospective owner would need to estimate what fuel supplies and prices would be like. Ideally, he should be trying to think about the behaviour of fuel prices and their availability over the period for which he will be owning the plant. He might, for instance, to be asking himself about likely developments in all primary sources of energy, namely coal, oil and gas, and nuclear; whether and how soon the investment in new coal fields is likely to result in increased supply, how that coal is likely to be processed, i.e. how it may be converted into secondary fuels (electricity, substitute natural gas, etc.); what the depletion policies for North Sea oil and gas are likely to be, and how soon therefore UK supplies may cease to provide a protective cushion against action by the OPEC countries; and generally what is to be the effect of all this on future fuel prices, that is to say how more steeply will the prices of the different fuels individually move than the general price rise in the economy due to inflation? He might even try to weigh up political influences, domestic and international.

However, only specialist organisations like the fuel industries carry energy analysts capable of formulating that kind of long term view, and the average domestic or organisational consumer cannot be expected even to try. Besides, if he did, his estimate would be too tentative to be a basis for an important investment decision. There are several ways out of this difficulty. The most popular perhaps is to assume that fuel prices will stay constant in real terms, which can be a serious distortion over the ten years or so that the plant will be in use. That is partly compensated by choosing a much shorter pay-back time, which may happen also to suit both the organisation's view of its cash resources, and possibly also the preference of financial markets for a rapidly moving business performance.

However, it can also result in undesirable fluctuations, as for instance with the sharp jumps in oil prices, upwards in 1973/74 and in 1979: behaviour was been essentially reactive, as consumers rushed to look for alternatives to oil. Domestically, local Gas Boards and heating contractors were swamped with enquiries about switching to gas, and for a time demand for gas may have outstripped supply. In industry, many firms hesitated over investment decisions which would continue their dependence on oil. The reactions were reversed when prices fell last year. Commentators treated the change as permanent; and even as showing the error of taking the previous higher levels as a guide for long term policy. Potential investments aimed at saving oil were cancelled or shelved, and the associated R&D was cut back. When we remember, as explained earlier, that the energy consumer's fuel choices commit him (because they are accompanied by the acquisition of the associated plant) to ten years or more, this shorter term approach is to be wondered at.

None of this is an argument for meticulous forecasting of fuel prices as part of the energy consumer's technique: given the shifting international influences, many of them political and therefore inherently impossible to predict, it could not be done. But to take today's fuel prices as constant, and to use a very short plant life span, seems just as wrong at the opposite extreme and unnecessarily defeatist. It gives no weight to the potential saving if fuel prices go up in real terms (more likely than a fall as the reserves are finite); in other words, additional capital spend to secure higher plant efficiency provides insurance against such rises, and should be given credit for it.

Provided a fair proportion of the market are informed purchasers ready to calculate total cost of ownership, the manufacturer has an incentive to design his products for best efficiency, and he asks essentially the same questions as his customers. He needs to judge future trends in fuel price and the way they may affect his business; but his assessment is complicated by his customer's limited ability or willingness to allow for those trends and the preference on the whole for short pay-backs calculated as if today's fuel prices were not going to change.

EFFICIENCY IMPROVEMENTS

There have been signs recently that procurement policies are becoming more thoughtful about efficiency and energy losses. The informed purchasers have always given them proper weight, and have provided the necessary encouragement to R&D into methods of reducing losses. In fact modern equipment is already pretty efficient, but that does not mean that further improvements are ruled out.

They can happen in three main ways: improved materials; better techniques of analysis; and a changed configuration. Better material properties are regularly being achieved, whether by new alloys or processes. They would translate into energy savings by for instance permitting higher maximum temperatures in a thermodynamic cycle, or eliminating auxiliary cooling; the recently announced increase in the minumum temperature for superconductivity is another example. Better techniques of analysis allow materials and flow passages to be disposed more effectively. They make it possible to reduce auxiliary power absorption through better design of cooling systems (of electrical machines, for instance); to improve understanding of test results on fluid machinery so that features causing friction loss can be identified and corrected; and to give better characterisation of flames to improve combustion efficiency; and so on.

As regards configuration, one good example is the variable speed solid state drive by means of which fans and pumps can now be driven by synchronous A.C. machines at optimum speed over the load range. Another major group of changes in configuration for better fuel economy embraces all the thermodynamic variants, from heat pumps to back end heat recovery, not forgetting combined heat and power and many others.

The general rule is that energy economy can almost always be improved if some extra capital cost is incurred; and the economics can be calculated to see if it is worthwhile, with appropriate assumptions about future fuel prices.

PLANT FOR THE ENERGY INDUSTRIES

The energy industries themselves constitute a special market where things are rather different. In the United Kingdom, they have great technical strength, which may often overshadow the resources of the manufacturers. They are equipped to analyse long term trends in the availability and price of primary energy resources, and are conspicuous in the national debate on energy policies. Precisely because they are technically sophisticated, these customers habitually assess total cost of ownership,

and will often initiate moves to enhance reliability or reduce energy
losses. Again using the example of electricity, they are responsible for
the whole chain of conversion from the original fuel, through combustion,
steam raising, turbine generator plant, transformers, switchgear,
transmission lines, etc., through to the customer's terminals, domestic or
industrial. At every one of these stages there is plant whose efficiency
must be less than 100%, so there are losses and potential for cost
effective improvements.

These customers readily discuss their future needs with their plant
suppliers, and since this is a field in which progress requires
engineering studies in depth, followed by prototype testing and
demonstration plants, the manufacturers are involved as a matter of
course. At the level of the individual pieces of plant, the scope for
efficiency improvements follows the pattern discussed earlier for the
general market: improved materials and better techniques. The difference
is that the customers, namely the energy industries themselves, take a
much more active part in promoting fundamental advances and in examining
how they might be embodied in product improvements, to the extent of
inviting plant manufacturers to participate in R&D programmes and of
putting their facilities at their disposal for demonstration purposes.
For example in the electrical industry, an improved burner or an improved
drive for an auxiliary would be tried out in an existing power station;
advances in transformers or switchgear in the distribution systems, and so
on.

Matters do not end with the individual pieces of plant. In each of the
key energy areas (oil exploration and enhanced recovery, deep mining of
coal, generation of electricity, recovery of gas, its distribution, and
perhaps its synthetic production), there is a continual search for better
system configuration and even for new concepts altogether. Collaboration
may involve three parties: plant manufacturers either alone or in
partnership, the energy industries in their role as customer, and
Government who may provide part-funding. Among current examples are
combined heat and power schemes for some of the major cities in the United
Kingdom, tidal power both at Severn and at Mersey, and wind generating
machines.

The technical resources of energy supply industries in export markets vary
widely. Some are equipped, like those in the United Kingdom, to analyse
plant proposals against a lifetime cost and to give credit accordingly for
better efficiencies. Others are still at the stage where first cost seems
the most critical factor, or may be required to treat it as such by
political direction or by their need for financial credits.

PUBLIC POLICY

It is worth noting in this connection, that the scope for more efficient
or improved methods of electricity generation is often a matter of keen
public comment. There seem to be several reasons: one is the
all-pervasive presence of electricity in the modern world, and the
tendency to equate national energy policy with policy for electricity;
another is the hightened interest in nuclear generation whose
environmental implications are considered to be of a special kind, which
has led to a concern with environmental aspects of electricity however
generated, and spurred the search for so-called "benign" forms of
electricity generation; yet another reason is the realisation that

classical methods of generation inescapably imply substantial losses (for fundamental thermodynamic reasons). It might seem that all this public debate is not a primary concern of the plant manufacturer, but to the extent that sentiment and misconception may override the technical realities and unsettle or distort the market for his products, he will try to take a hand in the debate and hope to deploy his experience and knowledge on the side of objectivity.

CONCLUSION

Energy is expensive, and invariably has an impact environmentally. It should therefore be used efficiently, even by those countries who do not need to import it. Energy plant ought therefore to be chosen partly for its efficiency. We have seen that the message reaching the plant manufacturer from his market is a confused one. His customers have to decide whether to buy his products, and if so which ones, i.e. what fuels they will be using and, according to their lights, what level of energy efficiency they can afford. The manufacturer in his turn has to get his products ready and because that takes time he cannot simply wait for the customer's specific investment decision: he must anticipate it, choose for himself his R&D priorities and decide where to put his effort on new product design.

The energy industries, who supply electricity, gas, coal, and oil to those same customers, are themselves a special group of customers with strong technical resources. The plant manufacturer must acquaint himself with their broad purchasing intentions and with the factors determining energy policy; he must also assess development trends, decide where to be involved in them, whether to form partnerships, and how to collaborate with his customers in prototypes and demonstration plants. All this advances the technology and should therefore have some benefit for the other parts of the market, at home and abroad.

The plant manufacturer will not neglect environmental aspects and the awakening interest of the public in energy matters generally.

Local Energy Planning in the UK: Past Experience and New Initiatives

A. Atkinson

ABSTRACT

Energy problems crowd into the headlines at ever more frequent intervals.
Major studies in North America and continental Europe have pointed to energy
efficiency programmes administered at the local level as providing a substantial part of the solution. However, local energy initiatives remain weak in
the UK. This paper analyses the structure of Britain's energy economy and the
strategy which has been pursued by the country's highly centralised energy
institutions. It continues by taking a look at the interest that has been
increasing at the local level to implement effective energy efficiency
measures, including the development of municipal heat distribution systems
(district heating - DH) that could make use of waste heat from electricity
generation (combined heat and power - CHP) and other sources, to reduce local
heating costs and the national energy bill. It goes on to note how so far
central agencies have done little to bring about effective action and appear,
rather, to have had a discouraging effect on these developments. The longer
term answer is seen as involving substantial decentralisation of power in the
energy field through institutional changes.

KEYWORDS

Local energy planning; energy economy; energy institutions; energy efficiency;
combined heat and power (CHP); decentralisation.

INTRODUCTION

Recently energy issues have come with increasing frequency to occupy the top
of the political agenda. One problem after another has claimed major media
attention, sometimes over extended periods. Items include:

o Determination of the Government to restructure the coal industry precipi-
tated the bitter miner's strike of 1984-1985, which was informed by a
determination to defend jobs and communities from imminent destruction;
since the collapse of the strike there have been more than 50,000
redundancies in the industry.

o A spell of cold weather in early 1986 revealed how badly heated many

11

people in Britain are in winter and how many old people are at risk of
dying of hypothermia; the situation was repeated in early 1987; the
solution of the moment was to provide warm blankets and 'cold weather sup-
plements' and there was little or no discussion of more permanent solutions

o The disaster at Chernobyl nuclear power station in the Soviet Union in Apri
   1986 demonstrated the environmental hazards of nuclear power; whilst this
   shifted public opinion and opposition political parties further towards an
   antipathy for the technology, neither the electricity supply industry nor
   the Government were prepared to admit that there might be better economic
   and technical solutions to our need for electricity supply.

o In early 1986 the international crude oil price collapsed; this was the
   latest in a series of wild fluctuations in oil prices since 1973, leaving
   many in authority confused and fatalistic concerning the possibility to
   make rational calculations about the most economic investment for the
   satisfaction of future energy needs; meanwhile the depletion of North Sea
   oil and gas resources in the coming years represents a general threat to
   the economic wellbeing of the country.

In general there has been a tendency to see all these as predominantly technological or financial problems, in some cases exacerbated by 'political motivation'. More considered judgement has, however, focussed upon existing insti
tutional structures as being in part responsible for generating the problems,
and then being badly adapted to heading off expected future problems. Major
studies by the United States National Research Council (Stern and Aronson,
1984) and the International Institute for Environment and Society, funded by
a number of national and international agencies (Joerges and Olsen, 1983),
have concluded that local institutions and community organisation - involving
a process of 'remunicipalisation' (Hennicke, 1985) - will have to play a major
role in the future energy economy if these problems are to be successfully
combatted. In many industrialised countries local utilities, often simply
local authority departments, have continued to play a significant role in the
national energy economy and are now being provided with additional resources
and responsibilities. In the United States and Scandinavian countries local
energy planning is now a statutory duty; in West Germany and France alternativ
possibilities for co-ordinated local activity in the energy field are obtainir
substantial backing from central governments. The European Commission (ECC,
1982) is also committed to 'far greater decentralisation of decision-making'
in the energy field.

The UK, on the other hand, with its extremely centralised energy institutions,
has been relatively untouched by these developments. Virtually no official or
academic recognition has been given to the possibility that local institutions
might play anything but a marginal role in the national energy economy. This
does not, however, mean that no local initiatives have been forthcoming, or
that initiatives in process of formation now might not at some future stage
provide the basis for more substantial initiatives in the medium-term future.
This paper therefore looks first at Britain's energy economy and institutions
and then reviews the local energy initiatives which have appeared in the UK ir
recent years, and assesses the prospects and possibilities for their future
development together with the problems which they confront.

THE UK ENERGY ECONOMY

In order to make sense out of the variety of local energy initiatives in the
UK it is necessary to obtain an overview of the energy economy and institutions which they seek to influence and change. This section thus provides a

sketch of salient dimensions of the UK energy economy first in technical
terms and then in terms of institutional and social questions.

The Uses of Energy

In the first instance, the concept of the energy economy is seen as involving
the question of the supply of fuels according to need. In detail, fuel supply
goes through various stages, the main ones being: extraction as primary fuel,
conversion and transmission, and end use. 'Conversion' includes oil refining,
electricity generation, coke and town gas production. In the UK this currently involves the loss of about 30% of primary energy, mainly in the form of
power station waste heat. There is further wastage by consumers after purchase, and whereas there are no statistics on this in the UK, in West Germany
this amounts to over 50% of end use energy and in the UK the situation is
likely to be similar. So useful consumption amounts to about 30% of primary
fuel. Technically there are quite straightforward ways of improving substantially on this figure and generally reducing energy demand. It is in this
area of 'demand management' that the major potential role for local energy
initiatives is seen.

The mix of primary fuels supplied through the UK energy economy has changed
relatively swiftly in recent years. Over the two decades following the
Second World War, UK energy use increased on average by about 2% per year.
Since the 1973 'energy crisis', use has fluctuated around an average of 330
million tons of coal equivalent (mtce) per year. Until the 1950s, coal pre-
dominated as the primary fuel, being steadily replaced across the 1960s by
oil, This, too, gave way as use of natural gas made rapid inroads in heating
markets after 1967. Currently coal and oil take about a third each of the
primary energy market, the former predominantly for generating electricity
and the latter for transport. Gas takes about a quarter of the market and
the remainder is taken by nuclear and hydro electricity.

Looking at use sectors to which end use energy is supplied, currently
industry, transport and the domestic sector each takes about 30% with the
remainder going to public administration and agriculture. In terms of energy
costs to these sectors, however, transport, as it uses high cost fuels, pays
over 40% of the national end use fuel bill with industry, using predominantly
low cost fuels, paying less than 20%. The differential price of fuels is
thus an important factor in the operation of the energy economy.

Looking at the evolution of fuel prices in the UK since 1970, one of the most
striking things is the way in which the nationalised fuel boards manipulated
prices in order to adjust and exploit market conditions to their best advantage and how even the current Government, ostensibly dedicated to laissez
faire, has altered fuel prices for its own ulterior ends. Thus uncertainties
concerning fuel prices due to international market conditions have been exacerbated rather than mitigated by UK energy institutions, making forward planning on the part of industrial and domestic consumers so much more difficult.
The problematic history of district heating in the UK is largely attributable
to this situation (Russell, 1986).

Nevertheless, in recent years, disregarding transport fuels, it is possible
to discern three distinct price bands for fuels: full price electricity
remains in a class of its own, over three times more expensive than other
regularly used fuels. Off-peak electricity also remains significantly more
expensive than other fuels, but because it is efficient to use, it remains
competitive with oil and gas in the domestic sector; until early 1986 burning
oil remained significantly more expensive than gas but this is no longer the

case. Finally, coal remains, in a further class, as a relatively cheap fuel dependent, however, on bulk burning and heat distribution technology (district heating) to make it efficient and attractive. Notwithstanding the possibilities of fuel industry and Government manipulation, this general situation is likely to hold well into the future, with coal (and perhaps refuse as a supplement) even increasing its price advantage.

Consumers have an overriding need to minimise the cost of energy for a given function. There is a tendency for certain uses to require a given fuel, but for many applications there is no very direct link between function and fuel type. Thus, transport, which consumes about 20% of energy used in the UK, requires 'high grade' and hence high cost fuels. About 7% of UK energy usage goes to power electronics, lighting and other 'essential electricity' applications and a further 12% goes to running motors in fixed applications which mainly also require high grade fuels. However, almost two thirds of energy use in the UK goes to provide heat and two thirds of this is low grade heat, predominantly for space and water heating. Over the whole of the latter market and much of the high grade heat market there is no technical restriction on what fuel can be used. The choice might therefore simply be one of adopting the cheapest fuel at any given moment.

However, the use of fuel for any purpose requires investment in conversion and distribution technology, and the use of cheaper fuels for heating applications (coal, refuse, heavy oil and the use of waste heat from industry and from power generation (CHP)) require substantial investment in infrastructure if their cheapness is to be exploited. The great success of gas in the domestic central heating market in recent years has been due to its convenience and relatively low capital cost to consumers; the fact that off-peak electricity has made significant headway in this same market, despite higher cost and lower convenience, is because capital costs to consumers are even lower than for gas. The fact that coal, and beyond that the use of waste heat, has failed to make significant inroads into this market in the UK, in contrast to many continental countries, lies in the requirement for considerable capital investment beyond the capacity of the individual consumer. The question then becomes: who will make the investment and carry out the work? This is no longer an economic but an institutional question.

Energy Institutions

The configuration of energy institutions varies considerably between countrie Amongst industrialised countries outside the Soviet block only France approaches the degree of severely rationalised centralisation found in the UK. Since the post war nationalisations, the UK fuel industries, with the exception of the oil industry, have comprised monolithic sectoral utilities beholden almost solely to the central Government. Although each possesses a regional tier (in the case of gas and electricity for distribution purposes and in the case of coal to organise extraction), this relates neither to thos of the other fuel boards - which would prevent integrated planning if this were called for - nor to areas used for other Government functions. The relationship between the boards and local authorities, as dimensions of the totality of the structure of national Government, is extremely tenuous and the boards are almost entirely unanswerable to local interests.

Working out of general terms of reference largely defined at their inception, the fuel industries have grown into immensely powerful organisations. The CEGB is amongst the most heavily capitalised organisations in the world and the gas and coal industries are massive organisations by national standards. One of the effects of amassing such power has been that the fuel organisation have, to a significant degree, come to usurp the Government in the determination of national energy policy: policy formation has become, to a large extent, a question of fuel board priorities and requirements rather than relating to the economic and social needs of the country in general. This situation became particularly clear when Wedgewood-Benn, as energy minister in the late 1970s, attempted to open national energy policy up to wider debate. The Green Paper issued at that time (Department of Energy, 1978) was little more than an attempt to reconcile the competing aspirations and demands of the fuel industries with scant attention paid to other potential energy sources and the potential for conservation or the broader question of satisfying specific social and economic needs. Current Government moves towards privatisation of the fuel industries as monolithic organisations, without effective regulatory structures found in the United States, renders even more tenuous the responsibility of the UK energy industries to satisfying any broader social and economic requirements.

It is of more than academic interest to note that the energy industries in this country, as throughout the industrialised world, started life in a decentralised fashion. Coal mining and distribution was developed by a large number of private organisations and private interests were also involved in the early development of gas and electricity undertakings. However, in the latter case, municipalities played the crucial role and, despite progressive legislative moves towards centralisation starting early in this century, it was only 40 years ago that municipalities were finally divested of their energy conversion and distribution role. Elsewhere throughout the industrialised world, it continues to be usual practice for there to be a complex of national, regional and local organisations involved in fuel supply. This has meant firstly that consumers have maintained in most cases relatively close political contact with the policies and activities of the energy industries, and secondly that skills and expertise in building and operating energy systems have remained attached to the local political structure. It is the loss of these in the UK which represents a basic weakness in the movement towards remunicipalisation of the energy economy by contrast with many other countries.

The configuration of national energy institutions has a powerful bearing on the choice of technical systems and investment patterns pursued. This in turn has a marked influence upon employment and other social impacts. A brief analysis of changing employment patterns in the UK energy industries illustrates this point. There are four distinct areas of employment concerned with the supply of energy: in extraction, in processing and conversion, in distribution and sales, and finally in the production and installation of energy technologies.

Employment in extraction is necessarily proximate to the resource. Following the First World War, the coal industry alone employed 10% of the UK workforce thereby supporting considerable coalfield communities; it currently employs less than 0.7% of the workforce. Whilst this reduction is in part due to contraction in production, it is predominantly a function of decisions on the part of British Coal (and the NCB before it) concerning which resources to exploit and how, technically, to exploit them (Burns, Newby and Winterton, 1984). Continuation of private exploitation of coal or some other decentralised organisation of coal exploitation would have reached very different decisions on these issues, with very different consequences for the workforce and economy.

Energy conversion facilities are the most footloose of energy industry installations and the policies of the CEGB since its creation in 1957 illustrate once again the relationship between institutions and wider social and economic impacts. The technical strategy of the CEGB has been to build large coal-fired and nuclear power stations on remote rural sites, in certain cases directly over coal resources, supplemented by small gas turbine units in urban areas which can be brought quickly on and off stream to cater for sudden fluctuations in electricity demand. Having inherited about 350 power stations at its inception, all in urban areas, by 1985 these were down to little more than 80 with the vast majority of electricity being generated in rural areas. Employment steadily declined: between 1975 and 1985, 18,000 jobs were lost, but there was a far more substantial loss in urban areas from which the power generating facilities were removed. In addition, there was a shift in the location of construction work: instead of using a regular urban workforce, supplemented by specialist engineering installers, large temporary workforces moved around the country and the delicate fabric of rural economies was in many cases significantly disrupted (Gwynedd County Council, 1976; The Planning Exchange, 1978). In countries where municipalities maintain responsibilities in electricity generation, the refurbishment and extension of existing urban power stations, often supplying heat as well as power, remains a significant facet of the national electricity industry.

The CEGB operates under terms of reference which require it to 'develop and maintain an efficient, co-ordinated and economical system of supply of electricity' which it has translated into a strategy involving large power stations which maximise electricity generation per unit fuel input and reduce staff costs. A very substantial case was put to the Sizewell Inquiry into the CEGB's application to build a new nuclear power station by the Greater London Council (GLC), demonstrating that it would be more economic for the CEGB to invest in urban CHP stations, which could supply both cheaper electricity and cheaper heat, than in rural nuclear stations. The CEGB disputed the findings and the Inspector, in his report (Layfield, 1987) dismissed the GLC case. This was not, however, on economic grounds, but on the grounds that the electricity supply industry was not prepared to adopt it and that no other institutions were immediately capable and prepared to construct CHP/DH systems. Whilst there is no practical example in the UK that might demonstrate whether or not the GLC case was correct, there are ample examples on the continent. In particular West Germany and Scandinavia present a wide variety of different arrangements of regional and municipal utilities that indicate quite clearly how municipalities can and do plan and build CHP/DH systems and also co-ordinate energy supply and demand management in general to the great advantage of consumers (Rüdig, 1986).

LOCAL ENERGY INITIATIVES

A great many kinds of energy initiative have appeared in UK local authority areas since the trigger of the 1973 'energy crisis'. Most of this effort has gone into reducing energy bills by introducing demand management through building insulation and investment in inexpensive technologies which increase energy efficiency for a given function. Some of these initiatives have, however, had energy as only one of their concerns and many have been quite ephemeral in terms of their actual or even potential impact upon the local energy economy. It therefore becomes necessary to impose some order upon their description so as to evaluate their worth as energy initiatives and a three part structure is adopted here. First, focus is placed upon work which local authorities have been doing to improve energy efficiency in their own activities, then those local initiatives which attempt to bring energy efficiency to all consumers are analysed, and finally the more substantial efforts to make an impact in the area of energy supply are looked at.

Energy Efficiency in Local Authorities

Local authorities in the UK act under the legal principle of ultra vires which disables them from undertaking substantial action which is not directed by the central Government. Hence, once divested of their responsibilities as energy utilities, they ceased to focus on energy as an issue that might be of any more than peripheral concern to them. In practice they are responsible for many decisions with respect to fuel use in their capacity as architects both for buildings used by themselves, including schools, community buildings and so on, and for council housing which, in a few cases, now represents the bulk of the local building stock. Particularly in the latter case, decisions have not always been taken in the best interests of tenants, as stringent capital restrictions have been applied, resulting in underheating or the installation of heating systems requiring the use of expensive fuels.

Following the first oil price rise of 1973, local authorities began to experience the consequences of their earlier capital decisions on heating and insulation, together with current poor management, in the form of considerable fuel bills. A few local authorities, in particular schools authorities such as Essex and Cheshire County Councils, rapidly organised a systematic effort to improve energy efficiency in their own buildings. In order to gain from the general experience, local authorities turned to their technical service organisation LAMSAC which carried out various studies to determine effective ways for local authorities to proceed and to indicate where they might look for good examples (LAMSAC, 1983). At the same time other local authority support organisations offered advice (STCELA, 1982; SOLACE, 1985).

The level and shape of response on the part of different local authorities remained extremely varied (Sheldrick, 1984) but in most cases some effort went into the following activities: fuel tariffs were scrutinised to take best advantage of alternative rates on offer; energy efficiency measures such as insulation, reduction in glazed area and improved heating controls were introduced into council buildings; council architects were required to take a more critical look at designs to minimise energy use in new buildings subject to capital restraints; lighting fixtures were replaced with more energy efficient ones; and council vehicles purchased and maintained with fuel efficiency in mind. Computerised systems for monitoring fuel use and targetting improvements were introduced and in some cases more elaborate centralised computer control systems were purchased. During 1986 the Audit Commission (1985) took up energy efficiency as one of the areas for development in local authorities and studies were undertaken in all local authorities, offering specific advice on the way to improve energy efficiency practice.

This effort called for some institutional framework and the results were as varied as were the technical responses to council energy problems. Generally a group of council officers came together to form an Energy Group to discuss mutual problems and disseminate experiences. In a few cases Energy Committees were formed to supervise efforts. Energy policies were adopted, generally restricted to committing the local authority to look after its own energy efficiency. Energy efficiency officers were hired in by many councils - and in the larger councils a whole unit or division was formed. Coming generally from an engineering background, these officers had to learn on the job and over a period of some ten years something of a new profession has been forged but with as yet no professional standards for evaluation of relative effectiveness. the Department of Energy did, however, provide some support to bring these new professionals (both from Government and industry) into group discussion at the local, regional and national levels.

However, the Audit Commission estimated that local authorities spend about 5% of the national energy bill, so that the 17% or so savings that it is estimated can be made in the short term in these local authority bills is clearly not going to make any major impact upon the national energy economy. Nevertheless, this effort has had the effect of developing significant expertise in the complex area of energy management and of alerting local authorities to the need to focus upon energy issues.

Community-Wide Energy Efficiency Initiatives

During the late 1970s, local authorities were encouraged by the central Government to take an interest in a further area of energy conservation work, namely in insulating council housing. However, special allocations for this purpose ceased in 1980 and whilst most local authorities continued some programme of insulation installation and heating system improvement as part of the general maintenance of council housing, rarely did this represent a high priority and very few local authorities currently have an overview of the thermal conditions of their own housing stock.

On the other hand, from the late 1970s on, a few councils attempted to address energy issues on a much broader front. This generally stemmed from pressure from a variety of local interests and over time an increasing number of local authorities has adopted a broader perspective. Some of these interests have found funding from sources other than the local council and have initiated local energy actions independent and ahead of the council.

The kinds of initiatives that have developed can be illustrated by reference to two exemplary cases: London and Newcastle. Following the publication of the 1978 Energy Green Paper referred to above, the GLC compiled a substantial energy policy document (GLC, 1978). This looked not only at the specific needs of London but provided a critique of the Green Paper by way of a preferred national energy policy, seen from the local authority perspective; it remains an important document, without parallel even today. Recognising that a more specific local energy strategy would require a more comprehensive energy data base, the GLC then undertook to produce one for London (GLC, 1981); again, this remains almost unique even today. In the final, unapproved revision of the Greater London Development Plan, a chapter on energy policy was incorporated (GLC, 1984).

The GLC also initiated public education and consultation work on energy matters as part of its Popular Planning initiative and provided funding for a number of local energy activities. These included the formation of the London Energy and Employment Network (LEEN) as one of a series of organisations aimed at developing technological innovation in London. LEEN originally brought together a number of environmental and educational organisations that had hitherto had plans for various energy initiatives but had had no funding; in time, membership broadened out to include many local organisations that had already been involved in building insulation work, giving energy advice to the elderly, and related social support activities. The organisation survived the abolition of the GLC, providing support for the formation of energy initiatives by London boroughs, organising seminars and issuing publications, and providing training, energy auditing and a number of other services on a consultancy basis. Further local independent energy planning initiatives were funded by the GLC in two of the boroughs and whilst these did not survive the abolition of the GLC, the councils concerned did continue to extend their commitment to community energy efficiency in line with the advocacy of these initiatives.

Interest in energy issues on the part of Newcastle City Council has been of a more pragmatic nature. Pressured on the one hand by trade unionists to take an interest in promoting employment in the ailing energy technology industry, which has a major concentration in the city, and on the other drawn into supporting work on house insulation initiated by Friends of the Earth in neighbouring Durham, the city adopted energy policies specific to these interests without undertaking any broader survey of energy use or needs in the area. Establishing an Energy Advice Unit in early 1979, Newcastle became the base for a network of insulation projects nationwide. In the city itself a number of insulation projects were established, together with a bulk-buying co-operative and a consultancy offshoot; subsequently, a shop-front Energy Information Centre was opened in the city centre. Branching out from this Newcastle base, a national organisation, Neighbourhood Energy Action (NEA), was established within the framework of the National Council for Voluntary Organisations (NCVO) to facilitate the formation of similar insulation projects across the country. In 1985 this became an independent charity, based in Newcastle, and had been largely responsible for the initiation of almost 300 insulation projects, supporting 5,000 jobs, by late 1986.

NEA projects provide jobs predominantly through Manpower Services Commission funding. They mainly install draughtproofing in the homes of those on supplementary benefits and, in some cases, also loft insulation partly paid for through the Homes Insulation Scheme (NEA was largely responsible for pushing the Government into adopting these funding schemes); a few also offer energy advice. Whilst addressing, fairly specifically, urgent social needs, both for employment and improvement in heating conditions amongst the elderly and disadvantaged, these projects are nevertheless limited in their scope relative to the broader possibilities for energy efficiency improvement in the UK building stock as a whole. Although it is true that they have demonstrated in many areas the possibilities for tackling energy efficiency problems, so that other local authorities have taken up funding for similar initiatives to those in Newcastle (in some cases with significant extensions, such as the Cardiff Energy Action City initiative and the activities of the Urban Centre for Appropriate Technology in Bristol), it cannot be said that any coherent or comprehensive approach to the local energy efficiency question is in sight.

Reviewing the field in general, local attempts to promote energy efficiency investments even up the levels which yield very short term economic gains, appear to be making very slow progress in the UK. There is no recent overview of the rate of progress; up to 1982 (Leach and Pellew, 1982) progress in the domestic sector was slow and there is no reason to believe that it has accelerated. Whilst in certain limited areas of industry and commerce major energy efficiency improvements have certainly been achieved through corporate strategies and with central Government assistance, information is non-existent on efforts concerning smaller organisations and premises, with virtually no advice or assistance to promote energy efficiency in these sectors.

Local Intervention in Energy Supply

It is something of a historic curio that in the late 1940s and early 1950s, before the newly-nationalised fuel boards formulated the strategy which subsequently led to the situation we find today, there was a good deal of interest in central Government and more particularly in a number of local authorities in building municipal combined heat and power (CHP/DH) systems (Russell, 1986). Failure of all these proposals, bar a small scheme built in Pimlico in London, was the complex result of a lack of any local remit, a lack of structured support from the central Government and the movement of the electrical supply industry towards a strategy within which CHP would have represented an inconvenience. Had these local authority initiatives been successful (it was only in the 1950s that the exemplary local heat supply systems in Sweden were initiated) then it can be postulated that the UK would now have an institutional framework much more appropriate to addressing present energy problems than the one we now possess.

Starting in the mid 1960s and peaking in the mid 1970s, local authorities did come to construct district heating systems, under Government encouragement, mainly as an adjunct to their remit to provide public housing; by 1983, there were more than 700 such systems in the UK, though not all of these were in municipal ownership (Orchard Partners, 1983). As already noted above, the impetus to construct these systems came not from energy policy considerations but as an attempt by the oil and coal industries to secure part of the heating market in the face of rapid inroads being made at that time by natural gas. The response of the gas industry was to offer a preferential tariff for district heating and so capture the market through this technology as well as through individual heating systems; once the battle had been clearly won, the preferential tariff was abandoned.

Meanwhile, local authorities had been induced to install technology which was often badly designed - there being inadequate standards and professional experience - and cheaply produced. The resulting problems experienced by consumers were exacerbated by the lack of willingness on the part of the local authorities themselves to provide a management and maintenance regime adequate to the sophistication of the technology and, either due to consumer pressure or from problems within the local authority organisation itself, a large number of local authorities came to reject the technology. At the time there was no realisation of the potential benefits that could be derived from the technology and only very recently have a number of local authorities come to appreciate the potential of this legacy.

These systems were never intended as means to greater energy efficiency but merely as means to distributing heat centrally generated from conventional fuels. However, interest in incineration of refuse arose amongst local authorities in the late 1960s and a small number of incinerators with heat recovery were built, two of which supplied heat to district heating systems. Legislation was passed in the early 1970s to facilitate the production of electricity and heat by this means and the distribution of heat by local authorities. However, by then refuse disposal responsibilities had passed to county authorities and although some experiments were subsequently undertaken to produce fuel pellets from refuse, no further local authority incinerators were built.

Following the 1973 'energy crisis', interest revived in certain circles in the potential energy savings to be derived from major investment in CHP/DH, and the Government initiated what became a rather leisurely programme of investigation into the possibilities of building urban-wide systems. By 1979 it had been ascertained that CHP might be a good thing and that some 40 urban areas possessed significant heat load potential (Marshall, 1979). By 1982 specific studies had assessed the technical and economic potential for nine inner city areas (Atkins, 1985) and detailed studies are currently being funded in three cities.

Meanwhile, an autonomous interest arose in the potential for urban-wide CHP/DH in a number of local authorities and the restricted nature of central Government initiatives led some, notably Newcastle, Sheffield and the GLC, to invest significantly in independent investigations into the local potential. The result of these studies was an indication that these systems contain great potential for employment and cheaper heat to consumers but there are severe financial and institutional impediments to their realisation; these are discussed in the final section below.

Besides this interest in urban-wide CHP/DH, many local authorities have recently focussed on other energy supply and conversion technologies as an extension of their in-house energy efficiency work or out of an interest in employment generation. In some cases both passive and active solar energy experiments have been conducted and many local authorities have now installed heat pumps or small CHP units in municipal pools. These have, however, remained isolated technology initiatives rather than being parts of a more general local energy strategy and, in urban areas, these will need to be evaluated in the light of possible integration into a local heat network (an urban-wide district heating system taking heat from power stations, industry and perhaps heat pump applications relating to sewage, solar and geothermal energy, as is being generally developed in Scandinavian local authority areas) as representing the core of an efficient local energy supply system.

Despite growing concern and investigation, practically no inroads have as yet been made into energy supply on the part of local initiatives. Only a few incinerators and pelletisation plants have brought a small proportion of the national refuse product into energy circulation and the torso of potential local heat distribution systems exist in a few urban areas in the form of fragmentary district heating systems. A few very small, isolated technology interventions have been made but local authorities and other local initiatives have yet to bring coherence and effectiveness to their efforts to increase efficiency and decrease the cost of energy to consumers through energy supply.

CONCLUSIONS

It was pointed out at the outset of this paper that major studies in North America and continental Europe have concluded that a vital contribution to combatting a growing array of energy problems can be made by increased local intervention in the energy economy. Many industrialised countries are thus moving towards a 'remunicipalisation' of the energy economy; although there are growing interests in following this trend at the local level in the UK, effective initiatives remain very restricted. There is a general argument that policies contrary to powerful interests often fail even to reach the political agenda (Crenson, 1971; Lukes, 1974). Energy efficiency is contrary to the interests of the fuel industries and the great power they wield in the UK has created a context in which coherent movement towards real improvements in energy efficiency organised at the most effective - namely the local - level, has been severely inhibited.

This line of argument can be made much more specific. Reference has already been made to the ultra vires principle under which local authorities operate in the UK. This simply confirms in legal terms a long tradition of central-ised Government in which local Government has played mainly a line agency role, lacking any vary substantial co-ordination or initiative at the local level. Attempts by local councils to break out of this mould in the past were generally suppressed by the central Government (Gyford, 1985) and recent attempts in this direction have been similarly handled, with the abolition of the GLC and metropolitan counties and by the assertion of greater central control over local Government spending.

In this way, growing interest in local energy initiatives has been channeled by the central Government into relatively ineffective activities: improving

22

energy efficiency in local authority buildings is important and assisting the most needy improve their heating conditions is vital, but these fail to make very significant inroads into the available potential to reduce energy consumption across the board. In the case of CHP/DH the confrontation is more direct, where the electricity supply industry has co-operated to the minimum required by the law and the central Government has done the minimum necessary to deflect the outright accusation that they are suppressing developments.

When it comes to the construction of CHP schemes, local authorities and interests are expected to find private finance, which requires at least a 10% real rate of return, and then compete in terms of energy sales with the electricity supply industry which makes investments on the basis of a 5% Government-stipulated test discount rate. Whilst CHP/DH schemes may still go ahead, testifying to the considerable immediate economic benefits inherent in such schemes (let alone the wider national benefits), much of the advantage will accrue to financiers rather than local consumers.

Local attempts to increase the scope of local energy activities, in particular involving increased legal and financial powers, are likely to continue to make slow progress under current conditions especially if these are pursued as isolated 'single issue' initiatives; it is all too easy for these to be delayed, emasculated and contained. Local interests must realise that this is part of a broader movement to shift the balance of power in favour of local areas and interests to address numbers of issues besides energy problems and that it is necessary to make common cause. Even a brief perusal of history indicates that this is not a party political matter but one more deeply rooted in national political traditions generally requiring new kinds of initiative. The development of local economic strategies by some local authorities indicates a certain line of movement in this direction, and the South East Economic Strategies Association is demonstrating possible changes in institutional approach to local and regional issues that include changes in energy institutions designed to shift focus away from energy supply operations to demand management, organised locally.

In the end one could envisage a structure of energy institutions in the UK which involves a division of responsibilities between local and regional energy agencies which would undertake integrated planning and the construction and operation of energy infrastructure on a departmental basis. At the national level all that would be necessary is certain research, policy (including standards) and co-ordinatory functions as support to the regions and to promote the national interest. Though this may sound quite radical in the British context, it would be doing little more than coming into line with the trends of developments elsewhere in the industrialised world and with a situation which is not dissimilar from that already prevailing in some countries.

ACKNOWLEDGEMENTS

Much of the material in this paper was collected in the course of research carried out for the Centre for Local Economic Strategies (CLES) and the South East Economic Development Association (SEEDS); it is used here with thanks.

REFERENCES

Atkins, W. S., and Partners (1982). CHP/DH Feasibility Programme Stage I. Summary Report and Recommendations for the Department of Energy. W. S. Atkins and Partners, Epsom, Surrey.

Audit Commission (1985). Saving Energy in Local Government Buildings. HMSO, London.
Burns, A., M. Newby, and J. Winterton (1984). WERG Report No. 6, Second Report on MINOS. Working Environment Research Group, University of Bradford, Bradford.
Crenson, M. A. (1971). The Un-Politics of Air Pollution: A Study of Non-Decisionmaking in the Cities. Johns Hopkins Press, Baltimore.
Department of Energy (1978). Energy Policy, A Consultative Document. Cmnd. 7101. HMSO, London.
ECC (1982). Community Investment in the Rational Use of Energy. Background Report ISEC/B19/82. European Communities Commission, Brussels.
GLC (1978). Energy Policy and London. Greater London Council, London.
GLC (1981). Energy Use in London. Greater London Council, London.
GLC (1984). Planning for the Future of London, The GLDP as Proposed to be Altered by the GLC. Greater London Council, London.
Gwynedd County Council (1976). The Impact of a Power Station on Gwynedd. Gwynedd County Council.
Gyford, J. (1985). The Politics of Local Socialism. George Allen and Unwin, London.
Hennicke, P. (1985). Die Energiewende ist Möglich. Für eine neue Energie-politik der Kommunen. Strategien für eine Recommunalisierung. Fischer, Frankfurt am Main.
Joerges, B., and M. E. Olsen (1983). The Process of Energy Conservation. International Institute for Environment and Society, Berlin.
LAMSAC (1983). Energy Management for Local Authorities - A Guidelines Report. Local Authorities Management Services and Computer Committee, Manchester.
Layfield, Sir F. (1987). Sizewell B Public Inquiry, Report by Sir Frank Layfield. HMSO, London.
Leach, G., and S. Pellew (1982). Energy Conservation in Housing. International Institute for Environment and Development, London.
Lukes, S. (1974). Power, A Radical View. Macmillan, London.
Marshall, W. (1979). Combined Heat and Electrical Power Generations in the UK. Energy Paper No. 35. HMSO, London.
Orchard Partners (1983). Present State and Future Evolution of the Heat Market in the Member States, Heat in Industry in the UK, District Heating in the UK, A Report Prepared for the Commission of the European Communities. Orchard Partners, London.
Rüdig, W. (1986). Energy Conservation and Electricity Utilities, A Comparative Analysis of Organizational Obstacles to CHP/DH. Energy Policy, April 1986, 104-116.
Russell, S. (1986). The Social Shaping of Energy Technology: Combined Heat and Power in Britain. Ph.D. thesis. Technology Policy Unit, University of Aston, Birmingham.
Sheldrick, B. (1984). Energy Saving in Local Authorities. Association for the Conservation of Energy, London.
SOLACE (1985). Energy Conservation II, Second Report. Society of Local Authority Chief Executives, Gloucester.
STCELA (1982). Guidelines for a Positive Local Authority Energy Policy. Second edition. Standing Technological Conference of European Local Authorities, London.
Stern, P. C., and E. Aronson (Eds.) (1984). Energy Use, The Human Dimension. W. H. Freeman, New York.
The Planning Exchange (1978). The Social Impact of Large Scale Industrial Developments. The Planning Exchange, Glasgow.

Energy Efficiency—the Barriers and the Opportunities

I. R. Bailey and J. W. Somerville-Smith

March Consulting Group, 33 King Street, Manchester M2 6AA, England

ABSTRACT

In a study carried out for the European Commission, energy use in the North West of England has been analysed to highlight the Region's energy profile, the scope for improving energy efficiency and the barriers to investment. The results of this work have then been used to put forward programmes for improving the effective use of energy in the study area.

KEYWORDS

Energy Planning; energy efficiency; barriers to investment; patterns of energy consumption.

INTRODUCTION

In July 1985, the Energy Directorate of the EEC appointed the March Consulting Group to study the use of energy in the North West of England. The study, one of 20 regional energy studies being undertaken throughout Europe at the present time, is the first of its type to be carried out in the UK. The project is due to be completed in July 1987.

The study has six specific objectives, namely:

- to establish the energy supply/demand pattern for the North West by fuel type and sector

- to investigate current levels of energy efficiency in each sector

- to establish the scope to accelerate implementation of energy efficiency measures

- to establish the scope for saving from alternative fuels and the use of waste heat

- to forecast energy demand patterns in the main sectors by the years 1990 and 2000

- to identify and stimulate action on key energy saving initiatives.

This paper provides an overview of the work carried out and discusses some of the initiatives that have been put forward. At the time of preparation, the study is four months from completion. Final results will therefore be presented at the conference.

THE NORTH WEST REGION

The North West Region of the UK comprises the counties of Cumbria, Lancashire and Cheshire as well as the Metropolitan Counties of Greater Manchester and Merseyside. This area, with its population of approximately 6.75 million people, accounts for around 14% of the UK's energy consumption. Its economy is very diverse, taking in a wide variety of industries from the hill farming of Cumbria to the energy intensive processes of North Cheshire, from the traditional industries of textiles, footwear, coal etc. to new high-technology activities in electronics and specialist engineering. The Region therefore encompasses a range of activities that is broadly representative of the UK economy as a whole.

In contrast to some of the studies being carried out in Europe, the North West does not have one industrial sector which dominates energy use in the Region. The chemicals industry is the largest user of energy in the North West accounting for 30% of industrial energy consumption.

In total, 16.5 million tonnes of oil equivalent were consumed in the study area in 1983/84 by final consumers, worth approximately £4.9 billion. Figure 1 details the fuels that are used to satisfy demand in the North West, the United Kingdom and the EEC. Oil is still the predominant fuel used despite attempts to reduce dependence. However, gas has become a more important fuel in the UK over the last few years and has replaced some oil and coal.

Analysis of where fuel is used reveals slight differences between geographical levels (Fig. 2). In the North West and the EEC, industry consumes the greatest proportion of energy (34% and 32% respectively), followed by the domestic sector, transport and commerce. In the UK, however, transport accounts for the highest portion of fuel used (29%) followed by industry, the domestic sector and commerce.

It is important, when assessing ways of improving energy efficiency, that one knows where energy use is focussed. This ensures that available funds are used to achieve the maximum impact.

INDUSTRIAL/COMMERCIAL SECTORS

To establish the scope for improved efficiency in the Region's industrial and commercial sectors, the top 400 energy users in the study area were identified. Together, they account for approximately 80% of the North West's industrial/commercial consumption. These companies were either interviewed by a consultant or were asked to fill in a questionnaire. The site visits normally lasted between 2 and 4 hours, this consisting of an interview which took approximately 1.5 hours and a tour of the site.

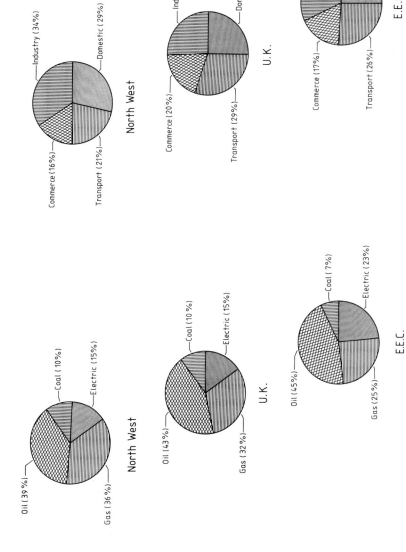

Fig 2 Main final users

North West

Industry (34%)

Domestic (29%)

Commerce (16%)

Transport (21%)

U.K.

Industry (26%)

Domestic (25%)

Commerce (20%)

Transport (29%)

E.E.C.

Industry (32%)

Domestic (25%)

Commerce (17%)

Transport (26%)

North West

Oil (39%)

Coal (10%)

Electric (15%)

Gas (36%)

U.K.

Oil (43%)

Coal (10%)

Electric (15%)

Gas (32%)

E.E.C.

Oil (45%)

Coal (7%)

Electric (23%)

Gas (25%)

Fig 1 Breakdown of final consumption
(by final user)

All the organisations interviewed had taken action to improve efficiency
through both better housekeeping and investment in large projects. However,
very few had monitored the level of savings achieved through these activities.

When assessing projects requiring investment, 85% of companies use simple payback as the principal criterion while 34% use discounted cash flow or internal
rate of return to supplement the decision. If investment in energy efficiency
is compared with other types of investment, 61% felt that measures were given
the same priority as other financial decisions whilst 29% recognised that
implementation of energy efficiency techniques were given a lower priority
than other measures. However, the payback criterion itself can discourage
energy efficiency investment. Simple payback is a straightforward, easily
applied and understood technique, but it is biased against long term projects.
The 5 year project life implied for most energy projects is short, although
many organisations are concerned with other uncertainties in the long term
including energy prices and changes to the production process.

Thus, in many organisations, there are good investment opportunities in energy
efficiency which could be rejected because a discounting method of appraisal
has not been applied.

Figure 3 highlights the main constraints to further investment. In the
public sector, financial restrictions are the greatest barrier for 80% of the
respondents. Lack of finance in the private sector does not, however, create
the greatest barrier to implementation. The main problem facing a high
proportion (50% of organisations interviewed) of companies is a shortage of
staff or management time to implement the measures identified. Contained
within this constraint, however, is a lack of the right skills to design,
specify and install the equipment.

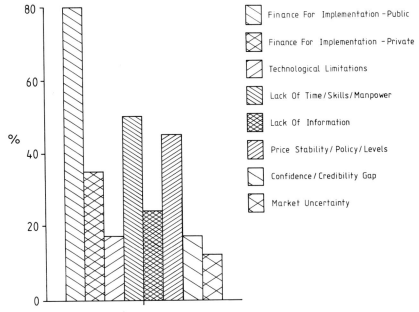

Fig 3 Barriers to investment

A cause of delay in 45% of the companies interviewed is uncertainty regarding the future price of fuels. Few firms felt confident that the paybacks quoted for a particular measure would actually occur in practice.

Interestingly, lack of information was only seen as a barrier to improved efficiency in 24% of the institutions contacted and an even lower proportion, 17% felt that technological limitations acted as a constraint. This may suggest that little further can be done to increase the effectiveness of energy usage. However, it is more likely that companies pay insufficient attention to energy matters to have any firm view about the future improvement of products.

Uncertainty about the future direction of the company and a disbelief that significant savings can be achieved also affected the take-up of measures.

DOMESTIC SECTOR

Owing to the nature of this sector, it is more difficult to target energy savings as there are no large individual users. A different form of analysis therefore has to be used.

There are approximately 2.5 million houses in North West England, each with a small and approximately equal energy bill (about £450). To predict how the energy is utilised and the scope for improving its efficient use, modelling techniques can be applied as these will average out the different consumption patterns.

The model used by March for the study was derived from a model formulated by the Building Research Establishment. This predicts energy usage in one household given specified parameters regarding the type and size of house and the measures already undertaken. The "North West Model" has taken this philosophy and extended it to predict usage in a large stock of housing.

The results arising from the research suggest that there is large scope for improvement in domestic energy efficiency especially through the more widespread use of cavity wall insulation and better heating controls. However, there are a number of barriers preventing these measures being taken, including:

- Finance. Householders rarely have the money to carry out major projects and, whilst building societies will generally lend money to carry out the work, little active promotion of this facility is carried out.

- Short Term Horizons. People tend to spend the money they have on items where they see an immediate benefit (e.g. video recorders, washing machines etc.). Energy efficiency improvements are therefore given a low priority.

- The Tenant/Landlord Relationship. Landlords tend to be reluctant to invest in measures when they are not responsible for energy consumption. Tenants, meanwhile, are unlikely to install products which will enhance the value of the property without providing them with a capital return.

- Low Income Households. People on low incomes or supplementary benefit are only able to afford the cheapest measures such as draughtproofing.

- Lack of Unbiased Advice. Owners either do not know who to approach for
  energy advice or feel that they cannot obtain objective assistance.

A number of activities need to be carried out in this sector to promote
awareness and encourage the public to look carefully at their energy usage.

TRANSPORT

The use of fuel in transport is very much a neglected area throughout Europe.
The vehicle manufacturers have been mainly responsible for the progress made
in this area through improved vehicle and engine design. However, further
significant savings can be achieved through better driving techniques and
more comprehensive monitoring systems. The application of programmes in two
companies with large fleets suggest that savings of up to 10% can be achieved.

Most of the companies interviewed during the study contract out their fleet
operations to haulage contractors. This helps to target improvements as a
high proportion of freight is conveyed by a small number of carriers who can
be approached individually.

March are currently establishing a Monitoring and Targeting system for fuel
usage in the Bus and Coach industry with a similar exercise being carried out
for British Rail on their traction energy. These systems will enable consumption to be compared against defined standards - an approach which is novel
in the transport sector.

ALTERNATIVE FUELS

As part of the study, the scope for replacing fossil fuels with alternative
energy forms has been investigated. There is already some experience in the
North West in the production of solid fuel from domestic waste and there are
a number of towns and cities in the Region which are large enough to support
a reclamation plant. However, plants operating in other areas are having
difficulty in reaching the necessary levels of reliability and price competitiveness.

In realising the potential reserves of mines and landfill gas, the study area
again has good examples of what can be achieved. Demonstration projects at
Cadbury Typhoo and Joseph Crosfields have shown that, under the right circumstances, waste gases can be used to meet base load energy requirements.

In assessing the future potential for plants of this type economics, reliability, marketing strategy and multi-product use must be seriously examined
at the concept stage.

In the Region there is a high level of industrial activity, with the potential
for the development of industrial waste burning. The heat demand and waste
availability must match, and the details of both (especially the properties
and quantities of waste available) must be carefully checked before expenditure is committed.

Overall, there is considerable scope for using alternative fuels in the
Region if the schemes are economically viable. However, reliability problems
and the expense of transporting the fuel have in the past limited its acceptance by consumers.

INITIATIVES

Having identified some of the major barriers to better energy efficiency in the North West and highlighted the areas with the most cost effective scope for improvement, a number of initiatives have been put forward to overcome some of the problems. These have been grouped under the sectors to which they relate.

Industry and Commerce

Energy Efficiency in individual sectors. The industrial interviews identified a wide variation in energy efficiency across the industrial sub-sectors. Following the success of Energy Efficiency Year, it is understood that the Department of Energy is investigating ways of ensuring that this benefit is maintained. One way of accomplishing this would be through targeted promotion in those sectors that provide the greatest scope for improvement. This could be done using case studies which demonstrate the savings that have been achieved by individual companies in the same industry.

Promotion could be carried out through industry journals, management journals and by buying advertising space in the Financial Times.

Overcoming the credibility gap. During the programme of interviews with main energy users, some firms were not aware of the level of savings that could still be achieved. These companies need a stimulus to examine their energy use in more detail and recognise that significant improvements can be made. This could be brought about through a short investigation (say 2 days) by an energy consultant together with a financial expert. Presentations to the board of directors would be given at both the beginning and end of the assignment. This would ensure that there is involvement at the highest level and that findings are put into the context of the whole business operation.

Changes to the energy survey scheme. The energy survey schemes promoted by the Energy Efficiency Office have generally worked very well in the past. However, the Department is aware that there is room for improvement, especially in raising the quality of the short surveys carried out. As a result of the interviews carried out for the study, March have been able to suggest amendments which the EEO is currently assessing.

Many companies in the Region felt they did not have the right expertise to implement measures identified during the surveys themselves. A scheme which provided assistance to companies requiring outside skills to carry out detailed design and project management for new schemes would not only increase the rate at which projects were implemented, but would also ensure that equipment is properly specified.

Contract energy management in the public sector. The main barrier to improved energy efficiency in the public sector is a lack of finance to carry out measures. The use of contract energy management, where a company will install equipment at a consumer's premises without cost and share the savings that accrue, is one way of overcoming this constraint.

The use of such a service has in the past been regarded by the Treasury as a form of leasing. This has meant that tight budgetary constraints have been applied which in turn has discouraged Authorities from taking action.

The Energy Efficiency Office (EEO) has held discussions with the Treasury and, as a result, the barriers perceived by the Authorities have been reduced.

More emphasis is now placed on the service/maintenance elements of the facility and some Authorities have recently started to implement measures in conjunction with Contract Energy Management Companies.

Marketing EEC demonstration projects. Very few companies in the Region are aware that the EEC provides support for demonstrating new technology which improves energy efficiency. The Commission has accepted the need to improve the marketing of the scheme, both to attract new projects and to increase the replication of technologies throughout Europe. Funds have been set aside to carry out more extensive promotion and work is to commence in mid-1987.

Domestic Sector

Local authority housing. The area where the tenant/landlord problem can be most easily overcome is in local authority housing as authorities hold the greatest stock of domestic property and therefore provide the greatest scope for improving efficiency. However, there are a number of constraints to investment in the public sector including:

- Finance. Authorities only have limited funds available to carry out work.

- Priorities. The money that is available has to be divided between a number of different departments, each of which has an urgent need for funds. Home improvement projects therefore have to compete against other projects for finance.

There needs to be greater co-ordination between departments within councils to ensure that priorities are properly established. To assist with this, the Energy Efficiency Office should provide guidelines for presenting the case for investment in Local Authorities.

Low income housing. The Neighbourhood Energy Action programme for improving energy efficiency in the homes of low income families is working well. Further work in this area should therefore be actively encouraged. Peter Walker, in his address at the National Energy Managers conference in December last year, recognised the need for action and intends to expand the programme from 336 projects to 460 projects by the end of 1987. These projects will provide warmth for the elderly, jobs for the unemployed and greater energy efficiency in thousands of homes across the country.

Energy advice centre. To help provide objective advice to the general public and promote energy efficiency in the home, an energy advice centre is currently being established in Manchester. This is similar in concept to the Newcastle Energy Information Centre which has been so successful over the last few years.

March has been involved in setting the objectives for the centre and in advising Manchester City Council on target markets. The centre has met with strong opposition from the fuel authorities who believe that it would duplicate areas already covered by them and would act as a pressure group working against their interests. The aims are therefore being reappraised to determine whether they can be reformulated to everyone's satisfaction.

Domestic heating controls. Few houses in the UK have comprehensive heating controls. However, before significant improvements can be made, the public need to be made aware of the savings to be made from installing good heating controls. The trades that serve the domestic sector (e.g. architects, plumbers etc.) should also be encouraged to specify more energy efficient systems. This could be done either through regulation or increased Government promotion and steps are currently being taken to assess this further.

Transport

The need for an energy efficiency programme in road transport is demonstrated in a recent European Commission document, "Rational use of energy in road, rail and inland waterway transport". Energy consumption in this sector rose by 25% between 1973 and 1984 whereas total final energy demand fell by 5% over the same period. The relative lack of emphasis placed in this area is borne out by the North West Study. Very few organisations have taken positive action to reduce vehicle fuel usage, even when they have done so within their buildings and processes. This is certainly also the case in the rest of the UK and, we understand, in other EEC countries. The fact that fuel consumption in transport accounts for around 25% of final demand in the North West, the UK and the EEC emphasises the need to extend the work already being carried out by March to other sectors of the industry.

Research and Development Projects

During the study, a number of organisations expressed a need for equipment which is not currently available on the market. These products included:

- accurate flow metering, especially for steam

- motor controllers

- speed control of pumps

- compact low energy lighting

- domestic heating controls

- cost effective, condensing boilers for the domestic market.

Accurate flow metering in industry was seen by many companies as a major requirement for improved efficiency. However, attention also needs to be paid to the promotion of good installation and maintenance practices. A manual which explained installation and maintenance principles for metering would not only have a large market but would also help to raise confidence in the metering already available.

CONCLUSIONS

By using statistical analysis to identify energy patterns in the North West of England, it has been possible to highlight the areas where effort is most needed to improve energy efficiency. As a result, it has been possible to:

- identify programmes in areas where the greatest benefit can be achieved at the lowest cost

- place sectoral initiatives in the context of the overall energy scene.

Studies of this type therefore make it possible to invest in measures with more confidence at both a regional and national level. This not only benefits the individual, but can also have a major impact on the economy as a whole.

where public £ are
S pront

Energy Efficiency Policy and Low Income Households

Brenda Boardman

Science Policy Research Unit, University of Sussex
Mantell Building, Falmer, Brighton, East Sussex BN1 9RF, UK

ABSTRACT

British energy efficiency policy for houses has focused on only two of the possible insulation measures and disregarded the importance of the heating system and fuel used. Most grants have gone to better-off families with mortgages, who also have the greatest access to the least expensive heating: gas-fired. The houses of the poor are, therefore, less energy efficient with little or no added insulation and heating systems that are more expensive to run. Low income tenants and owner-occupiers have limited opportunities to make the necessary capital expenditure, thus defining the need for government action if these homes are to be made more energy efficient.

Policy developments are seriously hampered by the lack of clarity in departmental responsibilities, particularly for the use of energy in the home, rather than for the building fabric. Another impediment is the narrow definition of cost effectiveness being used. However, developments in the Building Regulations and in a British Standard Code of Practice could provide an appropriate new method of defining energy efficiency for all dwellings. As the energy efficiency of a dwelling is dependent upon the soundness of the structure an integral part of a new approach should be to incorporate a minimum standard of energy efficiency into all building and environmental health requirements, such as the fitness standard for human habitation. A new system of enveloping grants, incorporating a wide range of energy efficiency measures, is proposed.

KEYWORDS

Energy efficiency, low-income households, cost of warmth, domestic heating, fuel poverty, energy policy, enveloping, fuel substitution.

INTRODUCTION

In order to cope on a low income, people purchase their needs as cheaply as possible: they buy inexpensive products, partly by shopping carefully, partly through substitution. When the commodity they want to buy is warmth, they have not got these options: they are dependent, in the short-term, on the heating system within the dwelling using fuel at a fixed price. Economies are only possible by using less of the product. To lower the cost of warmth, in the longer term, requires capital expenditure to increase energy efficiency through insulation or alterations to the heating system. In the absence of capital expenditure, the household is forced to buy expensive warmth, or go cold. Secondly, warmth is satisfying a physiological need and, thus, provides the same intrinsic benefit to everyone. Expensive warmth provides no additional value, on the contrary it represents a misuse of resources for the individual household and is avoided wherever possible. Therefore, to obtain adequate warmth a household must have sufficient money to cover either large running costs or to invest capital to obtain cheaper warmth. This paper investigates the present situation for low income families in Britain and the policy implications of ensuring that they can afford to be warm.

For about 30% of all households in the UK, at least three-quarters of their total income comes from the state in the form of pensions and benefits, and thus they are dependent upon the state for their standard of living. A similar number of households – and virtually the same population – are in receipt of the means-tested housing benefit, indicating that this proportion of the population are on low incomes, with limited capital or savings. This paper is, therefore, concerned with the 6.4 million households with the lowest incomes in 1986. At least half of the poor are pensioner households, and the whole group is in the following tenures (Department of Employment, 1985):

local authority tenants	57%
privately renting, unfurnished	8%
privately renting, furnished	4%
rent free	2%
own outright	24%
purchasing	5%

The first four categories listed above – 71% of all poor households – are in rented accommodation with limitations on the freedom to alter the dwelling and a disinclination to invest in the landlord's property, even if money is available. Another large, but overlapping, group is the elderly in old property, particularly the privately rented unfurnished sector and outright owners (32%).

The concern here is with the space heating needs of these low income households, with energy efficiency defined as the amount of warmth obtained for each unit of expenditure on fuel. Thus, the energy efficiency of a building encompasses the level of insulation, the technical efficiency of the heating appliance and the cost of the fuel used.

ENERGY EFFICIENCY POLICY

The concepts of energy conservation and energy efficiency have been gradually evolving, particularly since 1973, and the role of the cost of energy is becoming clarified. Some measures of energy efficiency can be made independently of the cost of the fuel used: for instance, 'miles per gallon' is appropriate for cars as the majority use petrol and the price of petrol is the same for all makes of car. Where the price of the fuel is important, most emphasis has been placed on economic pricing, so that consumers can take rational decisions and invest in cost-effective, energy efficiency measures:

"Economic energy pricing is a prerequisite for effective energy conservation policies" (IEA, 1987). However, economic pricing is concerned with the cost of a fuel, independently of other fuels. There is still little discussion about the importance of the relative prices of fuels, when a variety of fuels are used for the same purpose, as with domestic heating. British energy efficiency policy has not resolved this ambiguity over the role of energy costs.

The cost of energy is certainly recognized as an important factor in the domestic sector. For instance, in 1986, Energy Efficiency Year, the Department of Energy (DEn) coined the word 'monergy' and consumers were urged to consider both the use and the cost of energy: "Get more for your monergy". However, most advice concerned insulation measures with no reference to the actual fuel used, implying that the cost of the specific fuel used was irrelevant. This is treating energy efficiency in housing, as if it were the same as for cars, with the fuel price a constant. Thus, there is no attempt to focus assistance on those with the most expensive fuels and few policies to assist domestic consumers improve energy efficiency through fuel substitution.

This emphasis on building fabric and omission of consideration for the economic efficiency of heating systems is reflected in departmental responsibilities, as highlighted in November 1983 when the Energy Efficiency Office was created:
  "The Secretary of State for Energy will take on overall responsibility for promoting energy efficiency in buildings ....
  "The Department of the Environment ... will continue ... to be responsible for the energy efficiency of domestic buildings generally ....
  "The Secretary of State for Health and Social Security and the Secretaries of State for Energy and the Environment will continue to co-ordinate policy on energy efficiency with particular relevance to the disadvantaged" (Department of Energy, 1983).
The important difference between the energy efficiency **of** buildings and energy efficiency **in** buildings was not clarified then, and has not been apparent in subsequent policies. Instead, all the emphasis has been on insulation measures, so these ambiguous definitions probably indicate the continuing tension between the DEn and the Department of the Environment (DOE) over policy for the energy efficiency of dwellings. It is still not certain whether any department is concerned about the economically efficient use of energy in domestic heating systems.

BRITISH HOMES IN COMPARISON TO INTERNATIONAL STANDARDS

Before looking at the detail of low income homes in Britain, a comparison is made of the standards in British homes with those of other countries.

Using the present Building Regulations for new homes in different countries, a comparison can be made of the expected energy consumption in a standard home by an average family (Fig. 1). The Milton Keynes Energy Cost Indicator (MKECI) gives a measure of the cost of all energy use in a home, per square metre of floor area. Thus, fuel bills in a new home in the UK would be 75% higher than in a house, in the same location and of the same basic dimensions, but built to the standard of Finnish Building Regulations (173.0 ÷ 99.1) and providing the same level of comfort to the occupants. Figure 1 reveals that British homes are the most expensive to keep warm, because the present Building Regulations, dating from 1982, require a lower level of insulation than those of any of the other countries shown, and the faster rate of heat loss results in larger fuel bills to obtain the same standard of warmth.

38

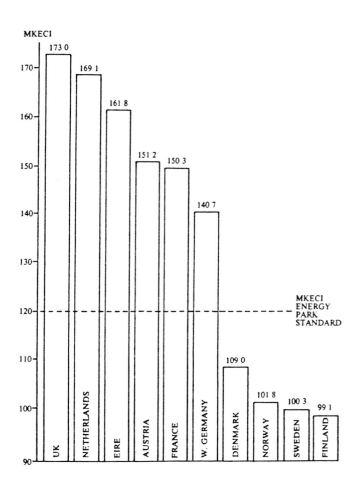

Source: Potter, 1985

Fig. 1   Comparison of Building Regulation Standards in ten European countries
using the Milton Keynes Energy Cost Index (MKECI)

That is the comparison for new buildings.   In existing dwellings the situation
appears to be broadly the same, with British homes requiring more energy to
achieve a given level of warmth than in other countries (Table 1).

TABLE 1 Energy Use for Heating and Resultant Temperatures, per Household,
         some International Comparisons

	Energy used for heating 1980 (kJ/m²/degree day)	Average indoor winter temperature (24-hour whole house mean, °C)
United Kingdom	230	14-16
Denmark	215	17-18
France	275	17-19
Germany	275	18-20
United States	275	19
Sweden	180	21
United Kingdom - future	310	18
	390	21

Source: Schipper and others, 1985, p3; future values - author's estimate

Thus, British homes, in general are less energy efficient than those in
countries with similar or colder climates, when climate is controlled for.
This disparity will continue whilst the present Building Regulations are in
force, and the standard suggested in the draft of the next set represents a
marginal improvement - there are still no proposals to have double glazing or
control ventilation losses.   There is a possibility that a British Standard
Code of Practice for Energy Efficiency in Domestic Buildings is introduced,
based on the MKECI, and that this would also be incorporated into the Building
Regulations.   If this occurs, reflects the cost of the fuel, and is applied to
both new buildings and extensive conversions, then the energy efficiency of
British homes will, for the first time, involve both the building fabric and the
heating system.

RANGE OF ENERGY EFFICIENCY WITHIN THE BRITISH HOUSING STOCK

The Building Regulations have been uniformly applied across the whole country
since 1965, and the level of thermal insulation required was increased
substantially in 1974 and in 1981.   Even now, the Building Regulations
stipulate the rate at which heat can be lost through the walls, roof and floors,
but no standards apply for heat loss through windows or as a result of
ventilation.   New buildings have to comply with the Building Regulations, but
lower standards are often permitted when existing buildings are converted.
Thus, only the 13% of the housing stock constructed since 1974 is known to have
some insulation in the original building fabric.   Of these, 1.9 million belong
to private individuals, and the remaining 1 million are in the public sector,
providing rented accommodation.   It is assumed that half these local authority
tenants are poor, by the present definition, as that is the national average for
this tenure.   The original building fabric of the other 87% of the housing
stock is similarly energy inefficient and can only be improved through the
addition of insulation.

Since 1978, about £350m has been spent by central or local government on
energy efficiency improvements to dwellings (Hansard 10.3.87; DOE, 1986a):
-    £140m by local authorities on draughtproofing and loft insulation in their
     own housing: as 50% of local authority tenants are poor (by this
     definition), a *pro rata* distribution means £70m has been to their benefit,
     though authorities were asked to give priority to the elderly and disabled
     (DOE Circular 23/78);

IEE-D

-   £170m as grant aid towards loft insulation, about a quarter of which has
    gone to low income households: the bulk has been given to private owners;
-   £40m to community insulation groups which draughtproof the homes of the
    disadvantaged, whilst providing part-time, temporary employment.
Therefore, £150m is the estimated capital investment in energy efficiency
improvements by government to low income households over nine years.   About
0.2m households have been draught-proofed only, 0.7m have had the loft
insulated, and 1.1m a mixture of these measures.   There is no evidence of more
extensive work being undertaken with government funds.   Nor is it known how
much has been done without the aid of grants, though the assumption is that
poor families need the assistance of grant aid.

Thus, of the poor households:
                2.0m have had loft insulation/draughtproofing/both
                0.5m live in property built since 1974
                3.9m live in older property, with no added insulation

Energy efficiency calculations include the type of heating system and fuel used.
In Britain, the cost of the delivered energy used in domestic dwellings varies
by a factor of over four: general tariff electricity costs £15.33/GJ and gas
£3.60/GJ.   Solid fuel and off-peak electricity prices are in between and other
forms of heating are used by few households in Britain.   The fuel used in the
home, therefore, strongly influences the cost of keeping warm.   The cheapest
method of heating is with gas, when the fixed costs of the appliance, standing
charges, maintenance and the efficiency of the appliance are taken into account.
It is not known what proportion of poor households use gas as the main form of
heating, though low income households are less likely to be connected to the
gas supply (66%) than the rest (75%) in 1983 (Bradshaw and others, 1987), so
that the poor have less access to the cheapest fuel than better-off households.

The most expensive forms of heating are those using on-peak electricity, such
as individual plug-in fires or ceiling heating, or technically inefficient
systems using off-peak electricity, such as underfloor heating.   At least 1.25m
families are dependent upon individual electric fires for their main source of
heating.   These individual fires are cheap to buy, and thus are predominantly
found in low income households – the classic example of limited capital
resulting in high running costs.   With the exception of night storage heaters,
most electric central heating is expensive to run and is found in 1.5m local
authority dwellings, affecting at least 0.75m low income households.   Thus, 2m
poor families are known to be using the most expensive forms of heating.

In Britain, there is a strong correlation between warm homes, high income, the
ownership of central heating (usually gas), and the presence of added,
particularly loft, insulation.   A non-centrally heated, privately rented home is
seven times more likely to have an uninsulated loft than the CH home of an
owner occupier.   The 40% of homes without CH are, on average, 3°C colder,
older, and more likely to be rented.   One factor is that many homes without CH
have few heating outlets: in a national survey, about a third of all rooms
(including the halls and landings) had no source of heating (Hunt and Gidman,
1982).   Thus, there is bunching of energy efficient measures in better-off
homes, and a dearth of them in low income ones.   This suggests that consumers
do respond to market forces where they have the legal right and finances to do
so, and confirms the need for Government policies to assist those on a low
income, particularly in rented accommodation.

A LOW INCOME ENERGY EFFICIENCY PROGRAMME

This programme, proposed in 1986 (Boardman), aims to make it possible for the poor to obtain affordable warmth. Therefore, the measures include fuel substitution and other improvements to the energy efficiency of the heating system, as well as insulation of the building fabric.

The conventional method of judging the cost effectiveness of energy conservation measures is to compare the value of the energy saved with the capital cost. There are no problems with this method where the home is adequately warm prior to the alterations. However, for the cold homes of low income families one of the main reasons for the programme is to enable them to be warmer. Thus, poor families may not save any money in running costs as a result of the energy efficiency programme.

Based on research of the effects of insulating cold homes (Boardman, 1987), the proposal is that the benefit of energy efficiency improvements is assessed as if the home had previously been warm. The monetary value of the energy thus saved provides the total value of the benefits to the low income household, whether taken as increased warmth or as fuel savings. This method allows the cost effectiveness of a measure to be judged, irrespective of the present temperature or level of expenditure on fuel within the household, and thus permits energy efficiency measures to be just as cost effective in cold homes as in warm ones. With present methods, the exclusion of the value of the additional warmth means that the Energy Efficiency Office and the Treasury consider that energy efficiency improvements to cold homes are difficult to justify on an economic basis (Macintyre, 1985).

Proceeding with this new method, the cost effectiveness of a proposed measure depends on the capital costs involved and the price of the fuel being used. Public sector investments have to be likely to achieve a minimum 5% real rate of return, before they can be considered against other objectives. The relationship between the life of the investment, the payback period in years and this required rate of return is given in Table 2.

TABLE 2   Payback required to meet a 5% real DCF rate of return (years)

Anticipated life of investment	Payback
5	4.3
10	7.7
15	10.4
20	12.5

Source: Department of Energy, 1982

Using figures provided by the Energy Efficiency Office (1986) at November 1985 prices, for 13 house types, different fuels and a variety of insulation measures, the options capable of achieving the 5% RRR has been calculated (Table 3).

TABLE 3   Energy efficiency measures able to achieve a 5% real rate of return
in relation to the fuel used for heating

	On-peak elec	Solid fuel	off-peak elec	gas
Cylinder cover	yes	yes	yes	yes
Draughtproofing (1)	yes	yes	yes	yes
Loft insulation	yes	yes	yes	yes
Cavity wall filled (2)	yes	yes	yes	yes
Solid wall - internal	yes	yes	most	old tce only*
Solid wall - external	yes	old tce only*	no	no
Double glazing (3)	no	no	no	no
Gas central heating	yes	yes	yes	yes
Night storage heaters	yes	yes	yes	no
Solid fuel central heating	yes	yes	yes	no

* only cost effective in older, terraced houses

Notes: (1) with a life of at least 6 years; (2) all methods of insulation;
(3) double or secondary glazing is only cost-effective, if expenditure is below
£700 and the product lasts 20 years.

Thus, in the most energy inefficient homes, such as those using general tariff
electricy, capital investment in many measures to improve the energy efficiency
is cost effective on the basis of accepted public sector investment criteria.
An assessment of the size of the total programme is given in Table 5.

TABLE 5 Cost effective capital investment in energy efficiency improvements
to low income households

Measure	Average cost £	Number of dwellings million	Total cost £m
Alter heating system (1)	750	4.3	3225
Add heating controls	200	1.5	300
Loft insulation (2)	200	3.0	600
Draughtproofing	75	5.0	375
Cavity wall	400	4.0	1600
Solid wall - internal	1500	0.6	900
Solid wall - external	3000	0.4	1200
Total			8200

Notes: (1) Includes fuel substitution and extension to inadequate systems;
(2) Includes topping up inadequate amounts.

Sources: Costs based on Energy Efficiency Office 1986; author's estimate for
numbers

This £8.2bn could be spent on low income homes and achieve a rate of return in
excess of 5%.   The rate at which the work could be undertaken depends upon a
variety of parameters, not least the ability of the manufacturers to supply the

materials and the speed with which the administrative framework is assembled. Initially, the value of the programme may have to be limited to about £400m pa, though rapid growth would be needed to ensure that all homes are treated within as short a timescale as possible - 20 years is too long.

The employment generated depends, partly, upon the design of the programme. For instance, a part-time, temporary employee of a community insulation project costs about £5,600 a year at 1987 prices (Hansard 30.3.87), though the level of productivity is low and cost per draughtproofing job completed is high. Full-time, permanent employees achieve a higher productivity and can undertake more skilled work, thus reducing the cost per energy efficiency measure. In 1984, the capital cost of a job created in energy conservation work in the European Community was calculated to be £11-17,000 (Hillman and Bollard, 1985). These gross costs are substantially reduced at a time of high unemployment by the saving in benefits and generation of taxes, which can be estimated at about £6,300-7,000 pa per registered unemployed person (Sinfield and Fraser, 1985). If the programme is assumed to take 10 years, the number of full time jobs created is in the range of:

     a minimum of 50,000 pa at £17,000 gross cost per job, and
     a maximum of 75,000 pa at £4,000 net cost per job (£11,000-7,000).

The total cost of the programme is further reduced by receipts from value added tax, which is levied on improvements to existing buildings. Employment creation in energy supply is considerably more costly.

The occupants have very limited capital resources and may even be in debt, so they are extremely unlikely to undertake the investment themselves. At the moment, they are obtaining poor value for their expenditure on heating - they have to buy expensive warmth. Thus, the state is obtaining poor value for the £2.2bn it provides annually for its dependents to spend on fuel. The Government could stimulate this investment in two different ways, either directly as part of a Government-funded national programme or through altering the responsibilities of building owners.

As demonstrated above, over the past nine years, the total capital expenditure on energy efficiency in low income households by government has been about £150m. Of the three programmes listed, the major one, for the poor, is the local authority investment programme, which has declined sharply in recent years owing to the Government's restrictions on housing finance. The other two programmes, covering loft insulation and draughtproofing only, are of limited effectiveness, although well-targeted on the poor. Therefore, a new approach is required to achieve the level of investment identified in Table 5.

About half of the potential improvement in energy efficiency in low income homes will come from alterations to the heating system. Because domestic heating systems generally commit the user to a specific fuel, obtaining a cheaper source of warmth requires the replacement of the heating appliance and, thus, fuel substitution. Even if there is no fuel substitution involved - as when open coal fires are replaced by enclosed room heaters - capital expenditure is still required. The DEn has never provided grants to achieve fuel substitution in the domestic sector, though, since 1981, £40 million of assistance has been allocated for industry to convert oil and (later) gas-fired plant to coal. The DOE set a precedent for fuel substitution in the domestic sector with the Clean Air Act 1956, which entitles householders to a grant of at least 70% of the cost of installing an approved alternative heating system. By 1983/84 the Exchequer had contributed £150m and local authorities a further £135m towards the cost of these replacement heating systems (DOE pers comm, March 1987). As this is a pollution control measure, the effect on energy efficiency in low income households cannot be calculated. The extension of this precedent, to give grants to low income households to improve heating systems in existing dwellings would be a major innovation in British energy efficiency policy, but a highly cost-effective use of resources.

In order for the Government to proceed with a Low Income Energy Efficiency Policy, a lead department would have to be identified. The DOE's responsibility for energy efficiency in domestic buildings has *de facto* included heating systems in the past, for instance when problems occurred with electric heating in local authority dwellings (DOE, 1978) and in relation to the Clean Air Acts, under which grants can be given for new domestic heating systems. However, the Department of Energy is, theoretically, responsible for both energy supply and energy demand, and the energy used for space heating in low income homes represents about 20% of all delivered domestic energy, or 6% of all energy. This split of concerns is seriously inhibiting the development of energy efficiency policies towards low income households.

There are also problems if energy efficiency improvements are to be made the responsibility of the owners of the building occupied by the low income family, concerning both the standard to be achieved and the need for grants. In order for the work to be done, it may be necessary to link financial assistance to the status of the occupant, rather than the owner. The minimum standard of housing, for owner-occupiers and landlords alike, is defined under the Housing Act 1957 and is known as the fitness standard. Any dwelling that fails to achieve this standard is "unfit for human habitation" and can be improved compulsorily. The fitness standard does not include any items referring to the ability to heat the property, it does not even require a heating system to be present. Even by this minimal standard, in 1985 in England, there were over 650,000 homes that were unfit, but occupied – 4% of the housing stock. Unfit properties are predominantly owner occupied (Hansard 15.1.86) and pre-1919 (DOE, 1982). In a Green Paper in May 1985 (Cmnd 9513) on grants for home improvements, including unfit dwellings, the Government stated:
"that a policy of raising the minimum standard to match rising social expectations would be inappropriate" (para 64).
Thus, the Conservative Government effectively ruled out using legislation as a way of raising the energy efficiency of the worst housing. This is unfortunate and it is hoped that when the White Paper or Bill is produced, the opportunity is taken to upgrade the standard and ensure that a home is only designated fit if it can be kept warm for a reasonable amount of money.

Under British law, landlords are responsible for repairs to the dwelling, including those relating to the provision of services (eg gas and electricity), though they cannot be required to undertake work that represents an improvement, such as adding insulation. There are no precedents in Britain for alternative methods of ensuring that rented accommodation achieves a minimal standard, such as requiring rented property to be certificated, as in some other countries. Similarly, the energy efficiency of rented property is not reflected in the rents that can be charged, so that landlords cannot recover their costs via higher rents, nor are high heating costs compensated for by low rents for the tenant. Thus, with the present legal and economic framework, there is no way of encouraging or requiring landlords to invest in energy efficiency improvements to their property.

Local authorities have, in the past, undertaken more energy efficiency improvements to their properties than private landlords, though this activity is declining through the restriction of funds by central government. Since 1983, local authority tenants have had the right to buy their homes, and about

750,000 have exercised this right. However, local authorities are now permitted to spend only 20% of the total proceeds from these sales in a year, so that £6bn is retained, on the instructions of the Treasury. The money is available, if the Government allow it to be used.

Local authorities have been able to use their resources to improve properties in other tenures, whether through home improvement grants or enveloping. The former of these is to be replaced by loans, under the Green Paper mentioned above, further reducing take-up amongst the poorest households. Enveloping is a useful initiative, developed by the City of Birmingham, primarily to improve the exterior fabric of whole streets at a time in Housing Action Areas. Typically, the properties treated are terraced, pre-1919 and solid walled, often occupied by elderly people who are unable to maintain their properties adequately. The work was free of charge to the owners and occupiers of the building, as it was hoped that it would stimulate further investment in the interior of the dwelling. Energy efficiency improvements, such as loft insulation, cannot usually be included, even when the roof is replaced, because of the financial restriction of £8,000 a dwelling (£10,000 in London), imposed by the Secretary of State for the Environment (Circular 26/84).

The omission of energy efficiency measures from enveloping schemes is unfortunate, as the cost of energy efficiency improvements diminishes substantially if they are incorporated with general repairs (DOE, 1986b) and is the only way to make some of the most expensive measures, such as solid wall insulation and double glazing, cost effective. Similarly, the unit cost is reduced through substantial economies of scale, for instance in bulk purchasing and administration. This may have little impact where the work is highly labour intensive such as draughtproofing, though the per dwelling cost of cavity fill can be halved by the greater productivity of equipment.

One of the advantages of a local authority initiative, such as enveloping, is that action on the worst houses in different regions of the country can take place simultaneously. At the moment, there is a considerable likelihood that many of the least energy efficient homes are also in a state of disrepair, though the overlap is not known. Thus, a programme to ensure that both improvements are undertaken together makes sound economic sense. As at present operated, enveloping is solely concerned with the building fabric and does not encompass the adequacy or economic efficiency of the heating system. If the scheme could be extended to ensure that the exterior fabric of the house was both in good condition and well insulated, and that the home included an efficient, inexpensive-to-run heating system, this could be the best policy for energy efficiency in low income homes in Britain.

The need for new initiatives towards the energy efficiency of low income households is particularly acute as the benefit system in Britain is being altered and simplified. Many of the payments made to counter the varying conditions of households, such as the use of an expensive heating fuel or the occupancy of a home that is difficult to heat, are being phased out. From April 1988, assistance will be related solely to the social characteristics of the family. The Department of Health and Social Security are implicitly assuming that all families live in equally energy efficient homes. It has been shown that the poor live in homes that are, generally, more expensive to heat than other, better-off families. The range amongst the disadvantaged is also considerable: a measured survey of local authority properties (EIK, 1980) found that the rate of heat loss varied by a factor of over five (120-620 W/°C). The cost of useful energy is difficult to estimate because of the lack of data on the efficiency of old appliances, though is certainly in the range

46

£5.00-£15.33/GJ, another factor of more than three. Theoretically, at least, a poorly insulated bungalow heated by on-peak electricity could cost 15 times more to keep warm than a small, well-insulated flat, with modern gas heating. As these properties are local authority owned, they could both be occupied by low income households who, in future, will be dependent upon benefits that ignore energy efficiency variations. Capital investment programmes need to be implemented immediately to counteract the increasing hardship that will be caused by these changes in the social security system.

CONCLUSIONS

Instead of the poor being able to buy warmth cheaply, they are forced to purchase an expensive product because of their lack of capital. Only central or local government can compensate, either with direct capital investment or indirectly through grants to improve the energy efficiency of low income homes. The standard to be achieved needs to include a wide range of insulation measures, and improvements to the economic efficiency of the heating system, if possible to reduce the cost of keeping warm below that of the average family. Two administrative mechanisms have been identified, that would work in unison. First, regulations concerning the minimum standard of fitness for existing buildings need to be upgraded, in conjunction with higher requirements in the Building Regulations for new buildings and conversions. Secondly, in inner city areas and on council estates, blocks of properties can be made energy efficient through an extension of the enveloping principle, to include insulation and heating systems. Local authorities have substantial capital receipts from the sale of council houses that they could use, if permitted, to finance the work. When fully operation, the programme would create between 50,000-75,000 jobs pa at a relatively low, net cost, because of the present high unemployment rate. Thus, the money and workforce are available. Many low income families have to choose between being cold, incurring fuel debts or depriving themselves of some other basic necessity. Evidence of deprivation and hardship are certainly present. All that is needed is the political will to take action and implement an energy efficiency policy to assist low income families suffering from fuel poverty.

ACKNOWLEDGEMENTS

The help of John Chesshire, Head of the Energy Group, and Gordon MacKerron, Senior Fellow in the Energy Group at the Science Policy Research Unit, University of Sussex, is gratefully acknowledged.

REFERENCES

Boardman, B. (1986). Fuel poverty: the need for a low-income energy efficiency programme. In A. Harrison, J. Gretton and G. MacKerron (Eds), *Energy UK 1986*, Policy Journals, Newbury, Berks.
Boardman, B. (1987). *Social, Economic and Technical Considerations for Fuel Poverty Policy*. DPhil thesis, in preparation.
Bradshaw, J., G. Hardman and S. Hutton (1986), *Expenditure on Fuels 1983*. National Gas Consumers' Council and Electricity Consumers' Council, London.
Cmnd 9513 (1985). *Home Improvement - A New Approach*. HMSO, London.
Department of Employment (1985). *Family Expenditure Survey 1984*. HMSO, London.
Department of Energy (1982). *Investment in Energy Supply and Energy Use*. DEn, London.

Department of Energy (1983). Extracts from the Government Observations on the Fifth Report from the Select Committee on Energy, Session 1981-82. DEn, London.

Department of the Environment (DOE) (1978). *Domestic Energy Notes 3.* Joint Working Party on Heating and Energy Conservation in Public Sector Housing. DOE, London.

Department of the Environment (DOE) (1986a). *Housing and Construction Statistics 1975-85.* HMSO, London.

Department of the Environment (DOE) (1986b). *Energy Efficient Renovation of Houses.* HMSO, London.

Department of the Environment (DOE) (1982). *English House Condition Survey 1981.* HMSO, London.

EIK (1980). *EIK Project, Report on Phase I.* City of Birmingham Housing Department and Department of the Environment, London.

Energy Efficiency Office (1986). *Cutting Home Energy Costs.* Separate editions for gas, electricity, solid fuel and oil. DEn, London.

Hillman, M. and A. Bollard (1985). *Less Fuel, More Jobs.* Policy Studies Institute, London.

Hunt, D. R. G. and M. I. Gidman (1982). A national field survey of house temperatures. *Building and Environment,* Vol 17, No 2, pp107-24.

International Energy Agency (IEA) (1987). *Energy Conservation in IEA Countries.* OECD, Paris.

Macintyre, W. (1985). Evidence to the Select Committee on Energy Inquiry into the Energy Efficiency Office, Eighth Report, Session 1984-85, 12 June. HMSO, London.

Potter, J. (1985). Evidence to the Select Committee on Energy Inquiry into the Energy Efficiency Office, Eighth Report, Session 1984-85, 19 June.

Schipper, L., S. Meyers, H. Kelly and Associates (1985). *Coming in from the cold: Energy-wise Housing in Sweden.* Seven Locks Press, Washington DC, USA.

Sinfield, A. and N. Fraser (1985). *The Real Cost of Unemployment.* Dept of Social Administration, University of Edinburgh.

The Development of a European Market for Contract Energy Management

Ian Brown

Association for the Conservation of Energy, 9 Sherlock Mews
London, W1M 3RH

INTRODUCTION

All available studies show that investment in energy efficiency within the European Community is not occurring at the optimum rate. The reasons for this are well documented and much discussed - barriers in the marketplace are preventing an adequate take up of energy efficiency opportunities.

Among these barriers are the lack of finance, or, as common, unwillingness to spend available finance on energy efficiency improvements, and a common barrier throughout Europe, a lack of credibility in energy saving technologies. The overcoming of these barriers may be assisted by the use of energy performance contracting, yet such an activity is at a very early stage of development in Europe.

In 1985 research was undertaken in the twelve countries of the EEC into the potential market for 'third party finance', or 'contract energy management' as it has become known in this country, (as defined below). This research was instigated because it was observed that the level of investment in energy efficiency equipment installed through the mechanism of performance contracting is considerably greater in North America than in Europe. This observation begged several questions, which the research set out, at least in part, to answer.

• Is there a market for contract energy management in Europe and how big is that market?

• Why has the concept not developed as fast (or indeed hardly at all) in Europe as in North America?

• What are the barriers preventing the growth of contract energy management?

• What actions can be taken (if any) to overcome these barriers?

Research results presented in this paper are the summary of well over one hundred personal interviews with relevant organisations and individuals throughout the twelve countries of the EEC.

DEFINITIONS

For the purpose of this study 'contract energy management' was defined as:-

The provision of the services of auditing, installation, operations, maintenance and financing on a 'turnkey' basis, with the cost of these services being contingent, either wholly or part, on the level of energy saving.

The current North American terminology of performance contracting can be used interchangably with the activity known in Europe as contract energy management.

DEMAND FOR CONTRACT ENERGY MANAGEMENT

The cost effective potential for energy saving in the European Community has been quoted as being 25% of present consumption, across all energy using sectors, by the year 2000.

However, the level of investment needed to bring about such savings is not a figure that has been predicted with any degree of accuracy. Nevertheless, using already published data, an attempt was made to estimate the potential level of investment in energy efficiency in Europe, but it should be stressed that the numbers quoted below should be treated as orders of magnitude rather than 'exact' figures.

Potential in the Building Sector

According to the European Commission study 'Towards a European Policy for the Rational Use of Energy in the Building Sector', the average investment cost per tonne of oil equivalent (TOE) saved each year for existing buildings is 1,300 ECUs for investments with an average simple payback of 3 years or less.

The same study estimated that 12% of the European Community's present energy consumption in the building sector (residential, commercial, industrial and public sector buildings) could be saved by investments with paybacks of 3 years or less.

Using such estimates as the basis for an estimate of the total potential investment in the building sector, based on the total final consumption in the European building sector of 270 million TOE (1984 being the most recent figures available), a 12% energy saving would equate to an investment need of some 42 billion ECUs.

Potential in the Industrial Sector

The potential for energy saving in the industrial sector has been estimated in a number of European countries, but estimates vary greatly according to the existing energy efficiency of the capital stock, and the methodology used for the estimate. The Netherlands has a target of 30% energy savings by the year 2000, a figure which was recently confirmed as economically feasible by a Dutch Government advisory committee.

Alternatively, a 1982 survey of the UK industry (Ref 3) concluded that the potential for energy saving investments with a payback of under 3 years was 14%. This figure is judged to be more realistic, on a European wide basis, and thus as a basis of estimating the market potential for energy saving investment through contract energy management (i.e. average paybacks no longer than 3 years), savings potential of 15% is assumed.

The French Energy Management Agency (Agence Francaise pour la Maitrise de l'Energie) calculated in 1985 (Ref 4) that in the industrial sector an investment of 1050 ECUs will be required to save 1 TOE, assuming that the investment has a payback of 2-3 years (the payback range necessary for contract energy management).

Conservation industry sources interviewed have confirmed the validity of this figure, and it is thus used to estimate the market potential for energy saving in the industrial sector.

Using the 1984 (latest figures available) Total Final Consumption in the industrial sector of the 12 European Countries of 279 million TOE as a basis for a market estimate, the 15% potential for energy saving would, using the 1,050 per TOE saved formula, equate to an investment of 44 billion ECUs.

Total Market Potential

Since transport was excluded from this study, the total potential market for contract energy management - being the total potential investment in energy saving projects (with a payback of 3 years or less) is the sum of the two sectors previously quoted - buildings and industrial - being a total potential market of $82 billion across the 12 countries of the European Economic Community.

Potential by Country

The aggregate market potential figures quoted above mask a wide variation between different countries.

Industrial Market:

The industrial market for contract energy management is most immediately promising in France, Italy, Spain and the United Kingdom. These countries all have relatively energy inefficient industrial sectors, where the concept of contract energy management could make a substantial impact on the level of investment in energy saving in the short term.

Among other EEC countries Denmark, Germany, and to a lesser extent, the Netherlands, have relatively limited potential for contract energy management in industry because of the substantial progress achieved in energy saving since the first oil crisis. In these countries most short payback investments have already been made.

There are particular problems restraining the ability of performance contracting to penetrate the industrial market in several countries - notably Belgium and Portugal.

In Belgium, the national government offers an incentive of a tax deduction of up to 20% of the value of an energy saving investment. This tax deduction is open to all industrial and commercial energy users. Under present rules this deduction can be claimed only if the investment is funded by the industrialist. If an energy saving investment is funded by an outside energy service company, then neither the energy service company nor the industrialist may claim the credit, thus putting contract energy management at a significant disadvantage.

In Portugal very high interest rates (currently 30%) and a general shortage of capital are barriers not only to performance contracting, but to investment in industry generally.

Residential Market:

The residential market in Europe, in common with North America, offers considerably less scope for contract energy management than other sectors because of the large number of relatively small investments involved and the major role played by occupancy levels and lifestyles in determining domestic energy use.

Contract energy management can approach viability only in multi-family dwellings where a central boiler plant is present. In single family dwellings or in multi-family dwellings where individual heaters are used it was judged that the concept was unlikely to be viable in the near future.

Multi-family dwellings likely to be a market for performance contracting are found on a significant scale in Italy, Spain and to a lesser extent France. In other countries multi-family housing forms a much smaller part of the housing stock - in the UK for example multi-family housing with central boiler plant accounts for less than 5% of the housing stock.

In Spain the mild climate and resultant short heating season together with the lack of cooling, imply low annual energy use for space conditioning, resulting in relatively long paybacks for energy saving investments. This problem, when allied to other difficulties of no tradition of multi-year contracts and the legal ability of a single tenant to block any capital investment, mean that performance contracting is unlikely to make rapid inroads into this sector in Spain.

In Italy the problems of the short heating season and long paybacks, a problem exacerbated by the highest percentage of oil fired space heating in Europe, imply little immediate market for contract energy management. Paybacks of 6 and 7 years are commonplace, and it was judged that such paybacks are uneconomic for the use of performance contracting.

Institutional and Public Sector Market:

A substantial market exists for contract energy management in institutional and public sector buildings in many European Countries, where restricted capital spending and lack of technical

expertise are commonplace problems. Unfortunately significant problems are likely to delay the introduction of contract energy management into one of its most promising sectors.

In theory a large potential market also exists in the public sector building stock of the UK, France, and to a lesser extent, Germany, Italy and Spain. Of these countries Spain is the only one that has taken any steps to introduce performance contracting in the public sector, and alone appears willing to show the necessary flexibility to successfully negotiate a performance contract.

In contrast, in the UK and Germany, and to a lesser extent in other countries, public procurement rules do not accommodate the performance element of a contract energy management investment, and certainly in Germany, Denmark and Ireland, public officials responsible indicated no willingness whatsoever to consider introducing the necessary flexibility to allow performance contracting.

The potential for energy saving in the UK public sector building stock is very considerable - 50% of all the UK's building stock (of all types) lies in the public sector, and savings of 20-25% of current consumption are economically feasible. This sector is seriously capital constrained, and significant opportunities for energy saving are not being addressed both because of the shortage of capital and also because of skill shortages.

However, despite interest in the use of performance contracting by the UK Department of Energy and also by local authorities, there is a significant barrier preventing the use of contract energy management in this sector. The UK Treasury has taken the view that such financing constitutes public sector borrowing, and is thus added to the 'Public Sector Borrowing Requirement' - equivalent to the US budget deficit. Because of very strict controls on spending, in order to keep the budget deficit low, public bodies face severe financial penalties - in the form of 'fines' - if spending rises above prescribed limits. Unfortunately the Treasury have ruled that contract energy management is counted as public sector spending in the year in which the contract is signed. It is obviously somewhat illogical to treat contract energy management as public sector spending but for wider reason of macro-economic policy the ruling persists.

The most immediately promising public sector markets are those of Spain (as previously discussed), Belgium and the Netherlands. These are countries where the public sector is capital and skill constrained, yet has shown, during interview, more interest and willingness in discussing the ability of contract energy management to aid investment in the public sector building stock.

In France and Italy, heat service contracts are widespread in this sector, and in this situation the attitude of these companies - with whom energy users have signed long term contracts - is critical. This issue is discussed in more detail later, but briefly the heat service companies have little or no economic motivation to reduce the quantity of heating fuel used, and as such are likely to be a barrier to any penetration of this sector by energy service companies offering performance contracts.

SUPPLIERS OF ENERGY SERVICES

Existing Suppliers

The Study revealed that there are only eight or nine companies operating in Europe whose activities can be defined as 'performance energy contracting'. It should be noted that heat service companies were judged to be outside this definition, for reasons discussed fully below. An examination of these existing energy service companies revealed a number of common characteristics.

(a)   No company was formed before 1984, the majority being formed in 1985. This shows the early stage of development of energy services in the Community.

(b)   All existing Escos are subsidiaries of parent companies, three of the eight being formed by multinational oil companies. No entrepreneurial Escos have yet been established.

(c)   Although most existing Escos claim to cover the public/institutional buildings sector, none has yet completed a contract for a government facility. The contract negotiation time has been so lengthy, and the bureaucratic obstacles so great, that no activity has yet taken place in this sector.

Potential Suppliers

If the European market is inadequately covered at present who could enter this business?

Consulting Engineers:

Twenty one consulting engineering practices throughout Europe were interviewed, and although all of these companies specialise in energy consultancy, less than a quarter were previously aware of the concept of performance contracting. Although expressing interest in the concept, there was near unanimity in the view that European engineers are extremely wary of entering the business of performance contracting because of a number of factors:

(a)   Increased Risk:

Consulting engineers are by nature risk averse, and are wary of any way of doing business that increases their financial and technical risk. Only two engineers interviewed indicated that they would consider taking their fee on a performance related basis while all were wary of accepting technical and financial risk of the equipment performing as predicted.

(b)   Professional Practice:

In some Member States engineers are prevented by their professional code from involvement in any 'commercial' enterprise. Engineers are wary of any overt involvement with any supplier or other service company - particularly if such involvement jeopardised their reputation with existing clients. Consulting engineers are not culturally accustomed to the concept of payment by results.

Engineers interviewed indicated that the concept was more complex than their traditional 'preferred' means of doing business, and thus less attractive.

(c)   Entry Cost:

The legal, administrative and marketing costs of establishing an energy service company are high, with the minimum viable figure estimated to be in the region of 300,000 in the first year.

Unless an Esco has parent company backing all early deals will need to be funded by equity alone. Few engineers possess the necessary capital to fund an operation.

(d)   Investment Funds:

In the absence of a loan or guarantee scheme, engineers would have considerable difficulty raising the necessary degree of bank loans to fund investment in energy saving, since they do not have sufficient capital or collateral.

Equipment Manufacturers:

Although many European energy efficiency equipment manufacturers have expressed cautious interest in the concept, few are likely to enter the business for a number of reasons:-

(a)   Manufacturers usually possess technical skill only in their own product sector.

(b)   Most Manufacturers are unwilling to hold products on the balance sheet until the end of a performance contract. Managements are usually under much pressure to maintain cash flow by keeping stocks on the company's balance sheet to the minimum.

(c)   Equipment manufacturers are very wary of upsetting existing business relationships, particularly with consulting engineers, whose influence on the purchase decision can be often crucial.

Heat Service Companies:

The provision of heat services, or 'Chauffage', is often quoted in North America as 'European Contract Energy Managment'. As previously mentioned, heat services have not been included in the definition of contract energy management techniques used in this study, because heat services, as operated in Europe, are concerned principally with the provision of heat. Although energy efficiency is an integral part of such operations, it is by no means the raison d'etre of heat service companies. An energy service company however exists to invest in energy efficiency improvements - not only to distribute or provide energy needs.

Heat service companies offer much scope to expand into the energy services area. They have the technical expertise in heating systems management, and, as established companies, are more likely to have access to capital than an entrepreneurial energy service company.

However, as these companies are contracted to supply a set level of heat, their incentive is to ensure that such heat is produced as efficiently as possible - but there is no incentive to see that it is used as efficiently as possible.

One unusually honest French chauffage company interviewed stated that the actual level of investment in energy saving by that and other French chauffage companies had been low, and that the primary source of profit for the company was the provision and distribution of heating oil.

For this reason this company was uninterested in a concept which implies investment in a package of measures to ensure the maximum possible energy savings. This conflict between heat service and energy efficiency is insufficiently appreciated, both in Europe and in the United States.

Utilities:

Utilities are a logical choice to act as energy service companies in Europe because of a number of factors, including access to capital; the close relationship to their existing business; some expertise in end use technologies; presence in the market and direct contact with energy users.

The attitude of the major European utilities, to the notion of performance contracting can be summed up as uniformly negative. Utilities throughout Europe do not regard energy saving as either a demand reducing tool, or as a possible business venture - two of the motivating factors which have caused North American utilities to promote or indeed enter the performance contracting business.

Gas and electricity utilities throughout Europe see their prime function as ensuring adequate supplies of their fuels. Demand management, as either a business venture as indeed a as 'supply' option is not considered within the remit of these supply industries.

However, the ability of energy efficiency to promote increased fuel sales through encouraging switching to efficient (and hence lower cost) use, particularly in the industrial sector, was very much a motivating factor to most of the utilities interviewed. The most characteristic sentiment expressed towards energy efficiency was that it is desirable only where it could be used to lower costs and thus maintain or even increase market share and sales, at the expense of other competing fuels.
A number of publicly owned European utilities indicated that they would face significant legal hurdles if they wished to 'diversify' into the energy services business. Most state owned European utilities are given a remit to provide adequate supplies at the lowest cost. No European utilities are currently active in the contract energy management business, and with one exception, none are currently planning such a venture.

The one exception is the monopoly British Gas Corporation, soon to be transferred from public to private ownership. It is known that this utility, at the specific urging of the House of Commons all party Energy Select Committee, is actively investigating the establishment of an energy services subsidiary.

AVAILABILITY OF FINANCE

Contract energy management is a highly capital intensive industry. Energy service companies have to meet the cost of marketing, administration and detailed energy audits in addition to the funds needed for the actual investment in energy saving equipment. Energy service companies must be well capitalised, and have access to low cost borrowing, to fund such investments.

In the United States small and medium sized Escos have obtained funding from private investors and local banks. In Europe those some sources of finance are not available to small Escos who lack parent company funding. The venture capital market in Europe is considerably less well developed than in the USA, and indeed there are far fewer sources of risk capital for start up ventures in Europe. European financial institutions interviewed were unanimous in their view that, in the absence of any risk reducing government or EEC scheme, or sufficient collateral, they would not make the necessary funds available. Energy saving investments themselves would not be accepted as sufficient security.

Since the degree and cost of borrowing is the single most important factor in determining the Esco's rate of return, European Escos need to have some access to funds, without which small, or even medium sized, potential Escos will simply be unable to enter the market.

RECOMMENDATIONS FOR FUTURE ACTION

Barriers to Entry

The barrier mentioned above - difficulty of raising finance, is only one of a number of barriers preventing potential energy service companies from entering the market. The principal barriers are:-

IEE-E

(a)   Risk:

The business and professional culture of Europe is more risk averse than that of the United States. Several interviewees indicated that as they were making satisfactory profits at present, they saw no reason to increase their risk to try to increase their business.

Consulting engineers are risk averse and are disinclined to accept the technical or financial risks for a project that contract energy management implies.

(b)   Significant Start Up Costs:

Because of the complexity of the contract, contract negotiations can be very lengthy - six months appears to be the absolute minimum feasible and 18-24 months is possible. This lengthy contract negotiation time implies a very high marketing and administrative start up cost for a possible Esco. This start up cost is estimated in Europe at a basic minimum of $400,000 in the first year.

(c)   Difficulty of Raising Capital:

In addition to the high start up costs which imply significant equity needs for an energy service company, such a company will need access to low cost borrowing if it can economically fund the level of energy saving investment needed for viability. Companies not backed by the resources of a major parent company foresee considerable difficulty in obtaining these funds without the necessary security.

(d)   Uncertain Energy Prices:

The recent dramatic fall in oil, and to a lesser extent natural gas and electricity prices, which has been seen in the United States, has been less pronounced in Europe.

US dollar denominated crude oil prices do not necessarily translate into a proportional fall in delivered local currency fuel oil. Further, falls in oil are not necessarily being matched in Europe by falls in other fuel prices - indeed energy service contracts may still have an expectation or static or even rising gas and electricity costs, usually supplied by a state owned monopoly supplier, which may be the predominant fuel in a performance contract.

Nevertheless, it is undoubtedly the case that the uncertain energy price picture, and the recent falls in the price of oil, have increased the risks for any potential European energy service company. However, in the opinion of those existing European energy service companies the effect of such energy prices will not lead to any dramatic curtailment of their potential market, for the reasons given above.

Possible Actions to Overcome Barriers

These barriers listed above are the cause of the supply problem for contract energy management in Europe. A question which should be asked however is, is the supply of energy services a problem which should be addressed, or should demand be stimulated, which will automatically lead to the growth of energy service companies?

It is certainly the case that knowledge of, and hence demand for, energy services is at a low level, and some of the actions recommended below in this paper are designed to assist in remedying that situation. However, the potential for energy services is very considerable, and the barriers to market entry so great that there is no certainty that companies will meet a demonstrated market need. Indeed in Spain the Basque Government spent three years trying to persuade potential energy service companies to enter the business in order to participate in several large performance contracting demonstration schemes. This effort was without success.

The potential for contract energy management in Europe will not be fulfilled until more suppliers of such services - small, medium and large - are available to meet this need. Therefore a number of actions are suggested which may assist in the development of this market both in the demand for, and supply of energy services.

(a)   Education:

Knowledge of the concept of contract energy management is very low both among energy users and indeed possible Escos in Europe. Less than a quarter of consulting engineers interviewed (and all of these specialise in energy efficiency) were aware of the concept.

It has therefore been recommended to the European Commission that they organise a series of seminars in European countries, and it is believed that a two-day European performance contracting conference will be held in 1987 under the sponsorship of the EEC.

(b)   Public/Institutional Buildings Sector:

In the public/institutional buildings sector there are a number of significant barriers to the use of contract energy management, including (i) public procurement processes (ii) very lengthy decision making processes and (iii) lack of motivation to save energy. These barriers must be addressed by European Governments if contract energy management is to make any progress in the public sector.

(c)   Pilot Projects:

In the United States pilot projects to demonstrate the worth of the contract energy management technique have been sponsored by Federal and State Governments. These schemes act not only to demonstrate the validity of the approach to energy users but also can actually encourage potential Escos to enter the market.

It has been recommended that the EEC organise pilot schemes in each sector (multi-family, industrial, commercial and public sector) throughout Europe.

(d)   Finance:

Comparing the contract energy management industry of the United States and the current supply of energy services in Europe (the starting point for this study), the provision of finance for small and medium sized Escos emerges as one of, if not the, key difference.

The difficulty European Escos face in borrowing to fund energy saving investments without the backing of a large parent company is a major problem in Europe. Several courses of action are possible to help overcome this barrier.

•   European Community Loans - European Community loans could be made to small and medium sized energy service companies under an existing EEC loan scheme, known as the New Community Investment, administered through the European Investment Bank. Such loans would both encourage the growth of small enterprises and would assist in the achievement of the Community's energy efficiency goals, two necessary predictions for this loan scheme.

•   EEC Loan Guarantees - EEC loan guarantees could assist small and medium sized Escos to achieve the high degree of low cost borrowing necessary. However, the administration of such a scheme by the Commission, on a project by project basis, would be both difficult and expensive.

•   Performance Insurance - Energy saving performance insurance policies guarantee a monetary level of energy savings, and hence the return to the investor in a contract energy management project. Such an option for assisting the financing of performance contracts differs from the two options above in that it is a private; rather than government option.

However, the disadvantage of performance insurance is its relative scarcity, and its considerable cost. Such insurance policies, if found, would cost between 3% and 5% of the total value of predicted savings throughout the life of the project.

(e)   Perceived Complexity

One of the problems affecting both the demand for, and the supply of energy services in Europe is the perceived complexity of the transaction. Energy users are wary of energy service contracts because this method of doing business is so different to the traditional procurement situations.

In order to assist such users the European Commission has contracted with the Association for the Conservation of Energy to produce sample contracts for energy performance contracting in both Industry and Buildings. These contracts are emphatically not model contracts - there is no contract that can be a model for all users in all situations. Nevertheless, the sample contracts, which have been written for Germany, France, Italy, Spain and the UK, can guide energy users through the issues involved and should encourage energy users to feel more confident in negotiating with an energy service company.

In conjunction with the sample contracts, an accompanying 'guidebook' has also been produced, which not only explains the intricacies of the sample contracts, but discusses how to procure such a contract; how to evaluate different proposals; discusses the various ways of structuring an energy service contract and the pros and cons of each; and how to manage such a contract.

Conclusions

This paper has argued that there is an enormous demand for energy performance contracting in industry, public sector buildings and in commercial buildings, throughout the countries of the European Community. The Community has set the ambitious target of increasing the energy efficiency of the community by 25% (relative to 1985 consumption) by 1995. Energy performance contracting can help to realise this goal.

However, there are many barriers to the wider use of energy performance contracting in Europe. On the demand side, energy users are either wholly unaware of this approach or are wary of it because of its novel nature and perceived complexity. On the supply side there are still insufficient energy service companies to fully exploit this large market.

The European Commission is acting to help overcome these barriers through the production of sample contracts and a guidebook to energy performance contracting which will simplify the transaction encouraging energy users to use this new approach. (Contract energy management is in its infancy in Europe, but the potential for substantial growth is there. I predict that with the help of the European Commission, individual governments and an active energy services sector, this approach will help us achieve an energy efficient Europe for the 1990's.

Updating Energy Policies in a Tanneries District

E. Carnevale, M. Tucci

Department of Energetics, University of Florence,
Florence, Italy

ABSTRACT

About one third of the whole Italian leather production is concentrated in a small area around Santa Croce sull'Arno (Pisa), where it accounts for most of the local industry. About 300 tanneries globally use 750,000,000 MJ/yr of thermal energy and 42,000 MWh/yr of electricity. We first analyzed the energy expenditure patterns of the three main processes, namely chrome tannage, vegetable tannage and fur dressing; we then estimated that a 10% to 25% energy saving could be obtained by technological updating in the production and use of low enthalpy steam, hot water and warm air for drying. Amortization plans have been drawn under various assumption; the investments required for a global saving of 1800 TEP/yr would pay back at most in three years, which could be reduced by half under existing legislation for promoting energy efficiency. The widespread implementation of electricity and heat cogeneration would afford even greater economies and benefits, which are expounded in the paper.

KEYWORDS

Leather industry; tanning; cogeneration; energy conservation.

INTRODUCTION

Energy conservation is a challenge to many people and a goal for everybody; it can only be based on knowledge, a complex knowledge ranging from physical laws to technical expertise, from economics to informatics. To us mechanical engineers the challenge is to pick up a real situation, to study and describe all its relevant aspects, the way they are and the way they could be. We have studied to this end a rather large, fragmented but homogeneous industry which has flourished in a geographically restricted area of Tuscany, the leather tanneries of the Santa Croce district. In this paper we describe the processes, we analyze the energy flows, we compare requirements with expenditures and we suggest ways to limit energy wastes.

PORTRAIT OF THE PRESENT ENERGETIC SITUATION IN THE DISTRICT

The recent worldwide yield of leather (F.A.O.,1982) is represented if Fig.1, which only considers industrial production, while a great amount of leather is still hand-crafted. Italy ranks first as a producer country, with about 36,000 people working in tanneries (ISTAT, 1981); this figure had increased by 33% in ten years. Italian production of leather is concentrated in some districts of five regions; Figure 2 shows the distributions of workers and of factories, and makes clear that most of the tanning is done in rather small plants. In fact the vast majority of tanneries employs from 7 to 30 workers.

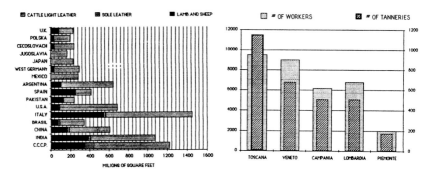

Fig. 1. Worldwide leather production.   Fig. 2. Italian leather industry.

The tendency to small size is particularly evident in the Tuscany leather area that we have surveyed; this gives a strong advantage to the industry, which must be quick to change and follow the fashion. In fact, the Santa Croce district produces high quality light leather for shoe uppers and garments and its many family owned firms are quite well suited to meet the variable and expanding needs of the market; on the side of energy conservation, this peculiarity is a hindrance to the optimization of the processes and equipments. The smallness of these plants also shows in the distribution of their thermal installed power (Fig. 3).

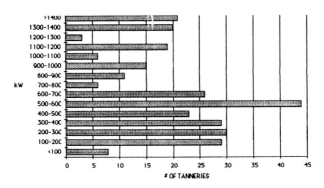

Fig. 3. Distribution of installed thermal power.

The total amount of thermal energy consumed in just one half of the district, belonging to the commune of S.Croce, is estimated in 750,000,000 MJ/year, while electrical energy reaches 42,000 MWh/year.

PROCESSES AND ENERGY SUPPLY

A short description of the main stages of the three principal processes of tanning is necessary for a better understanding of the various inlets of electrical and thermal energy (Fig. 4).

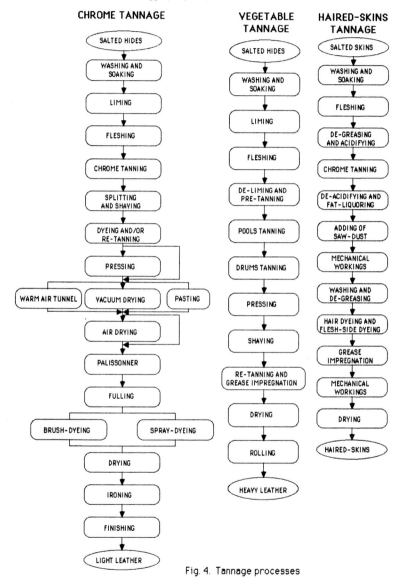

Fig. 4. Tannage processes

Chrome Tannage

Chrome tannage is the process used to obtain light leather for shoe uppers, fancy goods, gloves and garments. The raw material, represented by cattle hides, calf skins, goat and kid skins, sheep or lamb skins, is preserved in refrigerators, unless it is first salted and dried.

The tanning starts with a thorough washing and soaking of the hides, which are placed in revolving drums where an arrangement of pegs agitates the hides in a continuos flow of plain water (16-25°C) for about 20-24 hours. The actual time of rotation is 6-8 hours because of the pauses between cycles of alternating rotations. This phase only requires mechanical energy, obtained by electrical AC motors rated from 4 to 50 kW. There follows a liming phase which is done in similar drums at 26-30°C and lasts from 15 to 30 hours. The hides are then rolled between bladed cylinders to remove the flesh. AC motors used in these machines rate from 7 to 60 kW. The actual chrome tanning is done in smaller revolving drums with warmer water (35-40°C). The active part of this process lasts 2-4 hours and results in the so called "wet-blue" hides, which can be marketed as such. Afterwards the hide or skin is splitted in two parts, namely grain and crust; grain is the most valuable part of the hide, and it goes into fine leather goods. Crusts are treated to make low cost leathers, and have a market of their own. On both products shaving is then performed to obtain a uniform thickness. Splitting and shaving machines involve only electrical power (7-70 kW). Chemical treatments for dyeing and fat-liquoring are again performed in revolving drums with water at 60-70°C for 1.5 to 3 hours.

The product must then be dried, which is usually done in one of the following ways:
a) The wet-blue hide is spread onto a heated plate (60-80°C) under a vacuum bell, which enhances the temperature-dependent evaporation and hastens the initial drying, next completed in the open air.
b) The hides are nailed to frames, hung to transport rails and slowly driven through forced hot air ovens.
c) The hides are pasted to plates and driven through high temperature (60-70°C) tunnels.
Vacuum drying uses the least of thermal energy and little electricity; however it requires more complex machines and, by its nature, is a non-continuous process. Forced-air drying and pasting, on the contrary, may be run as continuous processes; both use a considerably greater amount of thermal power, and air forcing also needs a lot of electricity for the fans; however pasting is not suitable for the production of fine leathers. The next phase in the process is the pounding (palissonnage) which confers softness and pliability to the stiff dried hides, by passing them on a conveyor belt under the blows of a set of hammers. A variety of other mechanical operations may be done at this stage to obtain the requested characteristics, especially regarding the surface texture and appearance; they are mostly performed in roller machines or pressing plates, sometimes electrically heated. Fulling is done in small iron revolving drums, and is merely mechanical.

Application of covering mixtures, in order to obtain the right hue and surface smoothness, is done by means of automatic sprayers or brushes: a conveyor belt carries the hides through alternate spraying or brushing booths and drying tunnels. A great amount of thermal energy is used for drying and fixing the dyes over the hides. The heat is obtained by steam-air heat exchangers, or by infra-red radiating gas panels. Some electrical power is needed for moving the belt and spraying the pigments.

Vegetable Tannage

Vegetable tanning is performed to obtain heavy leather for soles, bags and

belts. The process starts with a washing and soaking phase to remove the
preserving salts. A liming is performed in revolving drums for about 24
hours, followed by fleshing and by deliming in water at 30-35°C in drums
revolving for 4-8 hours, occasionally for 24 hours. After an optional pre-
tanning lasting 3-4 hours in the same drums, the real tannage begins; the
hides, laid on rods, are submerged in vats or pools and periodically moved
into higher tannin concentrations; this stage lasts from 20 to 30 days. It
is possible to accomplish a fast vegetable tanning in revolving drums at 35-
35°C for 4-8 hours, but the quality of the leather is worse. After tanning,
hydraulic presses (10-20 kW) remove water and allow the recovery of tannin.
Shaving machines are used to uniform the thickness of the leather which then
undergoes a dry re-tanning in drums for 4 hours and a grease impregnation
for an hour. After roller pressing the leather is dried by passing in warm
(25-30°C) forced-air tunnels in about 24 hours, or in heated (35-40°C) lofts
for 7-15 days. The dried hides are then ironed in hydraulic presses and next
worked on rolling or hammering machines before being stocked.

Haired-Skins Tannage

This kind of tannage is milder than vegetable or chrome tannage, because the
hair is delicate. This is why liming is omitted and all the phases in
revolving drums are performed at a slower speed, in smaller drums or
agitating vats. Washing and soaking lasts 24 hours in almost steady drums.
Afterward grease is removed and the skins are acidified in other drums where
the chrome tannage takes place for 24 hours (5-6 hours only of actual
rotation). Solutions for neutralization (35-40°C for 1h 30') and fat-
liquoring (65°C for 2 h) are used in drums in the next step. Saw-dust is
added in fast revolving drums in order to soften the hair for as long as 24
hours, and then shaving and grinding are done in machines similar to those
used for leather tannage. At this point of the process the skins and the
hair are greasy and somewhat dirty, and so they need to be washed in
machines using solutions of water and perchloride (1-2 hours). Used perchloride is recovered by steam distillation at 160°C. Hair dyeing and flesh-side
dyeing are performed in drums for 2 hours at about 75°C, followed by grease
impregnation at 60-65°C. Afterwards the skins are hung from horizontal rods
and placed in lofts, where they dry in warm-air (35-40°C) for 15-24 hours.
Various mechanical operations may be done in order to recover softness
before stocking.

ENERGY EXPENDITURE PATTERNS AND WASTE

In order to figure the main patterns of energy input into the processes of
tannage, after a search for relevant information in the literature, we ran a
survey in the field; to this end we selected 15 tanneries to represent as
closely as feasible the whole range af the about 1,000 present in the
Tuscany leather area. We thank their managers for letting us gather informa-
tions on installed power ratings and for actully measure the power demanded
by single machines or by entire working processes. We also were granted
access to records of electricity, water and fuel consumption.

Some of our findings are summarized in the Tables 1 to 5; as expected we
found that the amount of both thermal and electrical energy involved in
every stage of the process has a great range of variation, partly dependent
on the different quality requirements for the leather in production. Some
treatments may be done at lower temperatures for longer periods in order to
obtain a better quality.

TABLE 1. Electrical Energy in Drumming

OPERATION	AC MOTORS POWER kW	ACTUAL AVER. OPERATING PW kW	ACTUAL AVG. REVOLVING TIME hours	AVG.ENERGY per PRODUCT UNIT kJ/ft²
WASHING, SOAKING AND LIMING	4-30	1.7-13.8	3	4.7
PRE-TANNING	9-56	4.6-28.8	2	10.1
CHROME-TANNING	9-56	4.0-25.2	8	34.9
VEGETABLE TANNING IN DRUMS	9-30	4.7-15.6	24	357.8
NEUTRALIZATION DYEING-GREASING RE-TANNING	9-45	5.4-26.8	2	9.7

TABLE 2. Thermal Energy in Drumming

OPERATION	NEED OF WATER liters/Kg	TEMPERATURE OF WATER °C	THERMAL THEORIC EQUIV. PER UNIT OF FINISHED LEATHER kJ/Kg
VARIOUS TREATMENTS WITH COLD WATER	35-70	TAP WATER TEMPERATURE = 12°C	0
WASHING, SOAKING AND LIMING	25-50	12-25	0-2720
DE-LIMING, TANNING NEUTRALIZATION DE-GREASING	35-70	30-40	2630-8200
FAT-LIQUORING DYEING RE-TANNING	5-10	50-60	788-2015

Data concerning electrical power of wet treatments in drums are listed in Table 1. The rated power of the AC motors is substantially higher than the measured power during operation, owing to their efficiency and their loading factor. The average energy spent in each phase per unit of product is based on the actual revolving time of the drums and the average load of hides or skins. The amounts of water and the corresponding temperature levels needed for each phase are listed in Table 2. Heat requirements assumed a basal temperature of 12°C, the weighted average of the water supply. The stated energy requirements do not consider the efficiency of the boilers nor the losses in the piping network and they are related to the unit of dried final product weight; we will consider the losses later, because they are usually due to the same factors.

TABLE 3. Mechanical Wet Operations

MACHINES	LEATHER	AC MOTORS POWER kW	ACTUAL AVG. OPERATING PW kW	AVG.ENERGY per PRODUCT UNIT kJ/ft²
FLESHING MACH.	HEAVY	30-60	17.6-35.3	31.6
FLESHING MACH.	LIGHT	7.5-22	5.3-15.9	15.1
SPLITTERS	AFTER LIMING	7.5-19	6.1-15.3	11.2
SPLITTERS	WET-BLUE	11-26	7.8-18.3	11.9
SHAVERS	HEAVY	30-75	17.4-43.1	39.6
SHAVERS	LIGHT	7.5-19	5.4-13.4	13.6
CONTINUOUS PRESSES	HEAVY	7.5-19	7.1-17.6	22.3
CONTINUOUS PRESSES	LIGHT	3.8-11	3.7-11.0	8.28
CONTINUOUS PRESSES +HEATED CYLINDERS	HEAVY	18.5-34	16.4-41	51.5
CONTINUOUS PRESSES +HEATED CYLINDERS	LIGHT	18.5-34	15.5-34.8	45.0

TABLE 4. Mechanical Dry Operations

MACHINES	AC MOTORS POWER kW	ACTUAL AVG. OPERATING P. kW	AVG.ENERGY per PRODUCT UNIT kJ/ft²
PALISSON	3.7-15	3.1-12.4	18.7
GRINDERS	3.7-45	3.4-41.4	44.64
CYLINDER IRONERS	11.2-37.5	10.3-34.5	32.4
PRESSES	11.2-45	10.3-40.9	61.0
SHAVERS	7.5-15	6.2-9.3	13.32
SPLITTERS	7.5-15	6.1-12.2	10.8
ROLLING MACHINES	4.5-6	3.8-5.1	8.6

Table 3 summarizes our findings on the mechanical operations on wet hides and skins. Here we had to consider separately two weight classes of leather, because of the great difference of powers involved. Table 4 collects homologous data on dry hides and skins.

The most widely used drying systems have been listed in Table 5, along with their estimated use of thermal and electrical energy; two extreme courses are contrasted: a strong preliminary pressing (to remove most of the water) versus a lighter pressing. The former uses less electricity and less heat than the latter, although the pressing is much more consuming.

Gas-panels and steam-air heat exchangers for drying the leather in automatic sprayers or brushers use similar amounts of thermal energy, although the former obtain the heat directly by fuel at almost 100% of efficiency. The power requested for drying 3,000 square feet per hour ranges from 100 to 250 kW and produces 70-90°C air. Electricity used for driving the conveyor belt, the gears and the compressors of the spraying guns, accounts for 10-20 kW, while power needed for the exhaust fans of the dyeing booths reaches 40-50 kW. The first process achieves a better quality for the finished leather, while the second one is quicker and requires shorter tunnels.

TABLE 5. Energy in Drying Process for 100 g of Finished Leather

DRYING SYSTEMS	AFTER STRONG PRESSING			AFTER LIGHT PRESSING		
	kJel	kJth	kJtot	kJel	kJth	kJtot
NAILING + WARM AIR	63.7	504.8	568.5	100.8	893.6	994.4
NAILING + DRIED AIR	190.8	116.0	306.8	355.0	116.0	471.0
VACUUM + NAILING	85.3	396.8	482.1	144.4	677.6	822.0
PASTING.	136.1	678.1	814.2	171.4	853.9	1025.3
VACUUM + ROOM TEMPERATURE AIR	139.3	10.0	149.3	188.6	10.0	198.6
VACUUM + WARM AIR	154.1	145.7	296.8	205.0	145.7	350.7
VACUUM + HIGH FREQ.	295.9	-	295.9	345.2	-	345.2

Similar data pertaining to a real process may be linked together and be graphically presented as in the flow-diagram of Fig. 5, which figures a process of chrome tannage as it is typically done in the district observed. The thickness of the arrows is proportional to the amount of energy used in each phase of the process for a unit of product; note that the scale is different between thermal (left side) and electrical energy (right side).The latter is mostly used in AC motors; its losses have been disregarded as non significant for our purposes. On the contrary, it is very important to compare the thermal requirements of each operation with what is actually spent. A rigorous thermodynamic analysis should be done on quantities like availability or exergy; however we preferred to deal with the heat quantities or enthalpies and the temperature of the supply. In so doing we emphasized the losses of thermal energy, starting from the burners and the boilers, through the piping to the heat-exchangers and, last but not least, passing the steam traps to reach again the boiler.

In our survey we found that most of the tanneries use steam for every heat requirement. For instance they obtain water at 30-40°C by mixing hot water (90-95°C) with cold water drawn from the wells. Water is warmed in boilers, also serving as storage tanks, by condensing steam. Similarly warm air is produced by steam-air heat exchangers. We measured an average efficiency for the burners and boilers of 85% (gas burners) and 80% (oil burners). We also found that most of the steam pipes are not insulated and the steam traps are not properly mantained, so they release vapours. We estimated such losses to average 15%, which lowers the overall efficiency to 70% or less.

The flow diagram clearly shows which phases use and waste more thermal energy: the primary drying, which requires low temperature air, and liming and tanning, which use water at 25-40°C. These are the phases where updating would be most rewarding.

66

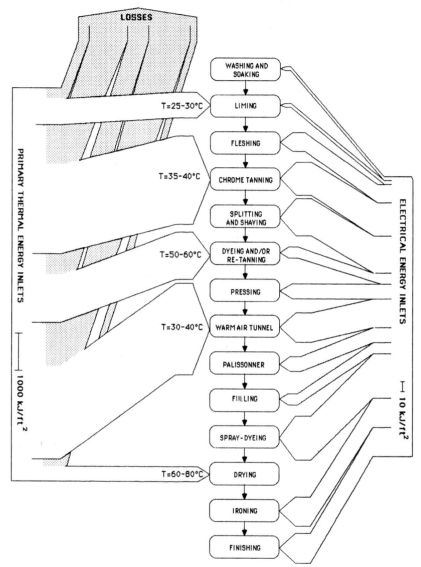

Fig. 5. Chrome tannage energy inlets

TUNING AND RATIONALIZATION OF THE EXISTING INSTALLATIONS

Some attention has been given by the manufacturers to increase the
mechanical efficiency of their machines (updated 'Y' sectors revolving

drums). Global thermal efficiency is still taken into little account.
A first step in energy saving could be made by a better control and maintenance of the existing equipment: the provision of automatic controllers for
burners and boilers and the use of natural gas to lower stack temperature,
may raise the combustion efficiency to 90-93%. Insulation of hot pipes and
prevention of vapour losses, by mantainance of the steam traps and pressuri-
zation of the hot well, may afford a further 10% increase in global
efficiency. Altogether, such simple measures might reduce by 10-15% the use
of fuel. This saving is consistent with the related expenses, as such
investments often return in less than one year.

IMPLEMENTATION OF NEW TECHNOLOGIES FOR HEAT AND ELECTRICITY PRODUCTION

Greater savings can be made by implementing well known technologies for low
enthalpy air and steam production. It is possible, without affecting the
operating characteristic of the various phases, to completely dispense with
the steam by directly heating either water or air by means of gas burners.
This reduces the irreversible processes like mixing waters with different
temperature, or using high LMTD heat exchangers or boilers.
The great amount of electrical energy used in tanneries and the fact that
heat requiring steps involve moderate temperatures suggest a close appraisal
of cogeneration. We will see later that use of natural gas is essential for
such updatings to the plants and machines.

Updating the Automatic Sprayers

Tanners of the district of S.Croce produce high quality leather, and thus
they prefer steam exchangers rather than infra-red gas panels, even if these
might be more economical; they do use them for pursuing high productivity.
The introduction of direct burning of natural gas in the drying booths is
not in conflict with quality, because both temperature and humidity of the
drying air can be easily modulated. A variety of safe burners ranging from
0 to 300 kW are available for installation in the existing machines in
place of the present steam-air exchangers. The overall efficiency of the
drying process can increase from the present 70% to almost 100%. For a
100mm automatic sprayer rating 240 kW one may use 6 burners of 40kW each,
which cost less than the equivalent infra-red gas-panels; 640,000 MJ per
year might thus be saved.

Any saving is counterbalanced by the loss in production occurring during the
substitution; given that global energy costs amount to a bare 5-8% of the
added value, a short standstill in production has a greater impact than a
large energy saving, specially now that gas is cheap. These considerations
do not apply to new installations, which certainly should be done as
suggested.

Use of Submerged Flame Boilers

From Fig. 5 we can see the importance of the amount of energy used in the
wet workings. The required temperature is quite low, ranging from 20 to
40°C, but all the heat is obtained by much richer fluid, like 120°C
condensing steam. As stated before this leads to irreversibility, which is
paid with a greater use of fuel. Up-to-date technologies heat water at 30-
40°C by means of submerged burners in water tanks. These equipments are well

68

tested and easily applicable to existing boilers: they can reach an overall efficiency greater than 105% at temperatures lower than 50°C. This performance is due to the condensation of the water produced by the combustion reaction; such heat is not included in the low heat value of the gas, which is used in all the thermodynamic calculations. For the process depicted, the saving on thermal energy expenditure for the wet operations can be cut down by as much as 18-22%.

Direct Burners in Drying Rooms or Tunnels

Similar cosiderations can be made about drying rooms and tunnels with more substantial reasons: the use of 120°C condensing steam to obtain 30-40°C air leads to irreversible exchange of heat. More correct use of thermal energy can be achieved using direct burners on the same principles described. Temperature and humidity level of the air can be easily regulated with automatic controllers provided with the burners. The benefit on energy conservation can be estimated in 25-30% fuel saving.

Heat and Electricity Cogeneration

The different processes of tanning involve both electrical and low temperature heat demand. However steam and gas turbines are out of question in our case, because of the small size of the plants considered (200-700 kW); we may only contemplate the use of 4-stroke gas engines, equipped with heat exchangers on the exhaust flue and on cooling and oiling circuits. As the ratio of electrical to thermal power is small, we must aim to satisfy the whole demand of electricity, and integrate heat production with conventional means.
A correct analysis of the problem starts with the study of the patterns of both thermal and electrical energy uptake from the public utilities Electricity and gas usage have been logged and plotted (Figs. 6 and 7) during two-weeks production relevant periods in several tanneries. The plots show a rather typical time course, with two daily peaks due to the discontinuous nature of several processes; moreover, heat and electricity peaks do not overlap, making it necessary to provide heat accumulators and auxiliary burners.

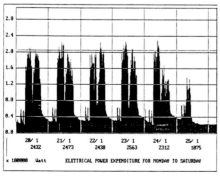
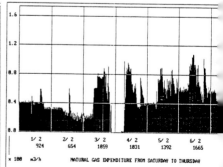

Fig. 6 Electricity expenditure.   Fig. 7 Natural gas expenditure.

he plant design for the cogenerating system applied to an actual haired-
kins tannery, is shown in Fig. 8.
cooling water and lubricating oil of the engine exchange heat with the water
o be used in wet processes, which is required at low level temperatures.
xhaust gas from the engine is mixed with fresh air and blown into the
rying rooms or tunnels along with air warmed by auxiliary direct burners.
hen dryers are not in use, the exhaust is used in to heat and store more
arm water. A 250 kW engine would cover the peak requirements, but would run
t low efficiency for most of the time. We therefore suggest the use of a
70 kW engine, with an integrating supply from the public utility to meet
he peak demand.

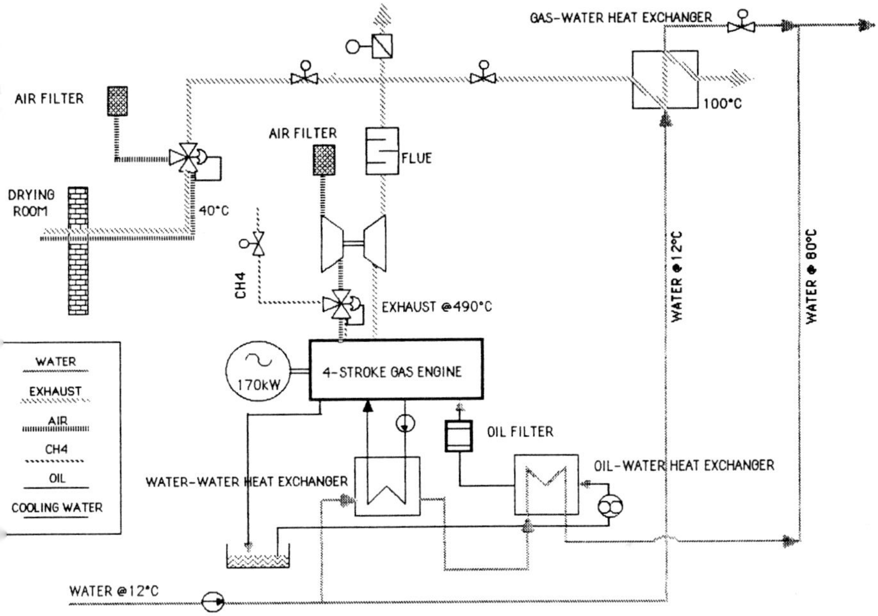

Fig. 8. Cogeneration system design

financial amortization plan was figured for the suggested design and it
ielded a return period of 3 years, strictly dependent on current local
tes for public utilities and for credit. This is a rather encouraging
sult; moreover, present Italian legislation is promoting energy
nservation by providing special credits at low rates for such investments;
is reduces return periods by a half, making these installations very
tractive.

timizing Pattern of Energy Uptake

sides the improvements of energy exploitation, reduction of losses, chan-
s in equipments and plants, it is possible to reap further benefits
rough the optimization of the patterns of energy uptake. The peculiar
ight given to the maximum contractual power in determining the rates of
ectrical public supply suggests the strategy of flattening the power
rve. The causes of the two daily peaks (see Fig. 6) are largely avoidable,
ing it possible to shift the timing of some operations. This policy might

achieve a 30% cut of the electricity bill and, on a larger scale, it would benefit the whole public utility system.

CONCLUSIONS

We have seen that it is very easy and economical to improve thermal energy efficiency by 10-15%, just by tuning and rationalizing existing equipments. This would result in a 150,000,000 MJ/year saving for the district of S.Croce. Greater savings could be made by implementing new technologies; the cost of their introduction in present factories may be high with respect to the benefits. In future plants such technologies must be widely used to improve the overall efficiency and to reduce air and water pollution. Especially cogeneration can achieve great efficiency because of the low temperature demanded in the tanning process.

REFERENCES

De Simone, G. (1977). La macerazione delle pelli. Cuoio Pelli Materie Concianti, 6.
De Simone, G. (1979). Preparazione delle pelli alla concia: Il rinverdimento. Cuoio Pelli Materie Concianti, 1.
De Simone, G. (1979). La concia delle pelli. Cuoio Pelli Materie Concianti, 3.
De Simone, G. (1980). Il piclaggio delle pelli. Cuoio Pelli Materie Concianti, 5.
Nardini, G. (1981). Studio di fattibilita' di un impianto di produzione di calore ed energia elettrica comprensoriale nella zona di Santa Croce sull'Arno. Unpublished.
Ramondetti, A. (1980). La rifinizione delle pelli. Cuoio Pelli Materie Concianti, 5.
Simoncini, A. (1983). Analisi dei consumi energetici nell'Industria Conciaria. Cuoio Pelli Materie Concianti, 4.
Simoncini, A. (1984). Risparmio di risorse nell'Industria Conciaria: Possibilita' di applicazione della cogenerazione. Cuoio Pelli Materie Concianti, 4.
Stratta, R. (1975). Corso di macchine e impianti per conceria. Levrotto Bella, Torino.

The Promotion of Energy Efficient Clothing

Keith Cornwell

Energy Technology Unit
Department of Mechanical Engineering,
Heriot-Watt University, Edinburgh EH14 4AS

ABSTRACT

A considerable saving of energy could result from the acceptance of slightly warmer clothing and the consequent lowering of indoor comfort temperature. The general trend towards higher indoor temperatures, and differences in heat energy use for work and domestic premises are examined and the level of saving is estimated.

Heat flow from the sedentary human body through the clothing is analysed and a new factor, the Garment Temperature Decrease (GTD) is proposed. The value of GTD (in $^{\circ}$C) represents the incremental drop in ambient temperature that the garment allows. The comfort temperature for a clothing ensemble is found by subtracting the sum of GTD values for all the garments from the nude base level of 27.6°C. Clothing styles and textiles are examined using this factor and some guidance towards producing more thermally effective, but still fashionable clothing, is given.

It is shown experimentally that Infra-red Thermal Scanning may be used to obtain GTD values. This could allow easy thermal labelling of garments as a first step towards promotion of general awareness about the warmth of clothing and energy efficiency.

KEYWORDS

Thermal comfort; comfort temperature; clothing, thermal properties; clo-values; thermal conductivity, textiles.

1. INTRODUCTION

Man evolved in hot climates and initially needed no clothing. His migration to temperate regions led to the need for protection from the environment. From earliest times he has used two layers of insulation, one being a building and the other clothing. This paper is essentially concerned with the division of the total temperature difference between these two layers.

72

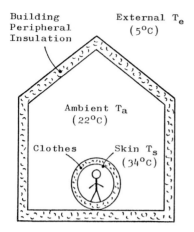

Building
Peripheral
Insulation

External T_e
($5^{\circ}C$)

Ambient T_a
($22^{\circ}C$)

Clothes

Skin T_s
($34^{\circ}C$)

Fig.1 The Two Insulation Layers.

Consider modern man thermally insulated
from the environment at $5^{\circ}C$ as shown
in Fig. 1. The heat energy consumed
is essentially a function of the intermediate temperature T_a. If he were to
wear a little more clothing so that T_a
decreased by say $2^{\circ}C$ then the energy
used for heating the building, if proportional to temperature difference,
would be decreased by 12%. This large
energy saving requires no effort apart
from turning down the thermostat, and
the payback period on the extra garment is usually only a month or two.

This potential energy saving is well
appreciated by those [eg. 1,2] involved with textiles and energy conservation. In this paper an attempt is
made to examine scientifically the
extent of saving, the effect on clothing and the way forward in promoting
this saving. It is concerned with
ordinary indoor clothing, not special
thermal garments, and with normal living conditions. Nomenclature is given
as an Appendix, and the term Ambient Temperature refers to the mean indoor
temperature.

2. SPACE HEATING AND COMFORT TEMPERATURE

2.1 Usage Pattern

This section is based entirely on heating in temperate climates. The UK has
a typical usage pattern and Table 1 gives the proportional distribution of
space heating and some other low temperature energy outlets. Figures are
based on [3] with some apportionment estimates by the author. Agricultural
buildings are included with Industrial, and institutional buildings with
Commercial.

Domestic heating accounts for about 60% of the total domestic delivered
energy consumption; about half going to central heating systems and half to
gas, coal and electric fires. Much of the cooking lighting and water heating
load is ultimately used for heating the building. As a conservative estimate
if 50% of the domestic cooking and lighting load and 10% of the water heating
is added to the space heating, the total energy used for heating human beings
is 30% of primary energy. Since 1980 the energy scene in the UK has remained
fairly static [4].

Reassessment on a per capita basis yields some interesting differences between work and home. In Table 2 the time ratio for Domestic and Total
Heating is based on a mean of 16 hours/day for 8 months/year while for
Commercial and Industrial it is based on 8 hours/day for 160 heating days
per year. This approximate assessment indicates that the energy per hour
used to heat a person at work is about 5 times that at home. This is presumably due to the large extent of factory and office space which is heated,

TABLE 1 Space Heating, Water Heating and Cooking Proportions of Total Energy Usage. (UK, 1980).

SECTOR	PRIMARY ENERGY %				DELIVERED ENERGY %				
	TOTAL	Space Heating	Water Heating	Cooking	Others*	TOTAL	Space Heating	Water Heating	Cooking	Others*
Domestic	30.0	14.6	6.3	2.7	6.4	28.0	16.7	6.2	2.4	2.7
Commercial	14.3	6.7	1.4	1.3	4.9	11.8	7.5	1.3	1.0	2.0
Industrial	36.1	3.7	1.3	0.6	?	35.2	4.8	1.6	0.8	?
Transport	19.6	-	-	-	-	25.0	-	-	-	-
Total	100	25.0	9.0	4.6	-	100	29.0	9.1	4.2	-

* Lighting and Appliances

TABLE 2 Space Heating Energy per Person (UK, 1980)

Sector	Primary Energy for Space Heating	Energy per Person	Total Heating Time Ratio	Heating Load per Person
Domestic	1233 PJ/yr	21.9* GJ/yr	45%	1.5 kW
Commercial	569	} 33.0°	15%	7.0 kW
Industrial	316			
Total	2118	37.6*	45%	2.65 kW

* Based on Total Population. O Based on Working Population

$PJ = 10^{15}$ J, $GJ = 10^{9}$ J

but unoccupied, and the higher ambient temperatures at work. The advantages to be gained by wearing slightly more clothing are therefore particularly relevant at work.

2.2 Comfort Temperature Trend

There has been a gradual rise in accepted ambient comfort temperature and a consequent decrease in clothing level over the last 50 years. Some corroborative evidence is given in Fig. 2. In the USA accepted temperature rose to a

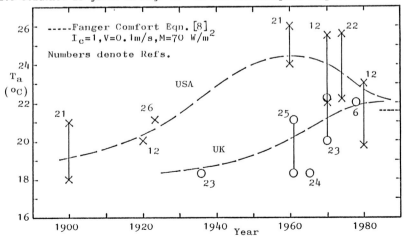

Fig.2 Variation of accepted Comfort Temperature T_a with Time.

high level in the 1960's when very low-cost energy was available while in Europe there has been a gradual rise. If the generally accepted level in offices of about 22°C [6] could be reduced to say, 20 to 21°C, the savings would be considerable.

2.3 Potential Savings from Ambient Temperature Reduction

Estimation from the average Degree Days for the UK over the period 1960 to 1980 [7] yields a mean effective external temperature of 8.75°C. This is the time-averaged value for the period when the heating is turned on. A mean of 2.8°C between indoor and outdoor temperature is covered by solar, human and other effects, leaving a mean ΔT of 9.45°C to be supplied by space heating when the indoor temperature is 21°C. If it is assumed (conservatively) that lineality exists between ΔT and energy use, this yields a saving in space heating of 10.6% per $^\circ$C temperature drop. From Section 2.1 this gives a primary energy saving of 3.2% per $^\circ$C. Compared to other schemes, this is an enormous potential saving at minimum capital cost (extra clothing) and effort (lowering the thermostat).

Expressed another way, a company with say 1000 employees paying a typical £6/GJ ($\sim$8.4 ECU/GJ) for its heat energy would, from Table 2, pay £198,000 ($\sim$280,000 ECU) per year to heat its staff. A reduction of 2°C would save 21.2% of the space heating cost, thus saving £41,600 ($\sim$60,000 ECU) per year.

3. COMFORT TEMPERATURE AND CLOTHING

3.1 Heat Flow through Clothing

Heat flow aspects of the human body are covered by Fanger [8] ASHRAE [9] and others and a concise account of the basics are given in Cornwell [10]. Essentially a thermal balance on the body includes the metabolic rate M, the physical work W, the heat flow Q_m due to mass exchange such as sweating and respiration and dry heat transfer Q_s from the skin through any clothing to the ambient surroundings. Thus:

$$M - W - Q_m - Q_s = 0$$

In this study we are concerned with Q_s, the dry heat transfer from the skin as it is this quantity which is most affected by the clothing. Evaporation loss through the clothing is low under the sedentary conditions in which humans spend most of their time indoors. It is assumed that there are no air layers between the clothing and it can be simplified to the situation shown in Fig. 3. The effects of evaporation, air layers and other factors are covered in the reviews of Slater [11] and Vigo & Hassenboehler [1].

Conduction and Convection equations for heat flow from the skin based on skin area A_s are:

$$q_s = Q_s/A_s \tag{1}$$

$$q_s = \left[\frac{k\,A_c}{t\,A_s}\right](T_s - T_c) \tag{2}$$

$$q_s = h_a \frac{A_c}{A_s} (T_c - T_a) \qquad (3)$$

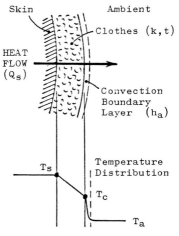

Fig.3 Clothing Heat Flow.

It is normal practice in clothing studies to use units of Clo or Tog where

$$1 \text{ Clo} = 0.155 \text{ m}^2 \text{K/W}$$

$$1 \text{ Tog} = 0.100 \text{ m}^2 \text{K/W}$$

and to rewrite (2) and (3) as

$$q_s = \frac{(T_s - T_c)}{0.155 \, I_c} \qquad (4)$$

$$q_s = \frac{f_c (T_c - T_a)}{0.155 \, I_a} \qquad (5)$$

Here I_c is the clothing resistance (in Clo), I_a is the air layer resistance (in Clo) and f_c is the area factor for the surface of the clothing, i.e. the ratio of outer clothing area to skin area.

It should be noted that these equations are not precise, but describe a model of the clothing in which the parameters represent average or effective values. Furthermore the terms within the square brackets of equation (2) are lumped together and equated to $1/I_c$ (here A_c is strictly the mean rather than the surface area of clothing) while in equation (5) the area ratio is retained and $h_a = 1/I_a$.

Equations (4) and (5) may be combined to give

$$q_s = \frac{(T_s - T_a)}{0.155 \, (I_c + I_a/f_c)} \qquad (6)$$

3.2 Sedentary Comfort Temperature

Equation (6) can be used to relate the ambient temperature to the clothing level. Sedentary conditions only are considered and are taken to mean seated and reading, at a keyboard or doing light office work. A large range of experimental work is summarised in ASHRAE [9] by the following relationships:

$$T_s = 25.8 + 0.267 \, T_a$$

for $0.6 < I_c < 1.0$

$$f_c = 1.0 + 0.2 I_c$$

$$h_a = h_c + h_r = 3.1 + 4.7 = 7.8 \text{ W/m}^2\text{K}$$

where h_c is the convective component at typical indoor air velocities of 0.1 to 0.15 m/s and h_r is the radiative component. Substitution into (6) yields T_a as a function of I_c and q_s.

The value of q_s can be found experimentally. Table 3 gives values from the data of ASHRAE [12] and the Fanger [8] comfort equation and the small variat gives some vindication of this approach.

TABLE 3 Estimation of q_s

I_c (Clo)	Metabolic Rate M (W/m^2)	T_a ($^\circ$C)	Ref.	q_s (W/m^2)
0.5	69.8(1.2Met)	24.4	12	40.5
1.0	"	21.2	"	39.2
1.5	"	18.2	"	41.6
0.5	69.7	24.6	8	40.0
1.0	"	21.4	"	38.5
1.5	"	18.4	"	41.1
			Mean	40.1 ± 1.6

Equation (6) then becomes on rearrangement:

$$35.2 - T_a = 8.48 \left[I_c + \frac{0.83}{\left(1 + 0.2 I_c \right)} \right] \tag{7}$$

3.3 Garment Temperature Decrease (GTD)

The analysis so far applies to complete ensembles of clothes. Individual garment resistances I_g do not add up to the ensemble value I_c owing to garme overlap and other factors. Experimental analysis [13] indicates that, as a average for men and women, for $0.4 < I_c < 1.2$,

$$I_c = 0.75 \sum I_g + 0.08 \tag{8}$$

The incremental effect of a garment on T_a can be found by substituting (8) into (7) and differentiating:

$$\frac{dT_a}{dI_g} = -6.36 + \left[\frac{1.056}{(1.016 + 0.15 \sum I_g)^2} \right]$$

$$= -5.53 \pm 2\% \text{ (for } 0.4 < I_c < 1.2)$$

If this tolerance is deemed acceptable we can write for an <u>individual garme</u>

$$\Delta T_{ag} = -5.53 I_g$$

where ΔT_{ag} is the difference in comfort temperature between wearing and not wearing the garment. Now if

$$\sum \Delta T_{ag} = -5.53 \sum I_g \qquad (9)$$

and, if T_{ao} is defined as the effective base comfort temperature with no garments, then

$$\sum \Delta T_{ag} = T_a - T_{ao}$$

and at $T_a - T_{ao} = 0$ equations (7), (8) and (9) yield $T_{ao} = 27.6^{\circ}C$. Although this datum is not strictly the nude comfort temperature owing to the restricted ranges of T_s and I_g, it is worth noting the close comparison with the nude value of $27.3^{\circ}C$ from ASHRAE Standard [12].

The value $-\Delta T_{ag}$ is a true indicator of the room temperature drop allowed by a garment and is henceforth termed the Garment Temperature Decrease (GTD).

GTD is the value (in $^{\circ}C$) by which the ambient temperature must be decreased when wearing the garment to give the same thermal comfort to that experienced without the garment under sedentary conditions.

It is shown in Section 4.1 that garment GTD's are typically in the range 0.1 to $3^{\circ}C$ and ensembles total 4 to $10^{\circ}C$. In summary the equations give:

$$T_a = 27.6 - \sum GTD \qquad (10)$$

and in terms of Clo values in the normal range $(0.4 < I_c < 1.2)$

$$GTD = 5.53 I_g \qquad (11)$$

and

$$\sum GTD = 7.37(I_c - 0.08) \qquad (12)$$

3.4 Level of Satisfaction with Comfort Temperature

Analysis leading to GTD is based on the idealisation of a comfort temperature acceptable to all subjects. The internal set point temperature of the body varies between individuals by up to $0.5^{\circ}C$ and over the daily cycle by up to $1^{\circ}C$, [14]. Coupled with this there are variations in metabolic rate and skin temperature which combine to produce considerable differences in comfort value.

Subjective assessment of comfort is usually based on a 7-point scale such as the Bedford Scale:

3	Much too warm	} Dissatisfied
2	Too warm	
1	Comfortably warm	
0	Comfortable	} Satisfied
-1	Comfortably cool	
-2	Too cool	} Dissatisfied
-3	Much too cool	

By analysis of the difference between the heat produced and lost under ambient conditions using the comfort equation and empirical data, Fanger [8] has developed the Predicted Mean Vote(PMV) as a measure of thermal sensation. This and other methods are reviewed by Humphreys [15]. Figure 4 shows the PMV and % dissatisfied on the above scale under the following typical sedentary conditions:

i)   Metabolic Rate = 69.7 W/m^2 (as Table 3)

ii) Air temp. = Mean Radient temp. = T_a

iii) Relative humidity = 50%

iv) Air velocity $<$ 0.1 m/s

v) I_c $\approx$ 1.0 (i.e. $\sum$ GTD $\approx$ 6.8 $^{\circ}$C)

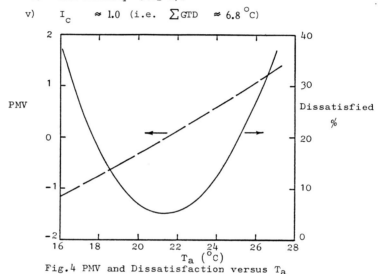

Fig.4 PMV and Dissatisfaction versus T_a

The prediction that under optimum conditions, 5% of subjects are dissatisfied has been experimentally confirmed [eg. 6]. It is worth noting, with energy saving in mind, the extension of this range to say 10% allows a drop in T_a of over 2°C.

4.   ASSESSMENT OF CLOTHING ENSEMBLES AND STYLES

4.1 GTD Values for Typical Clothes

Table 4 gives GTD values calculated from the I_g values given in [13] and included in [12]. Some slight adaptions for European clothing styles have been made and some items of declining popularity (such as long underwear) have been omitted. On the right, garment ensembles for various conditions and the resulting $\sum$ GTD and estimated ambient comfort temperatures T_a are given.

Clothing styles are highly individualised and the ensembles are those selected by the author as typical in the UK. Homes and offices have been selected because it is in these buildings that sedentary conditions requiring the highest T_a values occur. Some conclusions regarding this Table and the following Section are covered in Section 4.3.

TABLE 4   GTD Values (in °C) and Estimated Comfort Temperatures.

Garments	GTD Values (°C)	Home/Casual			Office			
		Light	Medium	Medium +	Light	Medium	Medium +	Heavy
WOMENS GARMENTS								
Bra & Panties	0.3	0.3	0.3	0.3	0.3	0.3	0.3	0.3
Slip (Half/Full)	0.7/1.1	-	-	-	-	0.7	0.7	1.1
Camisole (Half Vest)	1.0	-	-	1.0	-	-	-	1.0
Shoes, Stockings, Tights	0.1-0.4	0.1	0.2	0.3	0.3	0.3	0.4	0.4
Blouse	1.1-1.6	1.3	1.3	1.6	1.3	-	1.6	1.6
Skirt	0.6-1.2	-	-	-	0.8	-	1.2	-
Slacks	1.5-2.5	1.7	2.2	2.5	-	-	-	-
Dress	1.2-3.8	-	-	-	-	2.0	-	3.0
Jacket	0.9-2.1	-	-	-	-	}1.5	2.1	2.1
Woollen Sweater/Cardigan	1.0-2.1	-	1.5	2.1	-			
∑GTD		3.4	5.5	7.8	2.7	4.8	6.3	9.5
AMBIENT T_a (°C)		24.2	22.1	19.8	24.9	22.8	21.3	18.1
MENS GARMENTS								
Briefs	0.3	0.3	0.3	0.3	0.3	0.3	0.3	0.3
Vest (light-medium)	0.6-1.0	-	0.6	1.0	-	-	0.6	1.0
Shoes & Socks	0.3-0.5	0.3	0.3	0.4	0.3	0.3	0.3	0.4
Trousers	1.5-2.1	1.7	1.7	2.1	1.7	1.7	2.1	2.1
Shirt (sleeved)	1.2-1.6	-	1.2	1.6	1.2	1.2	1.2	1.6
T-shirt	0.7	0.7	-	-	-	-	-	-
Waistcoat	1.0	-	-	-	-	-	1.0	-
Jacket	1.2-2.7	-	-	-	-	2.0	2.0	2.7
Woollen Sweater	1.0-2.1	-	1.5	2.1	-	-	-	1.2
∑GTD		3.0	5.6	7.5	3.5	5.5	7.5	9.3
AMBIENT T_a (°C)		24.6	22.0	20.1	24.1	22.1	20.1	18.3

4.2   The Effect of Textiles on GTD

Measurements of heat flow through clothing textiles under a wide range of conditions are reviewed in [1, 11]. Important factors include thermal conductivity, air permeability, moisture retention, radiant transfer and the surface heat transfer coefficients. However, under our sendentary conditions, with low evaporation, sweating and air movement, the thermal resistance is primarily due to the dry heat transfer through the material itself.

A number of researchers have shown that this dry heat transfer resistance appears to be primarily dependent on the thickness t of the material and fairly insensitive to the particular fibre and textile structure. This is not surprising since the heat transfer is made up of components due to conduction and radiation through the air pockets and conduction through the fibres. As the volumetric content of textiles is mainly the air pockets, the scope for variation is therefore limited. The data on mean effective conductivity k_m of woven fabrics has been found to yield [2],

$$k_m = \frac{t}{\Delta T} \frac{Q}{A_c} \approx 0.0433 \quad \text{W/mK} \qquad (13)$$

If the heat flow through the material Q is equated to Q_s and the temperature difference ΔT to $(T_s - T_c)$, then use of equations (1) and (4) and substitution of f_c yields

$$t_m = 0.00671 \, I_c \, (1 + 0.2 \, I_c) \qquad (14)$$

Here t_m is the mean total thickness of woven fabric with k_m as given in equation (13). From equations (10) and (12)

$$I_c = \frac{27.6 - T_a}{7.36} + 0.08 \qquad (15)$$

Equation (14) and (15) combine to give $t_m = f(T_a)$ as shown in Fig. 5.

The high proportion of air pockets in the fabric also means that the effective conductivity is not strongly dependent on bulk density ϱ within the density range of normal clothing. Outside this range there is an increase in conductivity, both for low densities (due to radiation) and high densities (due to conduction through the fibres). O'Callghan and Probert [2] give $k = f(\varrho)$ and find minimum k at $\varrho = 90$ Kg/m^2. Use of their results and equations (2), (4) and (15) allows plotting of $\varrho = f(T_a)$ at given values of t as shown in Fig. 5

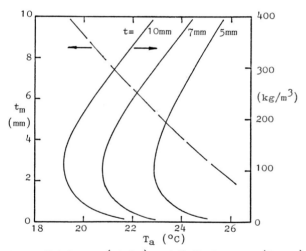

Fig.5 Mean Thickness (at k_m) and Bulk Density (Woven)

4.3 Discussion of Results and Clothing Style

Clothing not only fulfills the need for warmth, but also for image and style. Any advance to slightly warmer garments must be consistent with current fashions. In this Section some conclusions are drawn from the earlier results and the interaction with style is discussed. The following points can be made from Table 4:

i) The wide range of GTD values for many garments indicates that a considerable increase in Σ GTD can be achieved with no change of style by wearing similar garments of greater thermal resistance.

ii) 'Light' clothing is the minimum acceptable clothed level and yields ambients of 24°C, which is the normal setting for initiation of the cooling cycle of air conditioning systems.

iii) For casual clothing (worn in the home throughout the year, except in the summer) a decrease in T_a can be brought about in two ways. One is by increasing GTD values (essentially increasing thickness) and the other is by wearing extra underwear on the torso. This could be a short vest, under T-shirt, camisole or slip depending on sex and preference. So-called Thermal Wear is not necessary in surroundings above 19-20°C.

iv) Modern offices staffed by a mix of men and women pose a comfort problem. Men can adjust to the range 18-22°C very readily with only minor adjustment to style. (For example an underwear/shirt/trousers/jacket kit suitable for an office at 22°C could be increased by a light vest and thicker trousers for 21°C, by a thin woollen waistcoat for 20°C, by a thicker shirt and medium vest for 19°C and by a thicker jacket for 18°C). Women in offices have a basic kit (of say underwear/ half-slip/ skirt & blouse or light dress) which yields 24°C and a cardigan or jacket gives 22°C. Further decrease is only possible by wearing a camisole or trousers in place of a skirt. If stylish and fairly thick woollen wear were developed and promoted, then office temperatures could decrease down to the lower ranges possible for men. (This assumes $M = 70$ W/m^2; higher activity will decrease these requirements).

The conclusions regarding textiles are that fibre material and bulk density have a limited effect on thermal resistance and the thickness is the predominant factor. There is scope for development of chunky but light clothing with high GTD values. Scuffed or brushed surfaces effectively increase the thickness and are worth investigating. The total mass of clothing required can be estimated from $\rho\, t_m\, A_s$. For $T_a = 21$°C, $t_m = 7.8$ mm (from Figure 5) and ρ in the range 60 to 120 kg/m^3, the total clothing mass becomes 0.8 to 1.6 Kg. This range is corroborated by the data of Seppanen et al [16]. Many indoor clothes ensembles are at the upper end of this range, again indicating that equally thick but lighter clothing is feasible.

In work premises there are other factors which indirectly affect the clothing level and comfort temperature. Fishman and Pimbert [6] draw attention to the effect of external temperature on indoor clothing style and Langkilde [17] shows that ambients rather less than the comfort temperature can produce slightly faster work rates. Transient effects from seasonal acclimatisation [18] to hourly temperature drifts [19] can also influence conditions. There can be acceptance of completely unsuitable ambient temperatures for short duration exposure. In shops, for example, the low temperature dictated by energy saving requirements and customers in outdoor clothing is usually abandoned in favour of lightly-clad staff who are present for long periods. Light, fashionable clothing of higher GTD values for the staff could alleviate this problem.

The elderly are considered to require higher ambient temperatures while on the other hand they often by choice have home temperatures which are low. The following points are relevant:
i) It has been shown by Collins [20] and Fanger [8] that young and old require the same ambient temperature under the same conditions.
ii) The elderly tend to be more inactive than the mean sedentary level taken in this study. A drop from $M \approx 70$ kW/m (Table 3) to 58 kW/m^2 leads to a rise in T_a of 1.8°C (based on the comfort equation [8] at $I_c = 1.0$, $V = 0.1$ m/s).

iii) Clothing levels tend to be higher among the healthy elderly, partly due to this decrease in M and partly to a preference for heavier styles. This preference may be associated with thermal conditioning from earlier times, (see Fig. 2).

iv) Among the less healthy, the burden of carrying around 1½ kg of garments and the restriction of movement can favour less clothing and higher ambients. Homes for old people in care are therefore heated to above 22°C allowing $\sum$ GTD < 5.6°C. Here especially thick, light clothing needs to be promoted for both humanitarian and energy conservation reasons.

v) The thermal discrimination ability of the elderly is often impaired. Thermal perception is in the range 3-4°C, rather than 1-2°C for young subjects [20]. While this broadens the PMV and satisfaction bands (Fig. 4) it leads to the undesirable situation that the elderly may not appreciate when they are too cold. The body corrective mechanisms such as vasoconstriction and shivering are also less responsive, and in extreme cases this can lead to hypothermia. Promotion of a greater understanding of clothing and warmth and the use of low-cost temperature alarms is required in this area.

5. THE FUTURE; INFRA-RED SCANNING AND GRADING

If the warmth of ordinary clothing is to become of greater significance in the future, some simple means of measuring and grading the thermal resistance must be available . GTD fulfills the need for an easily understood scale, but its measurement using existing techniques is decidedly cumbersome. Current methods can be divided into those which measure the heat flow through the fabric (with or without moisture and convection layers) as reviewed in [1, 11] and those that estimate Clo values from a heated manikin [13, 16]. Infrared thermovision cameras have now been developed to a level where they can be used for accurate quantitative study. Body temperatures have been measured using this technique but detailed studies of garment resistances are not recorded.

The following exploratory experiments were undertaken on two similar sweaters one being Acrylic fibre (A) and the other Wool (B), manufactured by the Scottish College of Textiles (SCOT), Galashiels.

i) Samples of fabric from A and B were analysed using the British Standard method for measuring thermal resistance [27] on apparatus at SCOT.

ii) Samples were analysed by cladding an instrumented and heated dummy consisting of an aluminium cylinder (150 mm dia x 200 mm high) in each fabric and comparing the heat loss to that with no fabric.

iii) As ii), but with the surface temperature of fabric A measured using an infra-red camera kit (Aga Thermovision 782) supplied by Napier College of Science and Technology, Edinburgh.

iv) Infra-red thermal studies were made of the torso of a subject when nude and when wearing each garment in an environmental chamber at Heriot-Watt University, Edinburgh. Representative areas of the resulting thermographs, as shown in Fig. 6, were analysed to give area-averaged T_c and hence mean I_g. A summary of results is given in Table 5.

I values are expressed in Clo units and f_c is estimated from the areas. The results are in general agreement and in particular the infra-red method (iii) appears satisfactory. When worn by the subject, the GTD values (in °C) are higher because the sweaters were loose fitting and trapped air between the skin and garment. This is evident in the many cool fold areas in the thermograph, Fig. 6. Precise testing would require thermographs of garments on a standard heated manikin under prescribed conditions.

TABLE 5   Test Results on Garments A and B

| $(T_a = 21^\circ C)$ | GARMENT A | | | | GARMENT B | | | |
TEST	I_a	f_c	I_g	GTD ($^\circ C$)	I_a	f_c	I_g	GTD ($^\circ C$)
i	0.76	1	0.29	1.5	0.76	1	0.39	2.1
ii	0.80	1.05	0.30	1.6	0.80	1.05	0.37	2.0
iii	0.80	1.05	0.32	1.7	-	-	-	-
iv (Subject)	0.80	1.20	0.52	2.8	0.80	1.20	0.52	2.8

Fig.6 Infra-red Thermograph of Garment A.

If the objective of easy and reliable grading of clothes for indoor use is achieved, then the first step will have been taken on a path towards applying science to the selection of everyday garments for warmth. GTD values are needed on garment labels so that the purchaser is better informed and has his attention drawn to this aspect. Labelled garments would allow subjects (and especially the elderly) to dress according to the ambient temperature. It would also promote rational decisions on the level of this temperature for a balance between clothing and energy cost.

6.   CONCLUSIONS

. A slight increase in clothing level leading to a $1^\circ C$ drop in indoor temperatures would yield a 3% saving of primary energy.
. Garment Temperature Decrease, GTD, provides an easily used and understood index for thermally grading garments.
. Indoor temperatures, both at work and home, could be decreased by $1-2^\circ C$ without discomfort or major changes to clothing style.

84

ACKNOWLEDGEMENTS

Gratitude is expressed to the Scottish College of Textiles, Galashiels for th supply of garments and use of testing equipment and to Napier College of Science and Technology for the Infra-red camera work. The testing was conduc ed by Mr. John MacCafferty as his final-year undergraduate project.

REFERENCES

1.   Vigo, T.L. and C.B. Hassenboehler, Jr., (1979). A. Chem. Soc., Sym. Se: 107, 255-277.
2.   O'Callghan, P.W. and S.D. Probert, (1976). App. Energy, 2, 269-277.
3.   Hardcastle, R., (1984). The Pattern of Energy Use in the UK, 1980, H.M.S.O., London.
4.   Central Statistical Office (1987). Annual Abstract of Statistics, No. 123, H.M.S.O., London.
5.   Fanger, P.O. and O. Valbjorn, Eds., (1978). Indoor Climate, Proc. 1st. Int. Indoor Climate Sym., Dan. Building. Res. Inst., Copenhagen.
6.   Fishman, D.S. and S.L. Pimbert, (1978). Indoor Climate, Ref. 5, P. 677-
7.   Energy Efficiency Office, (1984). Degree Days, Fuel Eff. Booklet 7, H.M.S.O., London.
8.   Fanger, P.O., (1972). Thermal Comfort, McGraw-Hill, NY.
9.   ASHRAE Handbook, (1981). Fundamentals, ASHRAE, Atlanta.
10. Cornwell, K., (1977). The Flow of Heat, Van Nostrand-Reinhold, London.
11. Slater, K., (1977). Textile Progress, 9, 1-91.
12. ASHRAE Standard, (1981). Thermal Environmental Conditions for Human Occupancy, ASHRAE 55-1981.
13. Sprague C.H. and D.M. Munson, (1974). ASHRAE Trans., 80, 120-129.
14. Benginger, T.H., (1978). Indoor Climate, Ref. 5, p. 441-476.
15. Humphreys, M.A., (1978). Energy, Heating and Thermal Comfort, Construc tion Press, Lancaster, P. 237-265.
16. Seppanen, O., P.E. McNall, D.M. Munson and C.H. Sprague, (1972). ASHRA Trans., 78, 1, 120-131. (see Discussion).
17. Langkilde, G., (1978). Indoor Climate, Ref.5, p. 835-856.
18. Humphreys, M.A. (1978). Indoor Climate, Ref. 5, p. 699-713.
19. Berglund, L.G., (1978). Indoor Climate, Ref. 5, p. 507-525.
20. Collins, K.J., (1978). Indoor Climate, Ref. 5, p. 819-833.
21. Nevins, R.G., (1961). ASHRAE-ASME, Autumn Meeting, as reported in Ref.
22. ASHRAE Std., (1974), Thermal Environment Conditions for Human Occupancy ASHRAE-55-1974. (Data at 80% acceptance).
23. Kut, D., (1970). Warm Air Heating, Pergamon, Oxford.
24. I.H.V.E., (1965). I.H.V.E. Guide, London.
25. H.M.S.O., (1961), Parker-Morris Report, H.M.S.O., London.
26. Houghton, F.C. and C.P. Yaglou, (1923). ASHVE Trans., 29, 361-384.
27. British Standard, (1986). Determination of Thermal Resistances of Textiles, BS 4745: 1986, London.

APPENDIX - NOMENCLATURE

A_c	Area of Clothes (m^2)	Q_s	Dry Heat Flow from skin (W)
A_s	Area of Skin (1.77 m^2)	q_s	Heat Flux from skin (W/m^2)
f_c	A_c/A_s	t	Thickness (m)
GTD	Garment Temp. Decrease	T	Temperature ($^{\circ}C$)
h_a	Air heat trans. coef. ($W/m^2 K$)	ΔT_{ag}	Garment drop in T_a ($^{\circ}C$)
I_a	Thermal Resistance of air layer ($m^2 K/W$)	ϱ	Bulk Density (kg/m^3)
I_c	- of clothes ensemble	Suffixes	
I_g	- of garment alone	a	Ambient Indoor or Comfort
		ao	Nude Comfort
k	Th. Conductivity (W/mK)	c	Clothes
M	Metabolic Rate/A_s (W/m^2)	m	mean
		s	skin

AFME Programme for Energy Savings in Trade Centers

H. Despretz

Residential and Commercial Buildings Department, French Agency for Energy Managing (AFME) Sophia Antipolis, France

ABSTRACT

A general cooperation agreement between AFME and four major companies in the field of cash and carry should permit to save energy up to 30 % in trade centers. Text lists main fields of cooperation and presents results of some projects.

KEYWORDS

Energy saving, hypermarket, daylighting, waste incineration off-peak electricity, audit.

INTRODUCTION

The purpose of this operation is to convince directors of hypermarkets of the advantage of energy managing through demonstration operations and to urge competiting societies to share technical experiences.

HYPERMARKETS IN EUROPE

According to the International Self Service Organisation, an hypermarket is a warehouse of more than 2500 m^2 sales area, proposing a large selection of foods and goods, sold in self service.

Total Number of Hypermarkets and Evolution (1984)

Table 1 . Total Number, Area and Ratio

	Number	Sales area $(10^3 \ m^2)$	Hab/m^2
BE	81	561	17.6
DK	38	200	27.5
FR	492	2,800	19.4
DE	874	5,179	11.9
UK	345	1,226	44.8
NL	35	147	97.6
LU	1	8	46.2
IT	17	121	469
..			
EEC	>1,883	>10,242	

One can observe that national stocks are very different from one country to another but, according to the proportion of inhabitants per m^2, there is a large opportunity of construction of new stores.

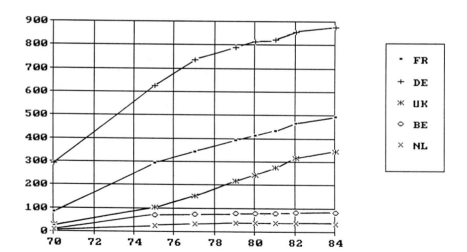

fig. 1. Evolution of construction through 1970-1984

Comparison of Hypermarkets in 3 Countries from the EEC

Structure of stock according to sale area

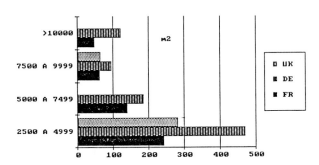

fig. 2. Structure of hypermarkets stock

As shown above hypermarkets are very different from one country to another; that may be explained by many factors: in Germany the non-sale areas (stores, laboratories, cold storages...) are smaller than in France and so are the sales areas wider. In England many warehouses are inside towns that accounts for their relative smallness.

Energy consumption: to make it easier we have expressed all consumptions in TOE by m^2 of sales area; the figures as known in 1984 are shown in table 2

table 2 Energy consumption in hypermarkets

	Cm (TOE/m^2)	TOE
FR	0.145	406,000
DE	0.09	476,000
UK	0.114	122,000
EEC	0.1	1,024,000

THE FRENCH PROBLEM

Two different points of view give light to the common interests that permitted the whole operation which will be described further on.

The National Point of View

Since 1984 the number of hypermarkets in France has raised up to 580 for a whole sales area of 3,260,000 m^2 and a total energy consumption of 500,000 TOE. There is but a problem of consumption and also of energy and power supplied. As illustrated underneath, electricity is the main type of energy used.

IEE-G

88

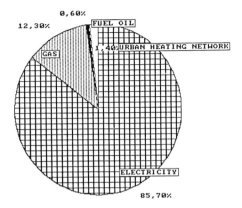

fig. 3. Distribution of consumed energies

Hypermarkets Point of View

It is merely difficult to argue on general financial aspects to
assess the interest of energy savings in an hypermarket; as a
matter of fact all hypermarkets are very different one from
another because of the so many situations: sales areas range up
from 2,500 m^2 to more than 10,000 m^2, energy consumption by m^2
is reciprocally proportional to sales area and depends upon
design and geographical location of buiding as well as on its
business. Cost of energy, turnover and profit margin also vary
in each center. Nevertheless if we take as an example a
5,700 m^2 sales area warehouse consuming 0.17 TOE/m^2
(2400 FF/TOE) with 60,000 FF/m^2.year turnover and 1 % profit
margin, we can find out that total energy consumption
represents 0.6 % of yearly turnover which is to compare with
profit margin. This explains the attention now payed to the
problem by cash and carry societies whose goal is cutting
energy cost by 30 % in trade centers

GENERAL AGREEMENT AFME-HYPERMARKETS

In 1983 four major companies (AUCHAN, CARREFOUR, CORA,
EUROMARCHE) owning one half of the total area signed a general
cooperation agreement with the AFME and grouped together in a
non profit association called AMEC (Association for Energy
Managing in Trade Centers).

Main Fields of Cooperation and Programme

- Energy audit methodology
- Energy auditing of all the hypermarkets built before 1976
- On-site experiments or innovative projects allowing prototype
  realisations and full size tests
- Exemplary projects (demonstration): in one existing and one
  new built establishment belonging to each of the four
  companies, to promote dissemination of proved techniques
  into this type of building.

- The contracting companies pledged to undertake the works specified following the energy audits within 4 years.
- Specialized educational and training programmes are to be organized.
- Technical datas on every operation will be published.

First Results at the End of the Third Year

- Energy audit method, and corresponding computer program ("DIAMEC") are available since beginning of 1986.
- 80 energy audits have been performed.
- Some demonstrations (1) have been carried out proving new techniques:
  . daylighting
  . waste incineration with pyrolitic effect
  . air conditionning control system
- In five hypermarkets (1) several techniques were combined (insulation, waste incineration, control system for HVAC or lighting, generators allowing off-peak electricity supply).
- Following the energy audits, the programme of works is currently in progress: for example AUCHAN Company invested 21 M.F.F. during 85 and 86 without any public subsidy.

PROJECTS AND RESULTS

Energy consumption in an hypermarket is distributed as shown underneath:

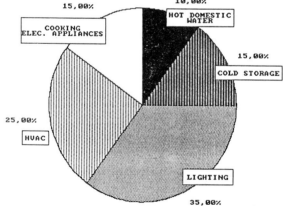

fig. 4. Distribution of energy consumption

So main efforts have beared on lighting and substitutions of energy.

Daylighting - AUCHAN AVIGNON

This hypermarket was built ten years or so ago. It had no openings neither in the façade nor in the roof. The sales hall

---

(1) Some of these actions were financially supported by AFME or EEC

(11,500 m^2) is illuminated by rows of 58 W fluorescent strip sources (installed power 29 W/m^2, horizontal illumination on floor level 900 Lux). The electric lighting has a feature allowing its subdivision into quarters (one row out of four).

Openings have been cut into the roof to provide illumination in the sales hall by daylighting with a contribution by electric lighting when natural one is insufficient. 105 translucent horizontal sheds (methyl methacrylate, diffused incidence transmission factor $t_{df}=0.5$) 8.5 m^2 area each have been installed. The 50 sheds placed above deep frozen goods display cases have been equipped with solar screens.

From july to december 85 a monitoring has been performed by CSTB NANTES (1). AT the rate of one scan every ten minutes illumination measurements have been made continuously: horizontal and vertical illuminance within the sales hall, horizontal lighting outdoors. The fraction of artificial lighting on line was also monitored.

Savings made in terms of electric energy for lighting purpose represents 65 % of the blind local consumption and the illumination levels were equal at least to those of the blind room for 60 % of time. Taking into account of the influence of the openings on the energy consumed in winter for heating and in summer for refrigerating proved no major effect on overall thermal balance.

Cost of transformation (including repainting of ceiling in white) is 125 FF/m^2(sales area) for roof modifications and 35 FF/m^2 for regulation.

As electricity for lighting in this hypermarket represents 42 % of total energy consumption, global economy is about 30 % and 1,000,000 FF/year.

Using Off-Peak Electricity Supply

In France development of thermal uses of electricity as led to a peak demand on a few unforseeable cold days. To cope with this power demand, EdF has to work up complementary generating plants what raises considerably prices of electricity. So as to limit such phenomenon EdF proposes to its customers to reduce (or even suppress) their electric consumption during 400 hours (22 days/year at the rate of 18 h/day - from 7 a.m. to 1 a.m. next day). As conterpart EdF offers a special tariff system called EJP (Effacement Jours de Pointe: Off Peak Electricity Supply).

Using EJP compels hypermakets managers:

- to buy new generating sets able to operate perfect
- in case of failure to stand more expensive tariff

---

(1) Building Scientific and Technical Center

To solve such problem AMEC :

- has launched a tender to generating sets manufacturers to obtain profitable sales prices as well as low cost servicing an emergency maintenance. Amec has selected two makers (SDMO and BERGERAT-MONNOYEUR) with whom they signed a "supplies agreement".
- has settled an emergency organisation to dispose without delay of mobile generating set.
- has trade the purchase of the generating sets by leasing
- has discussed an insurance contract with a group of firms (UAP, GAN, GFA, ALLIANZ and LLOYD CONTINENTAL) to limit financial risk in hypermarket up to 10 %

First results in warehouses are very satisfactory:

table 3 EJP tariff applied to four hypermarkets

	case 1	case 2	case 3	case 4
yearly electric consumption(kWh)	2,528,700	4,724,200	6,658,000	7,080,600
elect. bill(FF)	772,704	1,531,292	1,830,954	2,278,222
standing charge	314,349	409,445	917,318	551,480
TOTAL	**987,053**	**1,940,737**	**2,748,272**	**2,829,702**
EJP elect. bill	401,194	885,580	1,108,905	1,312,669
standing charge	120,260	143,306	385,560	195,863
TOTAL	**521,454**	**1,028,886**	**1,494,465**	**1,508,532**
variation (FF)	465,599	911,851	1,253,807	1,321,170
(%)	**47**	**47**	**46**	**47**

The EJP tariff allows to reduce the yearly electric bill by 50 % and to earn 30 to 35 % on overall cost (including fuel consumption and maintenance cost).

Waste Incineration EUROMARCHE LOMME

A large amount of refuse (2 up to 4 Tons/day according to the size of the buildng) is collected each day in hypermarkets - wrapping, cardboard, plastic, lost food..- and must be carried away. The interesting net calorific value of waste (3,000 to 4,000 kCal/kg) has induced hypermarket managers to study the energetic and financial aspects of their burning on site with waste heat recovery for heating of the premises (and/or cooling with an absorption heat-pump).

A 600 kg/hr waste boiler with pyrolytic effect has been installed in EUROMARCHE center in LOMME (north of France). This trade center is made up with an hypermarket (10,000 m² sales area), a self service restaurant and shops (about 10,000 m² more). Collected refuse amount to 104 Tons/month (26 days) with a N.C.V of 3,500 kCal/kg. A monitoring was carried up during heating season 85/86 and gave following results:

table 4 results of waste incineration survey

	consumed energy		produced energy
	kWh	FF	kWh
gas boilers	1,273,200	276,000	814,875
furnace burner	116,000	25,150	
elec. auxilliary	27,070	7,900	1,107,350
total		309,000	1,923,000

If heat production had been made by gas boilers overall consumption would have been:

C_{ons} = 1,923,000 x 0.64 = 3,005,000 kWh
C_{out} = 651,500 FF

Deviation = 350,000 FF

This fairly good results can yet be improved by a better maintenance and a more accurate regulation. One can see on that example that waste incineration allows a reduction of thermal consumption by more than 50 % and cutting energy bill by 15 to 20 %. Together with a fine regulation, especially modulating supply air flow as a fonction of occupancy, one can reach 25 to 30 % economy.

CONCLUSION

As shown before there are many opportunities of saving energy in trade centers. The whole operation with AMEC is a good example of dissemination of high performance techniques in commercial buildings which should be enlarged to supermarkets or other sectors as hotels. To intensify energy retrofits in existing buildings as well as in new constructions AFME is managing complementary actions, for instance:

- issue of a lighting guide in stores
- study of deep freeze appliances
- test of another computerized auditing method in small scale stores called "ENERMAG"
- innovation project of heating a supermarket through refrigeration system (for deep frozen foods) without boiler
- development of special waste incineration furnace

The results of these actions are (or will soon be) available to all those who, in EEC, are interested by the problem of energy consumption and managing in trade centers.

Renovation of Multi-Family Houses with Minimized Life-Cycle Cost

S-I. Gustafsson, B. G. Karlsson

*Institute of Technology, Division of Energy Systems,
S 581 83 Linköping, Sweden*

ABSTRACT

Our paper presents the OPERA - model, (OPtimal Energy Retrofit Advisory model), which enables the user to find the best retrofit strategy for each unique building. The model optimizes the implementation of both envelope, as well as heating equipment retrofits, in order to find the lowest possible Life-Cycle Cost, LCC, for the building studied. No other combination of insulation and installation measures thus can make the LCC lower.

Input to the model are e.g. the geometry and thermal status of the building, building and installation costs, the climate and prices or tariffs for energy deliveries.

The model can be characterized as an integer, nonlinear, mixed program and it is solved using both derivative and direct search methods.

Our results from calculations on both real and conceptual buildings show that a low operating cost heating system is essential to minimize the LCC. Such systems are e.g. district heating with a cost reflecting time-of-use tariff, or a bivalent oil-boiler heat pump system. These systems should be combined with cheap envelope retrofits e.g. weatherstripping and attic floor insulation. More expensive insulation retrofits e.g. external wall insulation or triple glazed windows can be a part of the optimal solution if a restoration has to made from other than energy conservation reasons. Only if a high running cost system e.g. electricity is installed in the building the exhaust air heat pump can be profitable.

KEYWORDS

Retrofitting; Life-Cycle Cost; Optimization; Energy Conservation; Heating equipment; Heat pump; District heating; Time-of-Use rates; HVAC-systems; Fenestration.

95

INTRODUCTION

Since April 1985 there is an on-going study, dealing with retrofitting of residences and the minimization of the building LCC, at the Institute of Technology in Linköping, Sweden. The project is co-sponsored by The Swedish Council for Building Research and the municipality of Malmö, Sweden. It was initiated due to the decrease in building new residences and instead emphasizing the restoration of the existent building stock. However, there were also major uncertainties how the retrofits should be combined in order to find the most profitable solution, both from the point of society and from the owner of the house.

Traditionally the envelope retrofit strategy was decided by the building contractor and the heating equipment strategy by the installation contractor. Of course both of them wanted an extensive strategy to be implemented. However, this was also encouraged by the Swedish subsidiary system, the capital cost for the building was, and still is, subsidized to approximately 3 % the first year. Almost all types of retrofits thus is profitable to implement for the owner of the building, the capital cost is lower than the inflation. From the societal point of view, however, this lead to sub-optimations, e.g. in district heated areas. The district heating utilities were and still is sometimes to a part heated by burning refuse, and the short range marginal cost for producing an extra kWh was in the summer as low as 0.10 SEK, (1 US$ = 7 SEK.) [Gustafsson, Karlsson, Sjöholm, 1986a]. The municipality had to get rid of the waste, and if they could not burn it they had to store it at the refuse dump. At the same time the building landlords could get grants and loans to install sun collectors in order to heat domestic hot water. It was obvious that something had to be done to prevent such misuse with the scarce resources of the nation.

A discussion in more detail can be found in [Gustafsson, Karlsson, 1986].

THE OPERA MODEL

Above we did describe our problem as an integer, nonlinear, mixed program, following the terminology in [Reklaitis, Ravindran, Ragsdell, 1983]. However, there are no commercial algorithms which can solve such problems with an absolute accuracy. We have thus elaborated the model so it is solved during the generation of the problem itself. The model is now implemented in a NORD 570 machine which solves the base case problem in about 30 seconds.

The program starts with the calculation of the existent LCC for the building. A retrofit is implemented, e.g. attic floor insulation, and the program elaborates the optimal extra insulation thickness using a derivative method. A new LCC is after that calculated and it is compared to the existent one. If this new LCC is lower than the original one the retrofit is selected, otherwise not. The program proceeds with a new retrofit, e.g. external wall insulation, and the procedure is repeated.

However, it is extremely difficult to find an optimal window construction. [Klems 1985, Benson and Tracy 1985, McCabe 1986, Anderson 1986]. Such retrofits are tested, using a number of different alternatives. The existing LCC is tested against the LCC for triple and new types of energy conserving glazing. The procedure is identical for the evaluation of

weatherstripping and exhaust air heat pumps. Either the measure is selected
or not.

Coming so far the program has found the optimal retrofit strategy for the
existent heating system in the building. This is now changed from e,g, an
oil-boiler to an electrical heated device. The evaluation will now start
from the beginning and a new optimal solution is found. This procedure is
now repeated for all the heating equipment possible to install in the
building. The model at this moment can work with oil-boilers, electricity,
district heating, heat pumps, T-O-U rates for district heating and
electricity and a bivalent oil-boiler heat pump system. Finally the lowest
LCC strategy has been found.

However, there are some uncertainties in choosing the proper discount
rates, optimization periods etc, and thus the program also evaluates the
LCC for about ten different discount rates, optimization periods from 10 to
50 years, up to 3 % annual escalation of the energy prices and 5 different
amounts of degree hours. This means that the optimal strategy can be
meticulously scrutinized in order to do a sensitivity analyze. If there are
more cases required, the FORTRAN code easily could be changed a little to
provide this. In [Gustafsson 1986 and Gustafsson and Karlsson, 1987b] the
theoretical background is presented and the model and how it works can be
studied in detail.

From the above discussion we think it is obvious that the main work has
been emphasized on generating the mathematical problem. Less have been done
to elaborate the optimization procedure. However, our method always finds
the true minimum point of the problem and that is exactely what we wanted
from the procedure to be used.

We also think it is obvious that we in the OPERA model have a perfect means
to evaluate different retrofits, if they are profitable, and how they shall
be combined in the best way. In the next part of our paper we will show
examples from two existing buildings in Malmö, one small with 18, and one
big containing 105 apartments.

CASE STUDIES

In Malmö we are co-operating with a group called the 7 - builders group,
which consists of 7 housebuilders aktive all over Sweden and in many other
countries. The 7 contractors are ABV, SKANSKA, BPA, JCC, Kullenberg, SIAB
and PEAB. The municipality is aktive through Malmö Energy and Heating
Utility. Due to this co-operation we have been enabled to calculate on two
real buildings owned by ABV and Svenska Riksbyggen AB. In a paper of this
length it is not fruitful to present all the input parameters, the
interested reader can contact us and we will send the total input file.
Instead we will show the retrofit strategies for a base case with 5 % real
discount rate, optimization time 30 years and assuming the heating system
is turned of if the outside temperature is higher than 17 °C. The desired
inside temperature is assumed to be 20 °C. It shall be noted here that
these buildings also are described in [Gustafsson and Karlsson, 1987a]
but the calculations then were done for an other base case. Furthermore no
sensitivity analyzis was made for different building or installation costs,
but instead for various economical parameters.

In table 1 and 2 we will present the base case and below we will discuss
how input changings will influence the optimal strategies. All the starting
values have been presented by ABV and SRAB, and no scientific study has
been elaborated in order to examine the reliability of those values. In
[Gustafsson 1986] however, there is a lot of references to Swedish
literature, concerning building and installation costs. Nevertheless it is
obvious that there is an immense lack of information about the costs for
different retrofits due to various buildings and so fourth, and we hope
that papers like this will initiate scientific studies about the topic. In
[Diamond and others, 1985] this is also emphasized.

TABLE 1 Optimal Retrofit Strategies due to Different Heating Systems. The
values in MSEK. ABV Building. 18 apartments

	Oil boiler	Electr boiler	Distr heat	Heat pump	Bivalent oil heat pump
LCC without envelope retrofits:	1.22	1.40	0.86	1.48	1.32
Savings:					
Attic floor insulation	----	----	----	0.02	----
Four glazed windows	0.02	0.04	----	0.09	0.04
Weatherstripping	0.09	0.12	0.06	0.16	0.11
New LCC	1.10	1.23	0.80	1.20	1.16

From the presented table it is obvious that the existing heating system,
district heating with a differential T-O-U rate, combined with
weatherstripping is the best retrofit strategy for this building. The
district heating with its low running cost, 0.21 SEK/kWh during the winter
and 0.10 SEK/kWh in the summer, makes almost all of the retrofits
unprofitable. The cheap weatherstripping, 200 SEK/window,door, was the only
one. The second best heating system is the oil-boiler, running cost 0.24
SEK/kWh, also combined with better windows.

In this case the four glazed windows had a better profitability then three
glazed ditto. The costs presented by ABV are 1300, 2250, 2650 SEK for two,
three, and four glazed windows respectively. The corresponding U-values
were 3.0, 2.0, 1.5 $W/m^2,°C$.

Expensive exhaust air heat pumps, approximately 100 000 SEK, never was
chosen by the computer. In [Gustafsson 1986] a minor study has been made
about air heat exchangers, which are very profitable in new buildings. This
study showed that it was very expensive to install new ducts, for the air
flow transported into the different dwellings, and thus this equipment
could not compete with the low energy prices nowadays, Feb. 1987.

The big building, will almost have a similar envelope retrofit strategy but
the best one also includes changing the heating system.

TABLE 2 Optimal Retrofit Strategies due to Different Heating Systems.
Values in MSEK. SRAB Building. 105 apartments

	Oil boiler	Electr boiler	Distr heat	Heat pump	Bivalent oil heat pump
LCC without envelope retrofits:	7.11	8.57	5.56	5.38	5.69
Savings:					
Three glazed windows	0.26	0.41	0.15	0.14	0.15
Weatherstripping	0.46	0.68	0.31	0.29	0.31
New LCC	6.40	7.48	5.09	4.95	5.23

Table 2 shows that three glazed windows and weatherstripping are profitable
for all the examined heating systems. In this case the heat pump system was
the cheapest which is not very expected. However, this can be explained by
low prices for heat pumps in this case, 150 000 + 3000 * P SEK where P is
the thermal power for the pump, and the fact that the existing building
does not have any chimney due to the existing district heating system.
Bivalent systems where the oil boiler is a part, of course will be less
competitable in such cases.

It is also important to note that the existing district heating system LCC
is very close to the cheapest one. Due to the uncertainties in all the
input parameters we assume that it would be preferable to keep the existing
heating system and only implement the envelope retrofit measures. Also in
this case the expensive retrofits are rejected by the program.

SENSITIVITY ANALYZIS

From table 2 above it is obvious that small changes in the input parameters
can change the most profitable solution. For insulation measures the
optimal thickness of extra insulation of course varies due to different
energy prices etc, but the most important is that at one special point the
profitability will vanish. The best solution will in such a case be to
avoid the extra insulation. This point can emerge for rather thick
insulations, more then 0.1 m, which is discussed more in [Gustafsson
Karlsson and Sjöholm, 1986b and in Gustafsson, Karlsson and Sjöholm,
1987].

In this paper we will instead emphasize the costs for the heating systems
dealt with during the optimization. These are described by a straight line
function:

$$C = A + B * P$$

where C is the total cost for the equipment and installation in the
building, A and B are two constants and P is the thermal power of the
heating device. In [Gustafsson 1986] the process to evaluate such
expressions is described in detail. The functions presented by the
housebuilders are:

TABLE 3 Heating Equipment Costs in SEK

	ABV	SRAB
Oil-boiler	30 000 + 200 * P	50 000 + 350 * P
Electr.boil.	25 000 + 150 * P	20 000 + 100 * P
Distr.heat	30 000 + 70 * P	100 000 + 400 * P
Heat pump	30 000 +6000 * P	150 000 +3000 * P

From table 3 it is obvious that there is big differences in the cost
functions especially for district heating equipment and heat pump ditto. It
is important to note that district heating is the existent system in the
two buildings which has many years of operating before it has to be changed
to a new system. The costs for the changing will thus not influence so
much, compared to the heat pump device which has a shorter life-cycle and
also is much more expensive. In table 4 and 5 we present the optimal
strategies when the costs for the heating equipment is ignored. The cost
functions are identical,

$$C = 10 + 10 * P$$

which means that only the different running costs for the systems will
influence the strategies.

TABLE 4 Optimal Retrofit Strategies due to Different Heating Systems.
ABV building, Heating equipment Cost Ignored.

	Oil boiler	Electr boiler	Distr heating	Heat pump	Bivalent oil heat pump
LCC without env.retrof.	1.10	1.30	0.85	0.47	0.57
Savings: Four glazed windows	0.02	0.04	----	----	----
Weather-stripping	0.09	0.12	0.06	0.01	0.03
New LCC	0.99	1.14	0.79	0.46	0.55

TABLE 5 Optimal Retrofit Strategies due to Different Heating Systems. The
values in MSEK. SRAB building, Heating equipment cost ignored.

	Oil boiler	Electr boiler	Distr heating	Heat pump	Bivalent oil-heat pump
LCC without envel.retro.	6.75	8.39	5.54	3.64	4.16
Savings: Three glazed windows	0.24	0.41	0.15	----	0.04
Weather-stripping	0.43	0.67	0.31	0.07	0.15
New LCC	6.09	7.31	5.03	3.57	3.97

Comparing Table 4 and 5 with 1 and 2 it is obvious that setting the heating
equipment cost to a very low value, almost free, will not change the
optimal envelope strategy very much for the cheapest heating systems. The
expensive ones, however, will loose some climate shield measures because
there is no money to earn on decreasing the thermal load of the building.
Of course the heat pump system is the cheapest, it delivers the energy for
0.10 SEK/kWh, and all the other systems to a higher price.

The important thing is that the situation is almost similar for heating
systems that cost 1000 + 1000 * P. We present this in table 6 and 7.

TABLE 6 Optimal Retrofit strategies due to Different Heating Systems. The
values in MSEK. ABV building, Heating Equiment Cost 1000 + 1000 * P
SEK.

	Oil-boiler	Electr boiler	Distr heating	Heat pump	Bivalent oil-heat pump
LCC without envel. retrof.	1.41	1.60	0.87	0.82	0.92
Savings: Four glazed windows	0.04	0.06	---	---	---
Weather-stripping	0.11	0.14	0.07	0.04	0.05
New LCC	1.27	1.40	0.80	0.78	0.87

TABLE 7 Optimal Retrofit Strategies due to Different Heating Systems. The
values in MSEK. SRAB building, Heating Equipment cost 1000 + 1000 *
P SEK.

	Oil boiler	Electr boiler	Distr heating	Heat pump	Bivalent oil-heat pump
LCC without envel. retrof.	8.07	9.72	5.59	4.48	5.41
Savings: Three glazed windows	0.30	0.47	0.16	0.03	0.09
Weather-stripping	0.51	0.76	0.32	0.15	0.23
New LCC	7.26	8.49	5.11	4.70	5.09

Table 6 and 7 shows that the heat pump still is the best alternative due to
its very low running cost. However, the existing heating system now gets
very close and if the heating equipment cost only gets a little higher the
best strategy will be to keep the existing system. The envelope retrofits
differs not much from the earlier ones.

This situation changes a lot in our next two tables where the heating
equipment cost is set to 10000 + 10000 * P, valid for expensive heat pumps.

TABLE 8 Optimal Retrofit Strategies due to Different Heating Systems. The
values in MSEK. ABV building. Heating Equipment Cost 10000 + 10000
* P SEK

	Oil boiler	Electr boiler	Distr heating	Heat pump	Bivalent oil Heat pump
LCC without envel. retrof.	4.24	4.27	1.05	4.00	4.02
Savings: Attic floor insulation	0.09	0.09	---	0.08	0.08
External wall insulation	0.11	0.12	---	0.10	0.09
Four glazed windows	0.20	0.19	0.03	0.19	0.18
Weather-stripping	0.28	0.28	0.10	0.27	0.26
New LCC	3.56	3.59	0.93	3.36	3.40

TABLE 9 Optimal Retrofit Strategies due to Different Heating Systems. The
values in MSEK. SRAB building, Heating Equipment Cost 10 000 +
10 000 * P SEK

	Oil boiler	Electr boiler	Distr heat	Heat pump	Bivalent oil-Heat pump
LCC without envel. retrof.	20.09	21.74	6.50	16.13	16.78
Savings:					
Attic floor insulation	0.15	0.26	----	----	----
External wall insulation	0.53	0.85	-----	-----	0.05
Three glazed windows	-----	-----	0.26	0.51	0.59
Four glazed windows	0.89	1.15	----	----	----
Weather-stripping	1.29	1.53	0.46	0.81	0.92
New LCC	17.24	17.94	5.78	14.82	15.22

From table 8 and 9 it is obvious that the existent heating system is the
most outstanding solution. The expensive heating systems will influence the
LCC very much, and a lot of envelope retrofits is generated by the program.
It is important to observe that exhaust air heat pumps never are chosen due
to small influence on the thermal load of the building. The tables above
also implies that low cost heating systems, lower than 1000 SEK/kW, only
will influence a little on the LCC and also on the chosen retrofit
strategy, the running cost is essential. For very expensive equipment the
opposite is true. In table 8 and 9 the LCC and savings tend to get closer
to each other for the different systems, the district heating of course
excluded.

This means that much work should be emphasized to find proper equipment
cost functions for the expensive facilities i.e. heat pumps, while less may
be done to the cheap equipment. Unfortunately, there is very little
experience, scientifically documented, about costs and life-cycles for heat
pumps, but, some references are [Gustafsson 1986, Pientka 1986 and Jaster
and Miller 1985].

The bivalent system has never been chosen by the computer. This can be
explained by the rather mild climate in Malmö and the use of 17 °C as the
inside temperature in the building. In the ABV case there is not enough
heat to produce in such a low running cost system, the existent heating
system will be the best. In the SRAB case the single system heat pump was
chosen due to the low running cost. The installation cost was also
acceptable. If the heat pump was more expensive the bivalent system gets
advantages but so does the existent heating system, a low cost system with
much less installation cost. In [Gustafsson and Karlsson 1987a], the same
building has been treated but with other base case parameters, e.g. a
different inside temperature. The bivalent system was in that case the
best.

104

REFERENCES

Anderson R.(1986) Natural convection research and solar building applications. Passive Solar Journal,3(1),33-76(1986), Marcel Dekker,Inc.

Benson D.K. and Tracy C.E. (1985) Evacuated window glazings for energy efficient buildings. 29th annual SPIE international technical symposium on optics and electro-optics, San Diego, California.

Diamond R. and others. (1985) Building energy retrofit research, multi-family sector. Applied Science Division, Lawrence Berkeley Laboratory, CA 94720, U.S.A.

Gustafsson S-I. (1986) Optimal energy retrofits on existent multi-family buildings. Thesis no 91, Institute of Technology, Linköping, Sweden.

Gustafsson S-I., Karlsson B.G. (1986) Why is life-cycle costing important when retrofitting buildings. Accepted for publication by The International Journal of Energy Research, U.K.

Gustafsson S-I., Karlsson B.G. (1987a) Minimization of the life-cycle cost when retrofitting buildings. To be published by ICBEM - 87, Lausanne, Switzerland

Gustafsson S-I., Karlsson B.G. (1987b) Bivalent heating systems, retrofits and minimized life-cycle costs for multi-family residences. Not yet published.

Gustafsson S-I., Karlsson B.G., Sjöholm B.H. (1986a). Differential rates for district heating and the influence on the optimal retrofit strategy for multi-family buildings. Accepted for publication by Journal of Heat Recovery Systems. U.K.

Gustafsson S-I., Karlsson B.G., Sjöholm B.H. (1986b). Renovation of dwellings - life-cycle costs. CIB - 86, Volume 9 p. 3886 - 3893, Washington D.C., U.S.A.

Gustafsson S-I., Karlsson B.G., Sjöholm B.H. (1987). Optimization of the retrofit strategy for a building in order to minimize its life-cycle cost. To be published by CIB - 87, Copenhagen, Denmark.

Jaster H., Miller R.S. (1985) Performance of air-source heat pumps. EPRI, Palo Alto, CA, U.S.A.

Klems J.H. (1985) Toward accurate prediction of comparative fenestration performance. Workshop on laboratory measurements of U-values of windows, Gaithersburg, MD, U.S.A.

McCabe M.E. (1986) Field measurement of thermal and solar/optical properties of insulating glass windows. ASHRAE symposium on low-e coatings and films, winter annual meeting, New York City. U.S.A.

Pientka K., Heat pump life and compressor survival in a northern climate. EPRI, Palo Alto, CA, U.S.A.

Reklaitis G.V., Ravindran A., Ragsdell K.M. (1983) Engineering optimization. John Wiley & Sons, Inc.

Solar Gains in Urban Renewal through Roofspace Systems

C. J. Hancock and J. Littler, Research in Building Group

Polytechnic of Central London, 35 Marylebone Road, London

ABSTRACT

This paper arises from a current demonstration project involving the construction and performance monitoring of 6 dwellings with roof constructions modified to admit solar gains which when convected into the house contribute to the heating load.
The application of this technology to re-roofing work within urban housing renovation and the effect on thermal comfort and ventilation requirements of the house are outlined in this paper.

KEYWORDS

Hybrid solar design, ventilation, housing renovation, thermal modelling.

INTRODUCTION

Solar roofspace systems are being studied as contributors to the heating and venting of dwellings where wall insulation is uneconomic, other problems eg condensation, occur and direct solar gain through windows is limited by overshading from closely spaced buildings.

In England more than 1 million dwellings are in need of substantial repair (1). Of these 980,000 were built before 1919 and of these more than 530000, 48% of all unfit dwellings, are two story houses within the high density inner urban areas. Almost all have poorly insulated solid walls, many have chronic condensation problems and others have discomforting draughts and ventilation heat loss .

Conventional energy saving programmes for older houses consist of a package of roof insulation, draught stripping and replacement double glazing. These will reduce the energy consumption and enhance thermal comfort. However problems remain when solid 9inch(215mm) brick walls remain uninsulated, due to the high cost of applying and protecting insulation, or the unwelcome disturbance to internal finishes or external appearance. Heat losses through such walls amount to 2.15W/m² °C. Typically for an end-terrace house when the outside temperature is 5°C this means an hourly loss of 2.6kWhrs.

Solid concrete and some suspended timber floors also prove difficult to insulate economically and without disturbance. Ground floor losses may remain higher than 0.95W/m²°C(solid),1.25W/m²°C(suspended) resulting in an average hourly loss exceeding 0.3kWhrs(solid-assuming an average ground temperature of 10°C) or 0.7kWhrs (suspended).

Low radient temperatures of uninsulated walls and floors perpetuate thermal discomfort and may prompt the occupier to compensate by adopting higher air temperatures for the heating system, leading to greater heat losses. Condensation on cold walls is a problem which may well be exacerbated by draught stripping. Local kitchen and bathroom extracts may alleviate the worst aspects if used purposefully, but unless the air circulation through the house is controlled, the risk of condensation remains.

All these problems are well understood by local authorities and housing associations. It is particularly difficult to carry out an adequate insulation and renovation package to cope with them, when tenants are to remain in occupation during the works.

Where the heating season of the house is not significantly reduced by insulation then solar radiation, strongest at the beginning and end of the heating season, can contribute to the comfort and heating load of the house. When the solar gain is distributed by controlled air circulation in the house, an improvement in the condensation risk is also probable. In urban areas with housing densities for two storey terraces upto 20 dwellings per acre (50 d p hectare) the opportunities for window solar gains are often limited by overshading or the use of curtains to preserve privacy. Roofs remain the principal unshaded surface for solar gain, and a high proportion of these have a southerly aspect particularly where back offshots make up 4 roof slopes per house.

EXTERNAL VIEW OF COMPLETED SOLAR ROOFSPACE SYSTEM

This advantage of an unobstructed roof is exploited by modifying the construction of a conventional pitched roof to form a solar apeture. The roof is recovered with translucent sheeting over a sealed roofspace. Solar warmed air is circulated through the dwelling with a simple fan and vents. The system uses normal joinery practice. Standard ventilation products may be integrated into the interior refurbishment.

Many older dwelling require reroofing as part of any renovation package. Where the solar roofspace system is substituted for a conventional replacement roof the real cost of the solar roof can be reduced.

OUTLINE DESCRIPTION OF THE SYSTEM

The principle of a solar roofspace was examined in a prototype incorporated into a cottage convertion near Bristol in 1983 . The energy performance was monitored by the Research in Building group and reported(2). The authors have now designed and built second generation roofspace systems,(6) in conjunction with the London Borough of Newham, on 6 Terrace houses in east London as part of a renovation and roof renewal contract. These systems are now in operation. Monitoring equipment is being installed to measure the performance of the systems over a period of one year.

Figure   illustrates the predicted contribution to the heating load, of a typical project house, from all sources. The figures were generated by a computer thermal simulation model SERI-RES (3), used extensively during the design process.(see House Performance Analysis). Solar gains occur through south facing windows(direct), circulated air(solar system) and reduced ceiling losses due to a warm roofspace(buffer effect). The solar system operates under thermostat control ,'topping up ' heat delivery from uncontrolled sources (occupants,cooking, direct solar). Any shortfall in heat supply is then made up by the heating system.

FIG. 1. **HEAT INPUTS TO THE HOUSE— YEAR TOTALS**
IMPROVED HOUSE WITH SOLAR ROOFSPACE

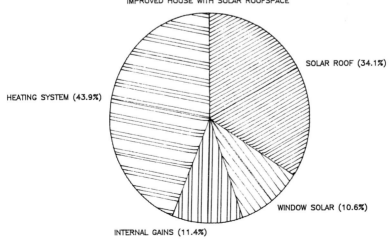

SOLAR ROOF (34.1%)

HEATING SYSTEM (43.9%)

WINDOW SOLAR (10.6%)

INTERNAL GAINS (11.4%)

Due to programme demands the project houses were selected within a contract
of houses with insulated cavity walls(U value 0.46W/m² °C)which will reduce
the the demand on the solar system. Never-the-less the simulated system
contributed 27% of the available solar gains during the period of operation,
some 3383Kwhrs per year, being 34% of the heating load.

The present design has been developed to meet the particular requirements of
tenanted houses namely an automatic control system that requires only a
simple selection switch for occupant's control. The roofspace fan delivers
air only when roofspace temerature exceeds a preset minimum and the house
temperature is below comfort setpoint. A Variable speed fan enables air
delivery tn match solar gain. Summertime overheating is checked with a
thermostat controlled fan and vent.
Due to variations in the house plans and householders requirement four
varients have been constructed showing the application to roof types
(duo-pitch, hip), air delivery systems (common supply duct or duct through
existing flue, axial or centrifugal fan), control systems (fan speed
controller, automatic or manual override).

PRELIMINARY APPRAISAL

A useful early assessment of the heating scope for solar gains may be made
by comparing a heat loss calculation for the house with average solar gains.
A manual degree day calculation or computer program ,eg Energy Designer (4)
or thermal model eg SERI-RES may be used to establish heat loads. Table 1
shows predicted heating for a typical unimproved project house. Heating to
20°C only in the evenings and early morning (Working household) was assumed.

TABLE 1

ESTIMATED ANNUAL HEATING LOAD FOR UNIMPROVED HOUSE			
(occurring upto set-point termperature, Sept-May)			
HEAT LOSSES	kWhrs	HEAT GAINS	kWhrs
Fabric	8528	Heating system	8899
Ventilation	3227	Internal gains	1180
		Solar(windows)	1676
Total	11755		11755

The potential heat gain to a solar roofspace was then examined,using
standard weather data(5) for the location (london), orientation (21°east of
south), tilt(30°) and glazed area(18m²). From October to May the daily
average radiation within the roofspace is 1.37kWhrs/m² giving more than
6500kWhrs of solar gain captured within the roofspace. Between 0.5-0.6 of
radiation is diffuse, favouring roof glazing even for low sun altitude in
overcast winter conditions. Only a portion of this gain will be useable as
heat in the house as peak solar gains will not coincide with peak heating
demand, and internal gains will displace solar gains. Later simulations
with SERI-RES were used to assess delivered heat. The design sought to
utilize as much of the 6600 Kwhrs solar gain.

PERFORMANCE SPECIFICATION

The main design requirements for a roofspace system in renovation work may
be summarized as;

The system needs to be simple to install by usual site craftsmanship, mimimising the number of trades involved to ease site co-ordination. Standard components and stock sizes should be used to reduce on-site fabrication and ensure a replicable system. The design must be adaptable to roof pitch ,construction, and limited space within the house. Translucent roof sheet need to be lightweight, able to be fixed to the existing roof structure and unbreakable. The user's controls need to be uncomplicated and robust. The solar system should be given priority over the house heating system to maximise fuel savings. The system should offer little maintenance liability with access to easily replaceable parts. The architecture of the system should compliment the style of the building.

It was decided that the roof construction would have a low thermal mass to deliver peak and instantaneous solar gains to the house. The mass of the house would be used for thermal storage. A single duct would be used to deliver air to the house. This would limit internal disruption and could be sized and insulated to reduce noise disturbance and required fan power even at high air delivery rates.

OPERATION OF THE SYSTEM

The Roofspace Solar System has a winter heating mode, and a summer self venting/cooling mode.

HEATING MODE

It was necessary to decide at the beginning of the design whether air input to the roofspace was to be drawn totally from ambient Fresh Air System, or recirculated from the house. The decision affected the construction of the roofspace, glazing, air vents and control systems. For a Fresh Air System ambient air is drawn in, heated to a margin above house air temperature, delivered to the house and allowed to exfiltrate through window vents. -Roofspace construction costs are lower in this design. Roof and glazing bars do not need to be air sealed and a permanent open vent (with one way flap) is the only other neccesary air path. A fan controller will be required to initiate (and if required-regulate) fan delivery.

For a Recirculation System the infiltration of ambient air to roofspace must be restricted. The roofspace enclosure is sealed. A draught sealed self-opening vent is needed if air cooling is required. A self-closing return air damper is required, capable of isolating house from roofspace. Dampers will be operated in conjunction with the fan controller.

The respective performances of these two strategies and their variants were examined. Table 2, extracted from this exercise, shows the relationship between air exchange rate (ach) and the proportion of heat gain delivered to the house(Qh/Qt). The exercise examined the interaction of two operating conditions in the short period whilst solar gain into the roofspace is increasing;
i) The condition to maintain a low roofspace temperature, by increasing air exchange, in order to limit roofspace heat losses to ambient. This was taken as 25ºC, the minimum consistant with comfortable air delivery to house.
ii) The condition of a rising roofspace temperature (air delivery temperature to the house) in order to multiply the heat transfer rate to the house.

The former condition when applied to the fresh air system shows a reducing proportion of roofspace gains delivered to the house. The latter condition which is the only reasonable operating mode for a recirculation system (within the limitations of acceptable air velocities), shows an increasing proportion of roofspace gains(Qh/Qt) delivered to the house. Within the assumptions of this calculation, the air delivery temperature floated between 33-35°C.

TABLE 2   Summary Analysis of Calculated Heat Transfer Rates for Four
Possible Heating Modes.

	:Solar Vetilation	:Fresh Air System:		Recirculation system		:
Solar	:Pre-heat	:	:	:	:	:
Rad'ion :		:	:FloatingTemp		:stabilized temp :
Qir :	Qh/Qt	: Qh/Qt	: Qh/Qt	: QH/Qt		:
						:
W/m² :	%	ach :	%	ach :	%	ach :	%	ach :
93 :	38	0.8 :	35	1.3 :	30	1.3 :	50	4.2 :
131 :	38	0.8 :	32	2.4 :	44	2.4 :	65	7.5 :
168 :	38	0.8 :	30	3.4 :	53	3.4 :	72	10.7 :
205 :	38	0.8 :	29	4.5 :	60	4.5 :	77	14.4 :

Key
    Qh/Qt -heat given up to house(Qh) as a proportion of the total heat
    gain into the roof(Qt)
    ach   -air change rate in house volumes per hour
    Qir   -incident solar radiation in the plane of the roof window

To optimise solar performance, the recirculation system was adopted in this project.   The challenge of finding practical and economic solutions to the problems outined above was taken up.

In different circumstances the operating condition "Solar Ventilation Pre-heat"may be adopted as the effective solution.   Baker(6) has shown that controlled infiltration wholly taken through a sunspace compensates for ventilation losses in highly insulated building.

RECIRCULATION HEATING

Warmed air is drawn from the ridge line of the roofspace through a variable speed fan and ducted to 3 rooms in the house.   House air is returned to the roofspace via a grille in the ceiling of the stairwell.   The solar system is activated when the house is below the comfort set-point temperature, normally adjusted to 22-25°C and the roofspace is above a minimum air delivery set-point temperature, normally adjusted to 5°C higher.   Delivery fan and sealing dampers remain open until either comfort temperature is exceeded or roof temperature falls below minimum.

The fan speed is controlled according to the roofspace temperature.   This minimises draughts from supply air and heat losses arising When ambient air is drawn into the depresssurised roofspace.   Air is introduced to the house at 100m³/hr and the rate rises by 3 steps to a maximum of 250m³/hr at a roofspace temperature of 35 deg C (these values are adjustable).   Air delivery to the lounge ceases once the lounge thermostat is satisfied, Thereafter air continues to circulate through the bedrooms until the whole house thermostat is satisfied.

The solar system is given preference over the house central heating by
automatically lowering the central heating set-point for the period of
activity of the solar system.to continue to gain solar heat at higher
temperatures each householder may select a higher house set-point to
"overheat " the house,normally by another 5°C. The thermal mass of the
house is charged and surface temperatures are raised during peak periods of
solar gains. This,"Unoccupied" mode is the only none automatic function of
the system.

COOLING MODE

In the cooling mode the roof fabric is protected from excessive temperatures
by a thermostat connected to a roof exhaust fan and motorized inlet damper
set into the roof eaves. 800m³/hr of air is drawn through to limit the roof
temperature to 60°C. As a back-up, in case of circuit failure, the
heating fan which is on a separate circuit maintains cooling and alerts the
householder. A smoke detector has been installed which monitors both the
house and the roofspace.

CONSTRUCTION

ROOFSPACE ENCLOSURE

A recirculation system requires a well sealed roofspace for optimum
performance. Builders details were developed to attempt a reasonable
standard of air tightness,eg plastic film interleaved at wall/roof
junctions, rubber seals to roof sheeting, draught seals to ventilation
dampers.
Infiltration tests using tracer gas decay analysis in a completed roofspace
showed air change rates lower than 1.4 air changes per hour. Considering
the high surface to volume for the space this demonstrates the success of
these measures.

The roof glazing is the major source of heat loss from the system and twin
wall sheet was therefore selected. Even so the glazing constitutes 94% of
the fabric UA value of the roofspace." Makrolon"is 10mm plycarbonate sheet
(U value 3.1W/m²deg C) and is an economic and practical covering. It's
other qualities met the requirements of being lightweight, unbreakable,
economic to fix in large sheets, complete with air sealed glazing bars, and
having a reasonable transmissivity at 0.73.

The roofspace was partitioned along the ridgeline with an insulated
plasterboard screen. This provided a cool north roofspace for water tanks
and household storage separated from the solar roof. Control and monitoring
electrical centres were also located here. Just below the ridge a shelf was
formed in the partition to act as an air collection plenum with baffled
inlets to ensure that air was drawn uniformly from the length of the ridge.
Smoke trace tests following completion show that this facility to balance
the intake of air is extremely important to raise collection efficiency.
With an open fan inlet, an airpath short circuit between recirculation inlet
and fan leaves air at the eaves and ridge relatively undisturbed . The
heating fan mounted at the head of the house duct draws air from this
plenum. All internal surfaces were insulated with 100 to 140mm of glass
fibre quilt, and then lined with film to provide a sealed, dust free
chamber.

112

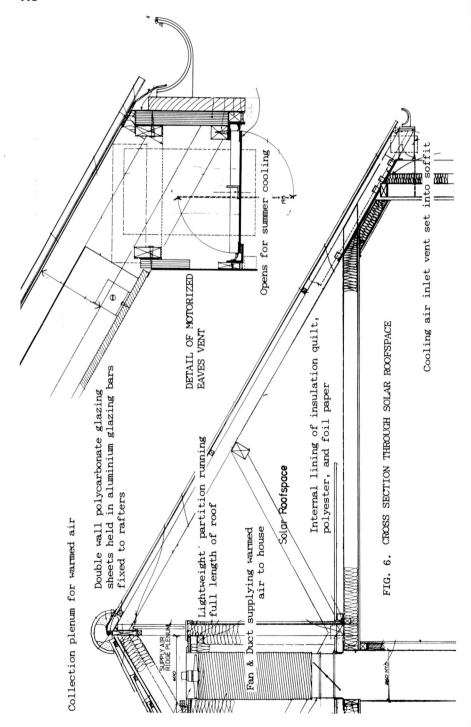

Collection plenum for warmed air

Double wall polycarbonate glazing
sheets held in aluminium glazing bars
fixed to rafters

Lightweight partition running
full length of roof

Fan & Duct supplying warmed
air to house

Solar Roofspace

Internal lining of insulation quilt,
polyester, and foil paper

FIG. 6. CROSS SECTION THROUGH SOLAR ROOFSPACE

DETAIL OF MOTORIZED
EAVES VENT

Opens for summer cooling

Cooling air inlet vent set into soffit

SUPPLY AIR
RIDGE PLENUM

AIR HANDLING

A single vertical duct of 0.06m² was designed to deliver air to the house. This was constructed in timber and plasterboard and located in the corner of a wall between bedrooms with sidewall delivery registers at low level into each bedroom and a ceiling mounted outlet into the lounge beneath. The duct contains an air filter and a shut off damper. The duct wall was insulated with accoustic lining to reduce noise transference, and the fan was connected to the head of the duct with flexible tube to isolate structure born sound.

Return air from the stairhead passes through another shut off damper which has been connected, via a flexible duct, to the solar roofspace.

For the heating flow a standard 70W axial flow fan(4 roofspaces) and an in line duct fan(2 roofspaces) were obtained from an economy range and for the cooling a 'through the roof' 80W centrifical fan and cowl unit was obtained from an industrial range. inexpensive shut off dampers were unobtainable off the shelf and so butterfly type dampers were fabricated to a standard design and operated with a remarkably low cost small electric motor. The eaves shut off damper, for cooling air inlet, was designed and fabricated to fit the house roof.

CONTROLS

standard inexpensive air thermostats with a range of 5-30⁰C and a 5⁰C set back were used in lounge and stairhead. Cylinder type thermostats with a 10-90⁰C range were most suitable for use in the roofspace. These were mounted on a heat sink plate within the shade of the plenum. A thyristor fan speed controller was developed to provide a monitoring facility with relays, and independantly adjustable speed settings.

An alternative micro processor based control, using temperature sensors would offer savings in wiring between discrete components. To be replicable, a commercially available inexpensive controller is required for this application.

VENTILATION AND CONDENSATION

In addition to heating, an equally significant feature of the roofspace system is the provision of air movement through the house. In a number of the unimproved project houses there were problems associated with inefficient ventilation. Condensation at some houses and draughts and exfiltration losses at others. Insulation, draught stripping and kitchen extract fans alleviated the worst of these problems. However the solar roofspace provides the facility to balance and control ventilation of principal rooms. Solar pre-heated ventilation air may be introduced in a draught free manner and circulated through rooms without excessive heat loss. One house occupier with a sinus complaint and a preference for wide open windows, has used the solar system instead to maintain ecomomic ventilation, even in winter. Another occupier has noted that her house is now free from dampness on bedroom and cupboard walls.

During air velocity tests on a completed house the roofspace system, on lowest fan speed, created air change rates of 1.7(lounge), 0.81(bedroom 1), 1.15(bedroom 2) with air warmer than the room. The overall delivery rate is adjustable and may be used to provide constant background ventilation.

114

HOUSE PERFORMANCE ANALYSIS

Simulation models, using SERI-RES, were constructed of the original
unimproved house, the house ugraded with insulation and double glazing, the
insulated house with solar roofspace. An extract from the performance
analysis(7) is given below.

HOUSE HEATING SYSTEM

Figure 2 shows comparative savings from insulation and solar roof gains.
As shown here, over the space heating season the insulation measures
save 23% (2055 kwhrs) of the demand on the conventional heating system.
Include the solar roofspace into the renovation work and the saving rises to
51% (a further 2294 Kwhrs).

The performance of insulation and solar gains are complementary over the
heating season. In January, insulation saves 26% and solar roof gains a
further 12% of heat load. In May insulation saves 13% and roof gains
another 74% of the heating load of the unimproved house.

FIG. 2. HEAT INPUTS TO THE HOUSE— YEAR TOTALS

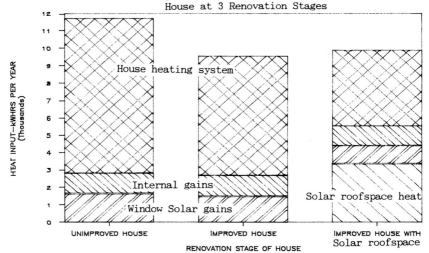

From fig 2, comparing unimproved to insulated house, whilst the heating
sytem load was reduced by 2055kWhrs, window gains were only reduced by
157Kwhrs(9% reduction) and internal gains by only 30kWhrs. This confirmed
that in both the unimproved and the improved house the heat need only rarely
fell below the level of the combined contribution of internal and window
gains. This implies that at both renovation stages there remains scope for
further solar gains.
    When the solar roof is added to the improved house then a major shift of
heat supply occurs. The roof system becomes the second principle source of
heat, supplying 34% of total compared to 44% of total from the heating
system. This proportion is an impressive demonstration of the potential of
this type of solar technology when used to optimum effect.
    The roof system displaces some of the direct solar (by 5% of its total
contribution) as the extra solar input shortens warm up time to reach the
set point temperature. The heating system input is reduced from 71% to 44%
of the total input (now 9909kWhrs per year).

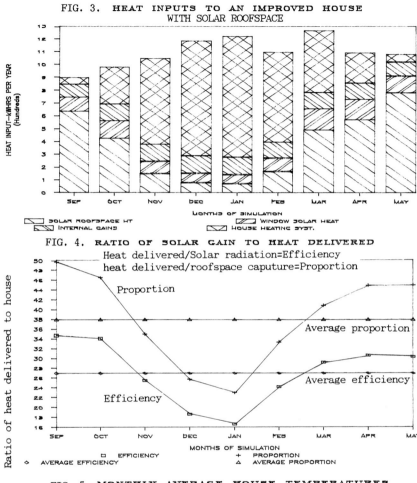

FIG. 3. HEAT INPUTS TO AN IMPROVED HOUSE
WITH SOLAR ROOFSPACE

FIG. 4. RATIO OF SOLAR GAIN TO HEAT DELIVERED
Heat delivered/Solar radiation=Efficiency
heat delivered/roofspace caputure=Proportion

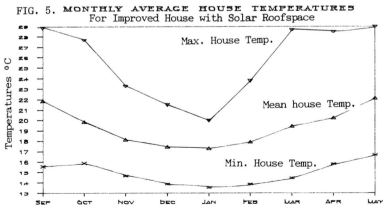

FIG. 5. MONTHLY AVERAGE HOUSE TEMPERATURES
For Improved House with Solar Roofspace

The total input is now 336Kwhrs more than the improved house. This is the extra heat delivered by the roofspace system to raise temperatures above the heating system setpoint   During "unoccupied" periods 10% extra useful heat input is gained by raising the setpoint of the roofspace system.

The proportional displacement of direct solar and house heating due to the inclusion of the roofspace system peaks at the beginning and end of the heating season. From fig 3, in May the roofspace contributes 72% to the total load, window gains 12% and the heating system 6%. By comparison in January, with low insolation levels and low solar altitude, solar gains through windows and the roof system are matched at 6% each of the total load. The heating system contribution in January has only been reduced from 84% to 77% of the total heat input to the roofspace house.

EFFICIENCY OF ROOFSPACE SOLAR SYSTEMS

The gross efficiency of the roofspace system in using available solar energy may be assessed by comparing the incident solar radiation in the plane of the roofspace glazing with the energy delivered to the house. Figure (4) shows that the convertion efficiency peaks during the early and late season high insolation periods reaching 35%, and falls to 17% in January. This gives an average efficiency of 27%.

COMFORT TEMPERATURES

The monthly simulated temperatures of the house are indicators of both the standard of comfort achieved in the simulation and the appropriatness of the selected control set points. Figure 5 shows that in the simulated"unoccupied" mode of the solar system, maximum temperatures averaged 28.5 deg C (for March April,May, September and October) indicating that this is a realistic mode in which to operate, that the set point is appropriate and that gross overheating of the house is not a major problem.

REFERENCES

(1) DOE, English House Condition Survey 1981 pt1, HSR10
(2) Energy Design Group. 1985 Woodbridge Cottage Solar roofspace Collector Final Report, UK D Energy, ETSU .
(3) Energy Designer (Energy Advisory Services Ltd,1986) is a variable base two zone degree day model derived from the BREDEM series of programs. From a description of the house fabric and services annual required heating will be calculated.
(4) Page,J. R Lebens,(eds) 1984, UK Data for solar energy Applications, ETSU S1134 .
(5) SERI-RES by Palmiter & Wheeling of Ecotope Group Seattle, now adopted b the UK passive solar programme, is a dynamic multi-zone simulation mode using finite difference and iteration over hourly intervals .
(6) Hancock,C., J.Littler, P.Ruyssevelt, P.Clegg, 1986, Hybrid Solar Roofspace systems in the UK, 11th conf pro. American Solar Energy Society,
(7) Hancock,C., J Littler, A Hybrid Solar Roofspace System-Performance Analysis, conf.pro. 1987 European Conference on Architecture.
(8) Baker,N.V., 1985, The Use of Passive Solar Gains for the Pre-Heating of Ventilation Air in Houses. ETSU report s1142 D En.

Heat Recovery by Heat Pipe Heat Exchangers to Improve Energy Efficiency in Farming and Industry

V. Hlavačka*, L. Macek**, F. Polášek*, P. Štulc*,
J. Valchář*, and L. L. Vasiliev***

* National Research Institute for Machine Design, Prague, Czechoslovakia
** State Commission on Technical and Investment Development,
Prague, Czechoslovakia
*** Heat and Mass Transfer Institute, Academy of Sciences,
Minsk, U.S.S.R.

ABSTRACT

The need for waste heat recovery in recent years yields a significant progress in design of heat transfer equipment. Among
others heat pipe heat exchangers (HPHE) have been used for a
large scale of applications. The proper prediction of their
performance requires theoretically and experimentally verified
data. The paper suggests an optimization procedure for the
HPHE thermal effectiveness under conditions occuring in ventilating and air-conditioning. Furthermore, some experience are
demonstrated that are obtained from measurements on HPHE´s in
farming and industry, concerning performance stability in
connection with fouling and other operation conditions.

KEYWORDS

Heat pipe heat exchangers, thermal effectiveness, optimization,
design, fouling, farming, industry, energy savings.

INTRODUCTION

Difficulties in fuel supply are felt still with increasing intensity. In terms of current prognoses one can hardly expect an improvement of this disquieting state of affairs in the near future. The finite nature of primary energy sources, in particular fossil fuels, has given to rise to a climate of energy consciousness which has injected additional impetus to the significant progress in the design of heat transfer equipment.

The early days of combustion air preheating for fuel fired furnaces showed that three essential features had to exist in order to generate consumer interest in heat recovery. First, a need for energy savings. Second, relatively low owning costs of waste heat utilizing equipment were necessary. Third, the recuperator had to have satisfactory reliability and low maintenance requirement. This continually applies in general for waste heat recovery in industry, farming and commercial or domestic sector.

The correct choice of heat exchangers and their sizing is probably the most important factor in designing an efficient and economic heat recovery system. From this point of view it seems to be a considerable problem to work out a proper method for an optimization of the recuperator thermal effectiveness.

There are many types of heat exchangers which are suitable for heat recovery systems. Recently, it has also come to a large application of heat pipes in the heat exchanger design.

A heat pipe heat exchanger (HPHE) can be considered as a special case of well known liquid coupled heat exchanger systems. Compared with this equipment the HPHE has a more compact configuration, the heat transfer by a heat carrier (a working fluid inside the heat pipe) takes place in boiling and condensing which, contrary to convection heat transfer for instance in duct flow, represents an obvious enhancement of the heat transfer. A tendency to augment the total heat transfer surface area occured at indirect heat transfer system is considerably reduced in the HPHE by several factors:

- the possibility to use very efficient finned surfaces on the
  outer side of cold and warm fluid streams;
- higher values of the heat transfer coefficient in the flow
  across the finned pipe bundle;
- easier achievement of pure counterflow.

These advantages make the HPHE particularly convenient for heat
exchange between gaseous fluids in heating, ventilating and
air-conditioning or in dryers. In this area the HPHE seems to
be more suitable with reference to thermal effectiveness, size,
manufacturing, easy cleaning and costs than many other types
of heat exchangers.

OPTIMIZATION OF RECUPERATOR THERMAL EFFECTIVENESS

The prevailing kind of the flow configuration for HPHE's should
be the counterflow. However, the parallel flow or cross-flow
may be desirable in some special cases.

The calculations for recuperator design are usually based on
the relationship between the thermal effectiveness

$$P = \frac{t_A'' - t_A'}{t_B' - t_A'}$$

and two dimensionless parameters, number of heat transfer units
$NTU = UA/C_A$ and heat capacity rate ratio $R = C_A/C_B$. The symbols
denote as follows:

t_A', t_A'' — inlet and outlet temperature of the fluid A, fresh
cold air ($^\circ$C)

t_B' — inlet temperature of the fluid B, waste warm air ($^\circ$C)

U — overall heat transfer coefficient ($W.m^{-2}.K^{-1}$)

A — total heat transfer surface area (m^2)

C_A, C_B — heat capacity rate of the fluid A and B ($W.K^{-1}$)

The recuperator thermal effectiveness can be expressed by a
generalized equation (Hlavačka, Valchář, Viktorin, 1980)

120

$$P = \frac{2}{1 + R + Z \coth \dfrac{Z}{2} NTU} \quad,$$ (1)

where

$$Z = \sqrt{(1 + R)^2 - 4p_p R}$$

and p_p is the flow factor: $p_p = 1$ for counterflow, $p_p = 0.82$ for cross-flow (NTU $\leqq$ 3) and $p_p = 0$ for parallel flow.

Solving Eq. (1) for NTU we obtain a relationship for heat exchanger total surface area calculations

$$NTU = \frac{1}{Z} \ln \frac{2 - (1 + R) P + \sqrt{Z} P}{2 - (1 + R) P - \sqrt{Z} P} \quad.$$ (2)

From this equation the following useful relationship may be derived

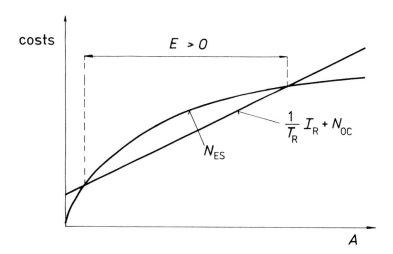

Fig. 1. Energy savings and investment and operational costs vs. heat transfer surface area

$$\frac{dNTU}{dP} = \left[1 - (1 + R)\ \bar{P} + p'_p\ RP^2 \right]^{-1} . \tag{3}$$

The annual energy savings N_{ES} expressed in X per year (X denotes a currency unit, e.g. $\mathcal{L}, \mathcal{P}, \ldots$) attained by a heat recovery system depend on the recuperator thermal effectiveness and increases with the surface area A as shown in Fig. 1. Capital costs of the heat recovery system I_R (in X) increases with the A as a rule linearly, i.e.

$$I_R = a + bA = a + b\ \frac{C_A}{U}\ NTU , \tag{4}$$

where b $(X.m^{-2})$ depends essentialy on the materials used in the equipment in question. Fig. 1 suggests that there may exist certain values of the thermal effectiveness giving a non-negative annual economic effect (X per year)

$$E = N_{ES} - \frac{1}{T_R}\ I_R - N_{OC} > 0 , \tag{5}$$

where T_R denotes the years of reduced life time of the heat recovery system (or of its main components) and N_{OC} (X per year) are annual operating and maintenance costs. In many countries special regulations apply when determining the items in Eq. (5). The annual economic effect E will reach its maximum when the "best" thermal effectiveness satisfies the following equation

$$\frac{dE}{dP} = \frac{dN_{ES}}{dP} - \frac{1}{T_R}\ \frac{dI_R}{dP} - \frac{dN_{OC}}{dP} \tag{6}$$

and results in E > 0.

The heat performance of a recuperator is given by

$$Q = C_A\ (t'_B - t'_A)\ P , \tag{7}$$

where the heat capacity rate of fresh air C_A and inlet temperature difference are assumed as annual mean values. Hence, the

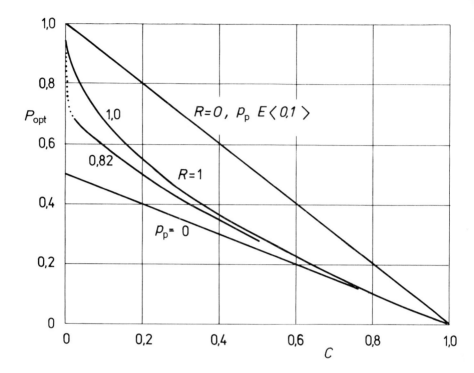

Fig. 2.   Optimum thermal effectiveness of a heat
exchanger

annual energy savings can be determined as follows:

$$N_{ES} = 31.54 \; 10^6 \; x \; Q \; N_E^+ \qquad (8)$$

Here, $x \in \langle 0,1 \rangle$ represents the relative annual operating
period, and N_E^+ is the unit energy or fuel price $(X.J^{-1})$.

Assuming that the operating costs N_{OC} are independent of the
recuperator thermal effectiveness, substituting from eqs. (3),
(4), (7), (8) in Eq. (6) gives a quadratic equation whose
solution for the optimum thermal effectiveness may be written
as

$$P_{opt} = \frac{1 + R - \sqrt{(1 + R)^2 - 4 \, p_p \, R \, (1 - C)}}{2 \, p_p \, R} \;, \qquad (9)$$

where

$$C = b \left[31.54 \ 10^6 \ x \ U \ (t_B' - t_A') \ T_R \ N_E^+ \right]^{-1}.$$

In the case of the above mentioned flow configurations, Fig. 2 shows the variation of P_{opt} with a dimensionless parameter C for $R = 0$ and $R = 1$. At low values of P_{opt}, the inequality (5) may be not fulfilled.

Similarly, as in the case of regenerators or regenerative loops, the overall heat transfer coefficient through a perfectly working heat pipe or in exchangers consisting of such pipes may be defined

$$U_o = \left[\frac{L_t}{L_c} \ \frac{1}{h_A} + \frac{A_1}{\pi \ d_i} \ \frac{s_W}{k_W} + \frac{L_t}{L_e} \ \frac{1}{h_B} + \frac{A_1}{\pi \ d_i} \ \frac{s_W}{k_W} \right]^{-1}.$$

The symbols denote:

L_t, L_c, L_e – total, condenser, evaporator heat pipe length (m)

h_A, h_B – heat transfer coefficient from the fluid A, B $(W.m^{-2}.K^{-1})$

A_1 – specific outer finned heat pipe surface area $(m^2.m^{-1})$

d_i – inner heat pipe diameter (m)

s_w – pipe wall thickness (m)

k_w – pipe wall thermal conductivity $(W.m^{-1}K^{-1})$

The overall heat transfer coefficient U for a real heat pipe is influenced by its internal thermal resistance R_i $(W^{-1}.K)$. Consequently, we obtain

$$U = U_o \ (1 + U_o A_1 L_t R_i)^{-1}.$$

Correct values of the R_i can be found in proper literature (Dunn, Reay, 1970; Chi 1970).

Note that they usually range from 0.005 to 0.05 $W^{-1}.K$.

The product UA for NTU calculations of the HPHE's is given by

$$UA = U_o \, (1 + U_o A_1 L_t R_i)^{-1} \cdot (L_c + L_e) \, A_1 N$$

where N is the total number of heat pipes in the HPHE.

HEAT PIPE HEAT EXCHANGERS MANUFACTURING IN CZECHOSLOVAKIA

The development of HPHE's for waste heat recovery was focussed on the most convenient economy. This led to the preference of gravitational heat pipes (closed two-phase thermosiphons). In working out the conception of the HPHE, the possibility of easy cleaning of the heat transfer surface was decisive.

It is thus typical for the whole layout that the two active heat exchanging zones, the warm (evaporator section) and the cold one (condenser section) are situated close above each other. Vertical heat pipes are placed in the guide grooves of the exchanger chamber. Each pipe is provided in the middle with a plastic separating lug. When assembled, these lugs serve as a continous baffle preventing the leakage of the fluids from one stream to the other. The dismountability of the heat pipe bundle is one of the main features of an efficient answer to the fouling problem and also permits to check the proper function of each pipe and to exchange the pipes which work unsatisfactorily.

Before the end of 1986 almost 900 heat recovery equipment with heat pipes were put into operation. The overwhelming part of the recuperators were produced by the enterprise Vzduchotechni Nove Mesto n.V. The heat pipes are made of aluminium tubes wit round rolled-out fins and also of bimetallic steel-aluminium tubes. The working fluid is mostly ammonia (HPHE of TH-type), only in some equipment water (TW-type) or freon were used.

OPERATIONAL EXPERIENCE WITH THE HEAT PIPE HEAT EXCHANGERS IN INDUSTRIAL AND AGRICULTURAL ESTABLISHMENTS

High concentration of domestic animals in common sheds of mass -production plants lay considerable demands on ventilation and air-conditioning of these buildings. The atmosphere of these

breeding spaces (temperature, humidity, concentration of
injurious matter and dust, temperature and velocity field)
affect not only the health condition, the weight increment of
the animals, consuption of fodder and its conversion, but also
the quality index of the final product and thus the economic
aspect of the enterprise.

In the agricultural animal mass-production the poultry
processing plants belong to the most exacting air-conditioning
installations due to their high claims on the program control
of temperature and moisture conditions, multiple air-exchange
and strict veterinary measures. The technical and economic
evaluation along with the operational experience have proved
that the HPHE are particularly convenient to utilize the
exchanged-air waste heat in poultry houses for fettering
broilers on deep litter systems, in cages or on slattered
floors, in layer rearing plants largely with cage technology
and in high-capacity hatcheries equipped with automatic
brooders with accurate thermal and humidity daily routine.

After extensive verifying tests and measurements it can be
said that in livestock housing the HPHE recovery is convenient
in prophylactic calf stations, in large-capacity calf-byres
with milk and especially vegetable alimentation where it helps
eliminate the projects of additional heat sources even in cool
mountainous areas and in old-style livestock-sheds where it
can significantly improve the microclimatic conditions preventing
the peripheral walls from getting wet and raising thus the
durability of the building.

SOME MORE EXAMPLES ILLUSTRATING THE OPERATION OF THE HEAT PIPE HEAT EXCHANGERS

The largest single investment project in Czechoslovakia was
the installation of 88 eigh-row recuperators comprising 10 560
two meters high heat pipes with ammonia filling in a new-built
factory to preheat fresh cold air for 11 lines curing and
drying artificial casings through waste heat of humid air from
the drier. The external air is drawn in by axial fans over
louvre boards and filters with a possibility of additional

induction of warm waste air in the winter period to prevent the icing of the condensation part of the heat pipes. The warm and humid air from the drying lines is sucked off by axial fans through the bottom part of the recuperators into the ambient surroundings. Measurements have shown (Hlavačka, Polášek, Štulc 1987) that the attained thermal effectiveness is as much as 65 % with average frontal velocity of 2.62 m.s^{-1}. The cut in annual heat consuption is as much as 23 000 MWh with an average of 7 400 operating hours per year.

Very effective is the application of the TWC-7 recuperators in a two-stage timber drier, type TWA II, although

1) the purchase costs of the HPHE system reached the double of the standard investment costs,

2) as a result of dirt deposits the thermal effectiveness has dropped from 48 to 38 percent and both recuperators have been unconveniently designed as co-current,

3) if the recuperators are redesigned and regular cleaning is provided for, the payback of the investment costs can be cut from the present 1.66 to 1.1 a year.

Appart from the most usual gas-gas type exchangers, there are also developed and employed heat pipe heat exchangers of the gas-liquid type.

Preheating of Fuel Gas through Wash Water

The fuel gas is used to heat pyrolytic furnaces and boilers on an ethylene unit. Natural gas, coming from the pipe line system beyond the bounds of the site, is mixed with fuel gas i.e. largely methane produced by the ethylene unit. Before mixing the fuel gas has temperature about 45 $^{\circ}$C, pressure 0.3 MPa and is burned, at flow rate of 22 to 30 t/h, without preheating in pyrolytic furnaces and boilers. Preheating by waste heat from wash water (80 $^{\circ}$C, flow rate 200 t/h) reduces its consuption. To preheat the fuel gas, an exchanger is used condisting of 40 ammonia heat pipes made of finned bimetallic tubes in 4 rows. The outer surface of the heat pipes in the wash water area is smooth while in the fuel gas section it is extended by circular radial fins. The thermal effectiveness

HVAC — heat pipe application

Better analysis tools — H

Fundamental — some example

Date: _____

To: _____

From: W.F. Kenney

of the exchanger designed in this way is as much as 70 %
according to operation conditions.

Heat Pipe Waste Gas Boiler

For the purposes of predominantly chemical industry a research
study including measurements was made of a small waste gas
boiler (540 kW) consisting of steel heat pipes to produce 0.2
$kg.s^{-1}$ of saturated steam with pressure 0.7 MPa by heat
recovery of waste gas having temperature 330 ^{o}C and flow rate
3 $kg.s^{-1}$. The heat pipes were finned on the combustion side
and smooth on the steam side. Because of high working temperatures the steel heat pipes in the first rows were filled with
diphenyl and in the remaining rows with water. Compared with
standard waste gas boilers, the heat pipe boiler is more
convenient from the constructional point of view. Its pressure
losses are as much as one order lower, the weight 40 % less
and the useful volume as much as 50 % smaller than the normal
waste gas boiler of the same performance.

CONCLUSION

The experience gained so far in running tests have indicated
that the HPHE is becoming an important feature in the
assortment of waste heat recovery equipment. Efficient extension of the heat tramsfer surface for both main fluid streams,
feasibility of the counterflow arrangement as well as easy
handling of the fouling or cleaning problem in a gas-gas heat
exchanger brings considerable advantages which, in comparison
with other types of heat exchangers, result in lower investment
and operating costs with equal thermal effectiveness. The
layout as well as the calculation for an HPHE design can be,
particularly in a case of gravitational heat pipes, considered
as sufficiently accomplished and creates favourable conditions
for proper designing of various systems intended to utilize
secondary energy sources.

The experience made so far in operating the HPHE´s offers good
prospects for their extensive introduction in both industry
and agriculture. The HPHE can be employed in many ways: not
only to recover heat, but to eliminate injurants and

adulterants from waste gas as well - e.g. in desulphurizing
combustion gas, deoxidizing nitrogen, removing organic solvents
etc.

From what we know so far from ten years´ operation of the
HPHE´s in agriculture we can say than in dusty working conditions of both animal and vegetable production their succesful
application is based mainly on these factors:

- the heat exchanging surface is easy to dismantle into their
  elements, the pipes being accessible after removing the side
  cover and releasing the quick-turn closures,

- where low-pressure axial fans are available, regular cleaning
  of the pipes can be done outside or even inside the recuperator (omitting the odd rows, a double-side nozzle connected
  with pressure steam can be inserted between the pipes),

- where medium pressure fans (axial or centrifugal) are working
  special wind-off filters can be placed in front of the
  airducts leading the contaminated air away to the recuperative exchanger, and/or large size frame filters can be built
  in with cleanable front wall or easily exchangeable filter
  elements.

REFERENCES

Hlavačka, V., J. Valchář, and Z. Viktorin (1980). Thermal
    Processes in Gas-Solid Particle Systems. SNTL Prague
    (in Czech).
Dunn, P. D., and D. A. Reay (1976). Heat Pipes. Pergamon
    Press, London.
Chi, S. W. (1976). Heat Pipe Theory and Practice. Hemisphere
    Publ. Corp., Washington.
Hlavačka, V., F. Polášek, and P. Štulc (1987). Heat Pipe Heat
    Exchangers for Heat Recovery of Low Grade Waste Heat of Air.
    Ed. ITMO AN B.S.S.R., Minsk (in Russian).
Zemánek, J. (1983). Energieanwendung, 32, No. 4, 26-31.

Ecological Building Project of Delft University of Technology

Hans Hubers/Ernest Israëls

Centre for International Cooperation and Appropriate Technology (CICAT)
Delft University of Technology, P.O. Box 5048, 2600 GA Delft,
The Netherlands

ABSTRACT

An integral and multidisciplinary approach to energy- and environmental technology led to a preliminary design of an egg-shaped half-underground ecological building. Feasibility studies showed an estimated energy use of only 12 % of the usual standard office (69 MWh/year). The running costs are estimated at standard office level (DF1 320,-/year). In principle there is no need for connection to the central drainage, gas, water and electricity. Rainwater is collected, the waste treatment system produces biogas (ca. 9 m³/day), electricity will come from a combined windpower/photovoltaic plant. A symbiosis is being proposed between an office building and food production (fish, vegetables, eggs). Depending on the fundraising the building could be finished in 1990

KEYWORDS

Ecological building; solar energy; environmental technology; underground building; waste treatment; building costs; design method.

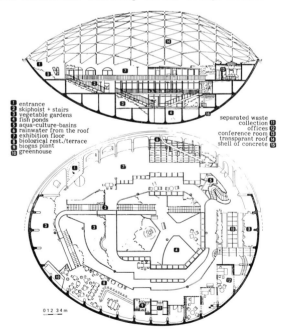

entrance **1**
skiphoist + stairs **2**
vegetable gardens **3**
fish ponds **4**
aqua-culture-basins **5**
rainwater from the roof **6**
exhibition floor **7**
biological rest./terrace **8**
biogas plant **9**
greenhouse **10**

separated waste
collection **11**
offices **12**
conference room **13**
transparent roof **14**
shell of concrete **15**

0 1 2 3 4 m

Fig. 1. section and floorplan

129

130

INTRODUCTION

Aims of the Project

The ECO-project wants to be a project of multidisciplinary research,
realisation and evaluation of energy saving and environmental
technology.
The preliminary design "the Egg" serves as a starting point. The
functions of this building will be:
- generation of future research;
- a test facility for this research with real users;
- demonstration object of the research for a large public;
- office of a Centre for Environmental Technology and Environmental
  Management.

Context

Building nowadays becomes more and more a multidisciplinary issue. A
building project effects the environment in many ways and the life of
men, animals and plants in it. Especially if the production of raw
materials, energy, food and other products that are used in buildings
are considered too.

The problems in the Western countries, where 20 % of the world
population consumes 80 % of the world resources, are well known:
pollution, economical crises, energy dependency and social problems
(allienation, vandalism etc.). These problems are interrelated: energy
production has economical effects, often pollutes the environment and if
people cannot understand and influence it, alienation will result.
This is the first argument for an integral approach of architecture. The
second argument is a purely economical one.
If, e.g. food production (fish, vegetables) is integrated in new
buildings, there is a good chance that food production becomes less
expensive (less energy, less transport, combined initial costs, combined
waste treatment etc.). At the same time the built environment becomes
more interesting, both physically (production of oxygen and humidity,
collection of dust and toxic elements) and psychologically (connection
with basic needs, pretext for social contacts). In ecology this is
called "symbiosis".

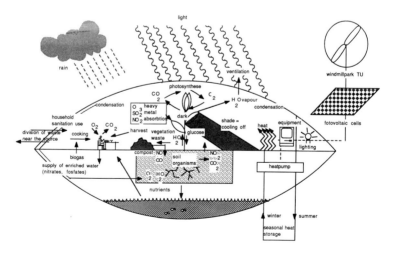

Fig. 2. global scheme of the cycles

Design proces

Ecology is a science which studies the relations of organisms with each other and their environment. It seems necessary to use the methods and results of this science in building. We therefore formulated the long and short term needs of everybody concerned with the project (also animals and plants) in all stages (from the production of raw materials until the recuperation of building materials after destruction). Using the combinative method (1) we tried to generate a complete list of criteria (fig. 3).

1	HEAT TRANSMISSION :	building envelope, isolation
2	DRAUGHT/WIND INFILTRATION :	chink length, building height
3	LOSS BY AIR-CONDITIONING :	contents of air-conditioned rooms
4	INTERNAL PRODUCTION :	persons, machineries, illumination
5	USE OF SOLAR ENERGY :	glass surface, SSE-front, shadow, efficiency
6	ENERGY CONTENTS OF BUILDING MATERIALS :	quantity, kind, residual value
7	ENERGY OF CONSTRUCTION :	construction time, materials, quantity
8	ENERGY OF MAINTENANCE :	kind, quantity
9	TOTAL ENERGY :	
10	RELATION COSMOSPHERE :	weather, seasons, sun, day light, spatial character
11	WIND HINDRANCE :	sheltered places outside, entrance, pyramide effect
12	WIND ENERGY :	foundations, turbulences, building height
13	RELATION ATMOSPHERE :	hearing wind, smells, seeing the sky, ventilation
14	USE OF WATER :	fishing, swimming, catch of rain water, storage of heat, humidity control, plants, light reflection, sanitary, evaporation heat (cooling)
15	RELATION HYDROSPHERE :	fountain, meeting places, reflexion (motion)
16	USE OF GROUND :	shadowing, possibilities of extension, spatial variation
17	RELATION LITHOSPHERE :	insertion in the location, territory, residences
18	FOOD PRODUCTION :	terraces, roofing, fishing pool, insectivora
19	DIVERSITY OF SPACES :	spatial variation, temporal rest, many gradual transition gradients, food, territories, shelter possibilities, natural enemies, no mono-cultures: illness and plaques
20	RELATION BIOSPHERE :	relaxation, astonishment, perceptive signals
21	HANDICAPPED PEOPLE :	parking places, stairs, safeguarding, escape routes
22	ADAPTABILITY :	flexible, variable, changeable, extensive, reduceable, additional potentiabilities, convertable, rebuildable
23	SPATIAL PLANNING AND ORGANIZATION :	relation between users and visitors, circulation, orientation, recognition, equivalence, social control
24	PRIVACY :	view, looking in, sound, communication, individual expression
25	SAFETY :	fire, burglary, violence, flodding, earthquake, assault, traffic, escape routes, illumination
26	ESTHETICS :	recognition, communication, composition, structure, contrast, proportion, scale, rythm, view
27	ENVIRONMENTAL JUSTIFIED MATERIALS :	little reinforced concrete, aluminium, plastics, radiating gypsum and chipboard, no asbest, preferably unbakened bricks, wood, reed, second-hand materials
28	TESTCASE :	glazed verandahs, solar windows, water walls, solar pools, solar collectors, heat storage, heat pumps, heat exchangers, isolation shutters, wind energy, earth heat, biogas, compost, green house, food storage
29	DESIGN AND CONSTRUCTION :	variation, nice materials, working method
30	COSTS :	macro-economical: unemployment, innovation, financial support, politics
		micro-economical: costs of initiation, exploitation, maintenance, demolition

Fig. 3. criteria list

n the preliminary design process we used several methods to stimulate creativity (2,3). With participation of the future users, we made a choise, on the base of a criteria-list, between four alternatives of the building lay-out. This was not only done by formulating advantages and disadvantages, but also by calculating points of appreciation. xperiments in this field showed that the use of this so-called SA-method (Potential Surface Analysis) often leads to discissions which re regretted afterwards. In fact the importance of criteria is very often related to the alternatives and cannot be expressed in the same numerical value. Decisions are best taken with both sides of the brains. During the discushions the calculated energy-use became the main selection criterium. Ultimately "the Egg" was chosen because of its potential attraction for a large public.

Fig. 4. artist's impression

DESCRIPTION OF THE DESIGN "THE EGG"

The typical building lay-out resulted from the sketches of an underground alternative. The constant earth temperature of ca. 10°C is an advantage in both summer and winter. However, a complete underground building does not take profit of passive solar energy and natural day light. Besides this, it might cause psychological problems for users of the building. So we introduced a large transparant roof at street level (40 x 50 x 10 m). This was also a good solution of the shading problem at the site. Thus maximal use is made of the diffuse radiation (ca. 60 % of the total radiation).
An upside down pyramide is created by starting the building lay-out at the lowest level with the offices, using them at the next ring as a gallery and so on to the street level. The enormous groundwater pressure (the site is in a polder) and the wish to limit the heat losing surface leads to a concrete shell. For reasons of balance between the weight of the building and the groundwater pressure, the top level is put above streetlevel. This also makes the shell visible and understandable.
Heating up buildings in the morning takes a lot of energy, therefore we located most offices in the morning sun. This resulted in the egg-shape and its orientation. Also the slope of the shell at one side becomes less steep. Here we planned the stairs and the gardens.
When entering, one has an immediate overview. There is no need for an info-desk. The restaurant is at street level to make seperate use in the evening and in week-ends possible. The meals (max. 240 daily) will partly consist of food produced in the building (fish, vegetables and maybe eggs). Preparation will be visible for the visitors, sitting at small tables under a pergola with grapes.
Waste from the garden and the kitchen is treated in the biogas plant (fig 13). Methane gas is used for cooking. So one of the biological cycles is closed. The effluent after treatment feeds the fish, and so another cycle is closed too. Rainwater from the roof is collected and used in the kitchen but also for toilet flushing, cleaning,

plant-watering, fish farming, heat storage etc. On top of the first
level the gardens also serve for cooling in summer.
A lot of space is reserved for demonstration and exhibition. Also the
galleries have this function, which is favourable for the gross/net
ratio of the floor area.
The materials that will be used should need little energy to produce,
little maintenance and must be re-used after destruction. For the
concrete shell these goals are difficult to meet. Also, for reasons of
fire safety, the floors probably will have to be of concrete. However,
research will be done into the use of roundwood and rammed earth. The
same goes for the separation walls and the inside supporting structure.

Another criteria for the choice of the materials is their acoustic
behaviour. Application of the law of Sabine (T = V/6A) shows that the
absorption coefficient should be a $\geq$ 0,57 to get a reverberation of less
than 1 second. The plastic films used in the roof are important in this
respect and also the inside gardens will be favourable. In the next
stage of the project (integral research, adaption of alternatives) more
detailed research will be done into these aspects. This again will make
clear that integral, multidisciplinary research makes sense. The second
stage of the project can start after the fundraising has been completed
successfully.

ORGANIZATION AND MANAGEMENT

Because Third World countries benefit most by changing attitudes in
western industrialized countries, the initiative for the ECO-project was
taken by the Centre for International Cooperation and Appropriate
Technology in 1980. In building more than 50 % of the money-, energy-
and material flow is used. As already stated before, a multidisciplinary
approach is necessary. However, without a specific object it is not
possible to make, e.g. urban planners and building physiciens work
together. Also without external financial support the research cannot be
done. Both statements were arguments to make a design, in order to get
things going.
Now a multidisciplinary working group of 8 departments from 3 faculties
of Delft University of Technology is carrying out the feasibility
studies. As a consequence of economizing by the Dutch government the
universities have to raise research funds themselves for over 25 % of
the total budget.
Therefore the steering committee (consisting of members of the
university board, representatives of the faculties of Civil Engineering
and Architecture, and the Directorate of Government Buildings) invited a
marketing bureau to investigate the fundraising possibilities for the
project. Results will be known by June of this year.
In the meantime the project group (consisting of the coordinator of
CICAT, the designer, technical coordinator and secretary), sponsored by
the Ministry of Urban Planning, Housing and Environment, is managing the
feasibility studies and the semi-integral research into new ideas. Many
building companies got involved. The plans received favourable publicity
in the Dutch media, professional as well as general.

FEASIBILITY STUDIES

Results

Economy. The final report of the graduate student about the costs is not
yet ready. Preliminary results are based on bids of building companies.
The initial costs will be ca. DFl 6 million. Considering the fact that
initial costs in general only make up $^1/_5$ to $^1/_8$ of the life cycle costs,
it makes more sense to calculate the annual costs. These are estimated
at ca. DFl 320,-/m^2 (gross. ex. ponds, real interest 5 %, including
correction for inflation 2 %), Januari 1985.
This is at the same level as average office buildings. It must and can
be optimized during the integral research.

134

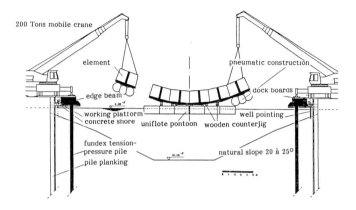

Fig. 5. one of the foundation methods

Foundation. At the site the street level is about 1 m. below sea level. In the polder the groundwater is kept artificially 1.5 m below streetlevel. Maximum difference between the highest and the lowest level was 0.9 m during the last few years. This means that if no foundation piles are used, the building will go up and down and soon break into pieces.
In the four building methods, investigated by graduate students in cooperation with four building companies, the forces during construction were found decisive for the dimensioning. One of the most important limitations was the prohibition of lowering the groundwater level. This would cause tilting of buildings next to the site. The result after more than one year of research is, that the most traditional method with a dry trench is less expensive (DFl 2 million). More interesting was the prefab solution, fig. 5, but sealing of the elements was too expensive. Digging of a wet trench of 10 m. deep, in the right form, was also found to be a rather expensive solution. Still, it has been shown that it is technically possible to make an underground egg-shaped concrete shell. There are some good ideas and possible combinations which have to be worked out in the future.

Fig. 6. wire lacing tool typical meeting of struts try out Lelystad

Roof. Here we investigate the possibilities of using Dutch roundwood. Timber poles need little energy (0,5 GJ/m^3) (6). The Netherlands produce only 10 % of their woodconsumption. Every year ca. 150.000 km^2 (^{2}/3 of the BRD) of tropical forest disappears (4). If no measures are being taken the world wood resources will be finished within 75 years ! If, on the contrary, the productive forests are enlarged with 30 %, the world's need of energy and wood can be provided in a closed biological cycle (5). So it certainly makes sense to use Dutch roundwood.

One difficulty to overcome is the problem of axial shrinkage cracks, due to drying of wood. Under supervision of dr.ir. P. Huybers (7) of DUT a wire lacing tool and several connection methods have been developed. (fig 6). Many tests took place in the Stevin Laboratory of the Department of Civil Engineering, DUT and in the first proto-type buildings in Lelystad (the Netherlands) and Winchester (UK). These tests on space-deck structures with poles of 100 mm diameter (Larch wood) resulted in an overall E-modulus of ca. 5100 N/mm^2.

Fig. 7. computer drawing of the roof structure

The research by two graduate students into application possibilities for the egg-shaped dome, showed:
- it was possible to write a computer program for the geometry, which gave adequate input for the program ICES-STRUDL (fig 7);
- a single-layer triangulated dome with 4 m. long poles of 200 mm diameter and four lacings was calculated;
- snap-through buckling under windload would be decisive for dimensioning;
Much discussion is possible about the interpretation of the test results and translation into this specific construction. If one is optimistic it will be possible to construct the dome within building regulations. So still more research has to be done. Everybody is convinced that a way to succeed will be found. Estimated costs in this stage of development are DFl 260.000,-.
The roof-cover was developed by a graduate student in a factory of plastic products. The roof will be covered by plastic film because this material needs 10 times less energy to produce than glass, does not break, is even more transparant for light (regrettably also for infra red !) and is lightweighted. It does not drip when burning. It can be recycled. Pollution during production must and can be stopped by better laws.
Triangular cushions with three layers of 'Hostaflon' (fluor film) were to be held by polycarbonate profiles. These profiles seemed to be too expensive (DFl 130,-/m^2 roof). Commercial alternatives already exist in aluminium but this material needs excessive an amount of energy to

136

produce (540 GJ/m³) (6). Theresearch is now aimed at larger cushions
with another geometry. At the same time the cushions must be used as
shading device and solar collector. In this way the profiles probably
become commercially interesting.

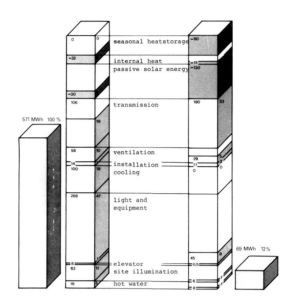

Fig. 8. standard office versus the Egg

Energy. One of the goals to achieve with this building is to minimize
the fosil fuel consumption and to avoid as much as possible the use of
supplementary installations. So the internal organization and lay-out is
very important for the future energy consumption.
The rooms and working area's are located under ground level to benefit
from the constant ground temperature. This temperature will be the mean
temperature in the rooms after some time. So insulation is only needed
where the building rises above the ground.
The hall is a buffer space. It functions not only as an entrance but
also as an integral part of the climatization system.By its height (20
m) it forms a very good thermal chimney, which keeps the ventilation
going, abolishing the need for mechanical ventilation. In the winter,
cool outside air flows down to the bottom of the hall, a heat exchanger
in the pond gives it a comfortable temperature. It rises up through
ducts to the rooms. From there to the hall and escapes through vents in
the top of the roof. These vents will be automatically steered according
to wind conditions and ventilation demand. In summer outside air flows
through the same ducts in the ground where it is cooled. Now it flows
directly into the rooms and again to the hall. During the days extra air
is introduced along the edge of the roof. The vents in the top are in
summer position with a larger aperture.
Calculations on micro-computer are made with the model of fig. 9.

An important mechanism in regulating the climate in this hall is
evaporation (from plants and fish ponds) and condensation. Of course the
roof will be constructed to "drain" the condensed water without
occasional "rainfall". The relative humidity, an important factor for
human beings and plants, will be controlled by the isolation qualities

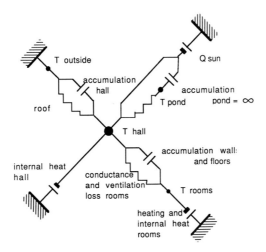

Fig. 9. analogon

on the roof. If necessary this can be changed easily by regulating the pression in the cushions.
Apart from the direct sun ligth the ponds will be the main heat source of the building in winter. The pond is heated by the sun and from the heat storage. In summer this storage is charged with solar heat recovered by the shadowing system in the roof. A pump nevertheless is needed but it can be efficiently powered by solar cells because power demand is exactly in line with power supply.
Another important role of the hall is the illumination of the rooms. Regulation of the light intensity takes place by the already mentioned shadowing system, without the heat loss, normally associated with daylighting. The advantage of this indirect way of illumination via the hall is more equal distribution, even in bright sunlight.
A detailed computer simulation with the SIBE program (Solar Irradiation Built Environment) of the DUT is executed at the moment.
Results of the energy feasibility study uptill now are:
- the U-value of the transparant film cushions is determined with a complex model to take the specific qualities of this film into account. The heat transfer by radiation, convection and conduction are seperately introduced in this model. Especially the transmittance of infra red is quite different from ordinary materials as glass.

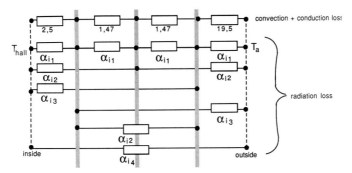

Fig. 10. calculation model for U-value of the cushions

138

- calculation of heat loss and comparison with a standard office. As we see in fig 8 the ecological building has a larger energy flow as the standard office. But it is free energy from the sun shining anyway. Another remarkable effect is minimization of the use of electric energy. Because the ecological building does not need mechanical cooling and uses more daylight.
- comfort conditions in average and extreme days. These will be the starting point for the detailed study and tuning of the integrated ventilation system.

hot summerday

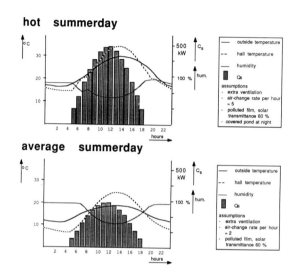

average summerday

average winterday

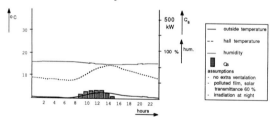

cold winterday

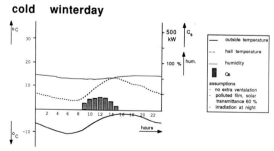

Fig. 11. temperature, humidity and radiation

<u>Flora & Fauna</u>.The effects of fish-farming and vegetable growing on the energy use and inside climate are multiple. For production of humidity next equations have been used:

$$G_{plants} = 2.0833 * 10^{-2} * A_{garden} * \frac{1 - \emptyset^*}{0.5} \quad [kg/h]$$

$$G_{pond} = k * A_{pond} [c_{max} - \emptyset^* * c_{max}^*] \quad [kg/h]$$

where: G = humidity production
      k = 20
      A = surface
      c_{max} = maximum concentration H_2O

      $\emptyset$ = relative humidity
      * = idex for hall

The energy flow through evaporation and condensation has not yet been introduced into the calculations. The research into selection of specimen and financial consequences just started. We hope to report about the feasibility of the food production next year.

<u>Waste treatment</u>. A choice had to be made between compost toilets and a biogas plant. First the quantity and quality of the input had to be determined. Especially the C/N ratio is important for biogas; this has to be ca. 30 and for compost ca. 25.
Since there would be too much compost for use in the building and not enough to sell, the biogas plant was chosen. Alternatives like burning, dumping, activated sludge, were also considered. Finally the scheme of fig 12 is worked out (fig 13).
In principle the system needs no connection to central drainage, gas, water and electricity. In spite of this, it will be done, but only for experimental reasons.
- input: rainwater, faeces, urine, garden and kitchen waste
- output: clean water, compost and ca. 9 m^3 biogas/day, enough for cooking.

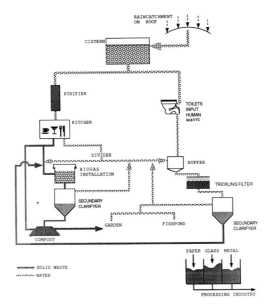

Fig. 12. scheme of waste treatment

140

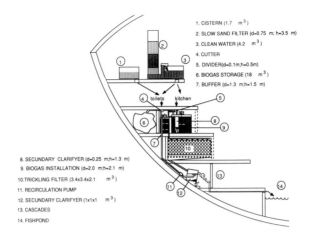

1. CISTERN (1.7   m^3)

2. SLOW SAND FILTER (d=0.75 m; h=3.5 m)

3. CLEAN WATER (4.2   m^3)

4. CUTTER

5. DIVIDER(d=0.1m;h=0.5m)

6. BIOGAS STORAGE (18   m^3)

7. BUFFER (d=1.3 m;h=1.5 m)

8. SECUNDARY CLARIFYER (d=0.25 m;h=1.3 m)

9. BIOGAS INSTALLATION (d=2.0 m;h=2.1 m)

10.TRICKLING FILTER (3.4x3.4x2.1   m^3)

11. RECIRCULATION PUMP

12. SECUNDARY CLARIFYER (1x1x1   m^3)

13. CASCADES

14. FISHPOND

Fig. 13. adaption in the building

FURTHER DEVELOPMENTS

The overall conclusion must be that most feasibility studies are
positive. In the integral research phase alternatives for foundation
have to be considered. A group of students already started to look into
the influence of form, depth and location on the costs and other
criteria. For the roof additional tests have to be done on the 200 mm
diameter poles with appropriate connections. Also windtunnel tests will
be necessary. Integral research has to be done into the roof-cover
cushions, where solar collectors, shading devices and ventilation
openings have great influence on the energy balance. Aspects of
condensation and acoustics have to be considered. The other subjects of
the feasibility studies (inside and outside gardens, fish farming, waste
treatment, participation and management) must be studied by the
multidisciplinary design group that will make the final design.
In order to survey the costs, initiatives are being taken to computerize
the initial and running costs with the aid of a cost database.
Financial and material support has to be obtained from software houses,
building companies, national and European programmes, and the universit
itself.
The latest development is the possibility to cooperate with a national
initiative for a 'science centre'. The roof, installations an
demonstration of the different ecological cycles are being considered
Probably after two years the whole could be taken down and reconstructe
at the site of the Ecological Building. This would really be a test fo
demontability, one of the properties of ecological design: a way o
re-use.

REFERENCES
Polak, B.M., Ontwerpmethodieken, DUT, Faculty of Civil Engineering
    Dept. Civil Management, 1981
Adams, J.L., Conceptual Blockbusting, San Fransisco, Freeman, 1974
Bono, E. de, Lateral Thinking, London, Wardlock 1970, Education Inst.
UNESCO, FAO, WWF, UNEP, IUCN, World Conservation Strategy, 1980
Wolterson, J.F., De vernieuwbare grondstof, The Haque, Staatsuitgeverij
    Ministry of Agriculture, Agraricultural series, 1977
Vissers, M., Energiebewust materialen kiezen, Bouwwereld 78 (1982)
    nr. 7, pp. 40 - 42
Huybers, dr.ir. P., Timber Pole Space Frames,Space Structures 2
    1986/1987, pp.77 - 86

Intelligent Electricity Metering

P. R. Hutt, MA, DIC, MIEE

Response Company Limited,
77 Wales Street, Winchester, Hampshire

ABSTRACT

Electricity accounts in many cases for over half of a Company's energy costs.

A new generation of flexible, solid-state lightweight, multifunction metering equipment is now available to put electricity consumption under control.

These new meters are highly effective in saving and accounting for energy either as individual controls on major supply points or as part of a centrally interrogated energy auditing system.

A number of examples of installations together with resultant cost savings will be described.

KEYWORDS

Electricity Meters; Electricity Supply; Energy Management; 3-Phase Meters; Power Measurement; Transmitting Meters; Pulsing Meters; Static Meters.

INTRODUCTION

In 1884 Ferraris invented the induction wheel meter. For the following century his principle of electricity measurement remained almost unchallenged. Meters based on the rotating aluminium disc proved to be quite adequate for the purposes of the electricity utilities. In fact they largely still do though attitudes are changing.

In the fifties and sixties electronic metering began to be researched. By 1980 a prototype electronic single phase meter was demonstrated. But it was not until 1984, exactly 100 years after Ferraris first conceived his meter that the first truly microelectronic meter for three-phase industrial supplies became available. In energy management the benefit to the consumer of energy is paramount and this paper explains how the microelectronic or 'intelligent' meter provides the means to minimise electrical energy charges rather than just to pay them.

THE FERRARIS METER

The Ferraris meter is a small electromechanical motor which rotates in sympathy with load power. Driven by its motor, a light aluminium disc, is a gear train which drives a set of indicating digits. These digits register the total amount of energy supplied.

The meter is heavy and cumbersome but is also rugged, reliable and reasonably accurate. Its main deficiency is the fact that it can only register one power parameter, it is also cumbersome to interface to electronic systems and intrinsically is incapable of indicating circuit instantaneous power. Its basic unit of measurement is a single disc rotation and this is indicative of consumption of a certain defined increment of energy. It can be configured to read reactive energy but is incapable of registering apparent, kVAh energy.

TRANSMITTING METERS

With the growth of the energy management industry the demand arose for meters which could interface to electronic systems. A meter was required which could produce a pulse output indicative of energy consumption. Various forms of optical electronic and magnetic sensors were developed which formed an add-on sensor generating a signal for each disc rotation.

The simplest meters produce an unconditioned electronic output pulse. This often represents quite an inconvenient quantity of energy. Figures such as 5 revolutions per 3 watt hour are common. So more sophisticated interfacing modules were developed to produce a pulse output in the form of a relay closure for a convenient increment of energy such as 1 unit or 1/10 of a unit

Electromechanical meters with pulse output form an uneasy compromise between electrical measurement and true electronic sensing. They have been termed 'transmitting' meters but only transmit in the simplest sense in that they produce a pulse which can be externally detected. They do not generate a true digital serial information signal.

ELECTRONIC METERS

It is self-evident that a quantity such as electricity flow should be measurable electronically. From the 1950's onwards electronic meters were developed utilising conventional electronic analogue circuitry. Many manufacturers developed circuitry for both single and three phase meters. Some meters have been designed to transmit pulses and some have been designed for the traditional domestic meter market.

The result has been a gradual evolution of electronic meters capable of metering both single and three phase supplies, but still, apart from pulse output, almost all have remained single function. Any tariff or communication function has been left to secondary devices.

ELECTRICITY CHARGING PARAMETERS

Industrial electricity tariffs are always structured by the supply authority to achieve the following objectives.

a) To charge on a unit basis for energy delivered to the customer.

b) To penalise the customer for inefficient delivery of energy, i.e. poor power factor.

c) To penalise the customer for irregular consumption of energy, i.e. peaky consumption.

COST REDUCTION TECHNIQUES

Mindful of the above charging strategy the customer must plan and regulate his consumption to minimise not only his total consumption of energy but also the way in which he consumes it.

The first requirement is to take as few KWh as possible. Electricity, on a thermal basis, is approximately four times the price of gas. Normal monitoring and conservation techniques will be applied in the same manner as for other energy sources. But in view of the price of electricity great care should be taken to measure it, audit it, sub-meter it and make individual users attributable for its conservation and consumption.

This creates the requirement for a meter which is simple to install, easy to interrogate and which can communicate both energy consumed and ideally also instantaneous rate of energy consumption, that is the instantaneous load power.

Poor power factor is penalised in a number of ways. Since conventional meters do not measure power factor directly, electricity boards must charge by installing two meters. A second meter measuring reactive kilowatt hours can indicate the average power factor over a period and this is often penalised if less than the magic figure of 0.9.

A meter which indicates power factor directly or measures both reactive as well as true power (or energy) can give the consumer instant monitoring of his consumption. Power factor correction measures can then be applied to minimise the cost penalty imposed by the board.

Peakiness in consumption is penalised on the basis of a demand charge. The demand charge is levied on the basis of the highest load taken in any half hour during the month. The basis can be KWh units or KVAh units depending on the area board. The advantage of the KVAh penalty is that it automatically also penalises for poor power factor.

The demand charge is often misunderstood. There is no penalty which is applied above a certain threshold. There is no threshold other than one that may be imposed by the customer on himself as a form of self-discipline. Demand charge is levied on the worst average KW or KVA loading in any half hour in the month. It is charged on the basis of pounds per KW or KVA starting at a zero base line.

A demand threshold set internally by the consumer maintains demand below a known level. By first examining consumption pattern he can determine a reasonable threshold for rate of energy consumption and determine to keep below it. This will minimise his demand charge which can often be as punitive as his units charge, particularly in winter months.

A meter which warns of excursion over a desired set point will significantly curtail demand costs.

THE CONSUMER'S NEEDS

Mindful of the sheer complexity of electricity charging, which has only been partly explained in the previous sections, it is clear that the consumer's interest can only be poorly served by the electromechanical meter and even the single function pulse-tramsitting meter is of limited use as a defence against unnecessarily high, and complex electricity bills.

Metering is performed within a consumer's premises at both the main supply point (or points) and increasingly at multiple sub-metering points throughout the building or production process.

Metering can be performed on a portable, quick-check, or instrument basis. It can be performed in a semi-permanent survey application. The latter often leads to a permanent sub-metering installation with perhaps fewer points metered than during a survey.

A meter is needed with the flexibility to perform the widest possible range of functions though in certain sites or applications not all these functions may be required. The meter needs to be comprehensive, portable, flexible and communicable if it is to meet all the requirements of today's energy management and measurement industry.

THE INTELLIGENT METER

In view of the wide ranging needs of electricity measurement it became clear in the early 80's that only a microelectronic approach could satisfy the design brief for an effective industrial supply meter.

Response Company undertook a programme of fundamental research and design which resulted in 1984 in the launch of a truly intelligent meter. Intelligent because for the first time a microprocessor was at the core of the design. The RESPONDER 3 with its intelligent integral microcontroller could be programmed to be a multifunction communicating meter and could as time proceeded be extended using more memory and programme to tackle further subsidiary tariff and metering tasks. A production instrument is shown in Fig. 1.

Fig. 1.   Instantaneous values of supply and load parameters
as well as energy consumption are displayed on a
Responder 3 electronic meter/transducer.

Some of the features of this 'Responder 3' meter are listed below.

- Class 1 performance (1% accuracy).
- Measures all the following parameters
  kW   kWh
  kVA   kVAh
  kVAr   kVArh
  Power Factor
- Communicates via opto-isolated and volt free relay
  pulses.
- Communicates via serial ASCII data signal.
- Infinite range of power handling via switch and soft-
  ware constant settings.
- Low 0.2 ohm burden means minimal size current trans-
  formers.
- Intelligence of meter utilised for installation self-
  test checks.
- Insensitive to vibration, position, environment.
- Instantaneous power functions allow quick testing
  calibration and customisation.

In addition to these benefits the inherently cheap mass production techniques
of printed circuit manufacture point to an economic solution to multiple
point metering as the volume energy management market matures.

BENEFITS OF THE INTELLIGENT METER

Resulting from the development of the intelligent meter there followed a
number of benefits hitherto unavailable and unassociated with metering.

Portable metering becomes simple and convenient due to position insensitivity
and low-burden clip round current transformer technology.

Local area networking of meters becomes possible whereby a number of meters
are all addressable via a single pair loop in a 'Christmas tree lights'
configuration.   A low cost computer addresses all meters via a loop controller to keep track of departmental or sub-building consumption figures.
Electricity auditing accounting and cost allocation becomes not only possible
but economic.

Control of electricity consumption to prevent undue penalties from power
factor and demand peaks now becomes a real-time preventative function rather
than an historic inquest.   Certain functions can be built into the meter
which hitherto were considered totally separate.   Demand alarms can be
raised by the meter itself.   The meter can conduct consumption logging. As
well as local area networking telephone modem communication allows wide area
networking of meters to a distant interrogating and managing computer.

And while the meter can communicate log and alarm it can also display on its
local counters and display panel instantaneous consumption parameters useful
not only in energy management but also in process and production control
and costing.

Finally while performing some or all of these functions ample intelligence
to drive a printer and signal to remote cost display panels.   Fig. 2 shows
an Interpreter printing 3-phase meter with its remote display panel.

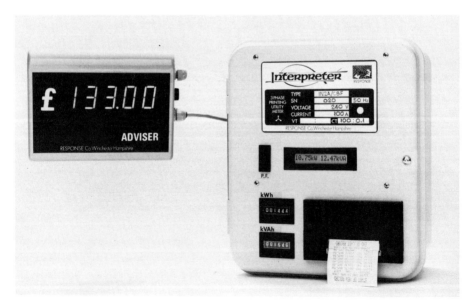

Fig. 2.   An Interpreter Printing 3-Phase Meter tracks
Maximum demand and displays billing figures
remotely.

ENERGY AWARENESS

One of the difficulties with metering is that most often the point of metering
is not the most convenient or accessible for regular reading. An intelligent
meter can communicate to remote displays and the RESPONDER 3 can drive a
number of different types of remote display panels.

It is one lesson of energy management that energy awareness is the first
step to reduction in energy bills. The first application of this principle
has been the development of an 'ADVISER' display panel. This consists of a
one inch high five digit display in sterling currency of the costs of electricity consumed to date in the current week or month. Instantaneous power
in KW can also be displayed remotely in an office more convenient than the
main supply switch panel so that appliances and plant need not be wastefully
left running. Power consumption becomes visible as illustrated by the
'Adviser' display in Fig. 2.

Finally in sympathy with the Energy Efficiency Office Monergy initiative a
large display panel, 'Monergiser' was developed. This display panel is fed
by KWh unit pulses and shows on its six inch high numerals the electricity
bill visible some 100 yards away from the factory gates. Fig. 3 shows the
Monergiser in action.

Fig. 3.   A Monergiser electricity cost display is an
          effective encouragement to conservation by the
          workforce.

CONCLUSIONS

The complexity of electricity tariffs for the industrial consumer are such
that the single function electromechanical meter even with pulse output is
inadequate for present day energy management systems.

A multifunction three phase industrial electricity meter is now available
which with tariff meter accuracy can measure all power and energy parameters
of electrical supply consumption.

This intelligent meter can be used in any monitoring measurement or control
application either as a portable instrument or as part of a building management system or as a part of a self-contained extended but local intelligent
metering system.

Energy awareness is a vital issue particularly for the invisible fuel of
electricity and displays of instantaneous power consumption on the meter
plus remote and easily visible displays of money and power consumption
provide an effective disincentive to wasting expensive energy resources.

The functionality of meters has now advanced to the point where the intelligent meter is to the induction wheel meter as is the word processor to the
mechanical typewriter. The secretary would no more choose a mechanical
typewriter than should the energy manager specify the mechanical meter.
One hundred years after the invention by Ferraris of the mechanical electricity meter a range of electronic intelligent meters has at last become
available to fill the needs of the 20th century energy manager.

Fuel Poverty as an Outcome Measure: A Comparative Study of Energy Policies in Norway and the United Kingdom

S. Hutton*, T. Braend** and L. Warren*

* Social Policy Research Unit, University of York, York, United Kingdom
** Council for Environmental Studies, University of Oslo,
Oslo, Norway

ABSTRACT

Whether low income families have difficulty in being warm at home or not can be considered as an outcome measure of the effectiveness of policies for housing, energy, and income maintenance. This study aims to compare such policies in Norway and the United Kingdom. As background, evidence of fuel poverty in a wider international context is presented. To conclude, policies from Norway which might help to relieve fuel poverty in Britain are set out.

KEYWORDS

Fuel poverty; comparative; Norway; United Kingdom; income maintenance; housing; energy policy

INTRODUCTION

Whether low-income families have difficulty in being warm at home or not can be thought of as one outcome measure of the effectiveness of policies for housing, energy and income maintenance. This study aims to compare government policy in these areas particularly energy, in the United Kingdom and Norway and assess their effectiveness in enabling low-income households to be warm at home. As a background to this it considers whether the United Kingdom is unique in its concern for fuel poverty - the inability to afford a warm home. It is a collaborative study by the Social Policy Research Unit, University of York and the Council for Environmental Studies, University of Oslo. At this stage we are waiting for further results from Norway which will be presented at the conference.

BACKGROUND

It is ironic that it is in energy-rich Britain that lobbying organisations have highlighted the problem of fuel poverty. Britain has indigenous supplies of coal, and natural gas, also nuclear power and the climate is temperate. Fuel poverty is difficult to define precisely but shows itself in three ways:

(1)   Fuel debt and disconnection: Over 100,00 consumers are disconnected from the

electricity supply each year and Berthoud (1981) has shown that the majority of these households are in hardship.

(2)   Cold homes: Although no large-scale national survey of temperatures in the home exists some evidence is available from Wicks' (1972) survey which showed that 55 per cent of the sample of elderly households had living room temperatures below the minimum required in shops and offices. A later survey (Hunt and Gidman, 1982) showed that households were likely to be living in cool conditions if their homes were built before 1919 or if they were on low incomes. An evaluation of local energy projects in 1983 (Hutton and others, 1985) showed that the mainly elderly clients had average living room temperature of 16°C. The DHSS and Department of Energy recommended temperatures for the elderly at home is 21°C.

(3)   Seasonal variation in mortality and cases of hypothermia: In Britain the death rate of the over-sixties increases by 20 per cent in winter compared with summer which is considerably greater than in Norway when the climate is much more severe. (Bradshaw and Hutton, 1986)

If these problems do not exist in other countries, it is interesting to know why. What combination of policies and circumstances have averted the difficulty? Although there has been comparative work in the separate fields of housing, energy and income maintenance policies nothing has been done on how these interact to affect the warmth of low income households.

In this country the policies to alleviate fuel poverty are mainly within the income maintenance scheme through heating additions in supplementary benefit and the single payment allowed for draughtstripping materials. Within housing policy, higher standards of heating equipment and insulation have developed through the building regulations for new properties and grants have been made available for loft insulation and lagging hot water tanks within the Homes Insulation Scheme for existing properties. By insisting on high price energy, however, energy policy has hindered low income households' ability to be warm at home. As something of a counter to this the Department of Energy has supported a voluntary sector initiative which delivers simple low cost energy efficiency measures to low income households via local energy conservation schemes.

RESEARCH DESIGN

To provide some background to a study of fuel poverty in an international context we wrote to contacts we had in different countries to establish what information was available on fuel poverty. Letters were sent to contacts in countries in approximately the same latitude as Britain - Canada, Holland, Germany, Sweden, France and the USA. Australia was also included because of some interesting work being done there. Information was asked for on: knowledge of fuel poverty; fuel debt and disconnection; cold homes; and seasonal mortality statistics.

A study of fuel poverty as an outcome measure requires detailed information on relevant policies in the areas of housing, energy supply and efficiency and income. This phase of the study focuses on only two countries - Norway and the UK and a visit to Norway provided the opportunity to talk to researchers and practitioners in these fields.

Finally a small survey in each of Britain and Norway was undertaken to illustrate the effects of these policies on the ability of low income households to keep warm. Contact had been made with Tore Braend at the Council for Environmental Studies, University of Oslo who collaborated on the survey of households in Oslo.

RESULTS

International background The response to our letters to personal contacts in the
USA and Australia and on personal recommendation in the other countries was
disappointing but some rather hard-won information was gained.

Recognition of fuel poverty as a problem It was clear from the replies that the UK
is not alone in its concern about fuel poverty - in the USA and Australia
considerable research has been undertaken and a number of policies developed for
its relief. The following quotation is illustrative:

> 'If a major low income energy assistance program is not established by
> this winter, the health and safety of thousands of low income people
> and the elderly will be jeopardised.' (American Public Welfare
> Association, 1979)

The 'Lessons from America' series (Brown, 1986) includes a report on policies for
low income households which quotes the statistic that the poor in America spend 25-
40 per cent of income on fuel costs compared with 5-10 per cent for the average of
all households. The average for the United Kingdom is around six per cent for the
average of all households.

A report from Australia outlines the problem and sets out a policy agenda.
(Department of Industry, Technology and Resources, 1986)

Replies from Denmark, Sweden and Germany while agreeing that fuel poverty was
probably not non-existent felt it was probably not so extensive as in Britain. The
reasons given were better housing in Denmark and a variety of circumstances in
Sweden. In the towns in Sweden low income households live in flats and the cost of
heating is included in the rent and in the country firewood is free and very cheap;
the social welfare system is better and finally heating is essential so houses are
generally centrally heated and well insulated. The respondent in Germany did not
give many reasons for his claim except that there are legal restrictions on
disconnection.

Fuel debt and disconnection A surprising result from the information received on
debt and disconnection from Holland, Norway, the USA and Australia was that
disconnections in Oslo were at a considerably higher rate than in the other
countries (1 per cent). The UK and USA (Warkov and Ferree, 1984) and one supplier
in Victoria, Australia (Deasy and Montero, 1983) had disconnection rates of 0.06
per cent and Holland 0.04 per cent. Legal restrictions on disconnection were
reported from Canada and Germany.

Cold homes Information on temperatures in homes in Britain is not comprehensive or
very readily available so we were hoping that as a result of this comparative study
we would be able to point to another country which had satisfactorily acquired such
data. This was not to be, however, as we only received information of any sort
from Holland. This described similar temperatures found in small scale studies in
Britain and a similar concern in the press about low temperatures in the homes of
the elderly.

Seasonal variation in mortality Denmark and Australia sent information on seasonal
variation in mortality. In Denmark in 1985 deaths per month were 21 per cent
greater in the winter than the summer and in Australia they were 28 per cent
greater. This compares with 20 per cent for the elderly in the UK.
A great deal of research has been done on international comparison of the variation
in season mortality, and a paper by Alderson (1985) which describes this work
presents the following table of the coefficients of variation in mortality over the

152

months of the year for different countries:

TABLE 1 Coefficient of variation in mortality over the months of the year

France	0.06
Italy	0.11
Scotland	0.12
Sweden	0.06
United States	0.05
England and Wales	0.13

The table shows that England and Wales and Scotland have much higher seasonal variation than Sweden, the United States or France. Italy also has high seasonal variation. Some of the work reviewed suggests that poverty may be an intervening variable between changes in temperature and death, which may be borne out by the position of the United Kingdom and Italy in the economic league table. (Treasury, 1987) They rank 13 and 14 of the developed industrial countries in gross domestic product per head compared with USA, Sweden and France which rank 1, 5 and 7.

One of the reasons suggested for reduced seasonal variation in mortality is the greater use of central heating in the USA, Canada and Sweden. Keatinge (1986) questions this hypothesis on the basis of a study of seasonal death rates of elderly people living in some housing association centrally heated sheltered accommodation. Death rates rose in winter by a similar ratio to that in the general population. He noted that all but one switched off the heating in the bedroom or slept with the bedroom window open and all who were fit enough walked for up to four miles or waited for buses for up to 40 minutes. He suggested that the beneficial effects on mortality of high indoor temperatures were balanced by the British love of 'fresh air'. Other research by Collins (1986) suggests that temperatures in the home below $15^{o}C$ for long periods may result in increased morbidity among the elderly. Further work by Mant and Gray (1986) suggests that the following threshold temperatures might be appropriate for the literature on the maintenance of health: a minimum of $12^{o}C$ and a mean whole house temperature of $16^{o}C$. These are relevant to temperatures observed in the household interviews.

COMPARISON OF POLICIES AND CIRCUMSTANCES IN NORWAY AND BRITAIN

Energy Both countries have substantial indigenous supplies of energy and have adopted the policy of not subsidising energy prices which are based on long run marginal costs. Electricity is the cheapest fuel in Norway and priced differently by each of the mainly municipal generating boards. These prices do not vary a great deal, however. Ninety-five per cent of electricity is hydroelectricity but this source is not to be expanded for environmental reasons and in about ten years Norway will have to find other sources. All gas is exported - mainly to Britain and Germany and coal is only used in industry but oil is used to heat some homes although many have changed from the more expensive oil to electricity in the past ten years. The oil should last five to ten years depending on the price and it has been decided that there should be no nuclear power before the year 2000. The other fuel used in the home is firewood, the price of which is very variable. Many people go to the forests and cut it for themselves when it is free - they are helping to thin the trees. More than half of firewood is got this way. Firewood sold in Oslo is however very expensive and many elderly people living in older housing still use it. Some of them receive free firewood in the winter from a voluntary service - Aksjon Ved.

In 1983 the price of electricity was 26.8 ore per kwh which at the 1983 exchange rate was 2.460p compared with 5.098p per kwh in the United Kingdom (arithmetic average for all boards for 1983). Thus electricity is almost twice as expensive in

Britain but peak rate electricity is not the most often used fuel for heating in Britain. Comparing household expenditure on fuel shows that on average Norwegians paid £5.15 per week (for the years 1978-79) from the 1983 Norwegian Social Survey) which was 4.2 per cent of total expenditure. For the same period of the Family Expenditure Survey gives average fuel expenditure of £4.80 which was 5.9 per cent of total expenditure. Thus fuel expenditure is greater in Norway but perhaps less of a problem because income is greater. Total expenditure for the same period was £114.64 per week compared with £94.17 per week in the UK.

The energy efficiency section of the Department of Energy in Norway is small and similar to that in the UK. The Department of the Environment and Natural Resources in Oslo, however, is making decisions about the relative merits of investment in energy efficiency compared with investment in energy generation which includes the cost to the environment of further hydroelectric schemes. An example of the comparison made (and of the involvement of a fuel utility in energy conservation) is the target set by the electricity supply board for Oslo, Oslolysverker, of reducing consumption by 15 per cent by the turn of the century. It offers a free energy audit to consumers and auditors are taught to stress the home improvement and comfort aspects rather than pure energy savings.

Other energy efficiency measures are administered by the State Housing Bank which finances most housing associations. The Husbank offers low interest loans and grants for repair, insulation and changing heating systems to households where at least one person is more than 60 years old and the accommodation is more than 30 years old. The UK Homes Insulation Scheme is moving in this direction by targetting on households in receipt of means-tested benefits but does not cover such comprehensive measures.

Building regulations are considerably more stringent in efficient use of energy in Norway than in Britain. The same house built in the same location to British building regulations would cost 70 per cent more to heat than the same house built to Norwegian building regulation standards. (House of Commons, 1986). Norway however still has problems with existing housing as we shall see in the next section.

Housing   It has been an aim of housing policy in Norway that as many people as possible should own their own dwelling. In this and in the proportions of owner-occupiers it is similar to the UK. There are major differences, however, first in the use of the State Housing Bank to finance the building of homes to a modest but adequate standard and second in the large number of homes owned wholly or partly by housing associations. Thus there is only a small specialised local authority housing sector in Norway. In 1980 19 per cent of households in Norway were in housing association accommodation and 26 per cent were private tenants compared with 12 per cent private tenants and 33 per cent local authority tenants in the United Kingdom.

In Oslo however many more live in housing association accommodation than nationally. The proportions are 45 per cent housing co-operatives, 38 per cent private tenants and 17 per cent owner occupiers. Standards in housing co-operative dwellings although not luxurious are good. For example, many housing co-operatives in Oslo are linked to a larger administrative agency, OBOS, which is very aware of the energy efficiency aspects of its work. Energy consumption is monitored and savings made through changing boilers district heating schemes. Once occupants loans have been paid off, they are encouraged to maintain similar levels of payments so that energy efficiency measures can be undertaken.

Although 64 per cent of housing has been constructed since 1948 in Norway surprising numbers of older houses lack basic amenities - 13 per cent of dwellings

built before 1900 have an outside toilet and 22 per cent of single elderly
households have neither bath nor toilet. A further surprising statistic is that
over a quarter of Norwegian homes and more than half those in blocks of flats still
have single glazing. The interviews in Oslo corcborate these statistics.

Income support in Norway The Norwegian system of income maintenance is similar to
that in the UK. The National Insurance Scheme introduced in Norway in 1967 is
based on contributions over a 40 year period and covers old age retirement,
sickness, physical defect, pregnancy, unemployment, disablement and loss of
supporter. The retirement pension is paid at the age of 67 years and an earnings
related supplement is payable.

Those who have had no income from earnings and been unable to contribute are helped
by the Social Assistance Scheme which performs a similar function to that of
supplementary benefit (SB) in the UK but is a municipal scheme and rates of benefit
vary throughout the country. For example, in Oslo if housing and energy costs are
over a certain amount a subsidy is paid. Most of this subsidy is for fuel expenses
because rents in the older properties in central Oslo are very low. Normally the
claimant brings the fuel bill to the local office but if the bill is not available
NOK 100 per month was allowed in 1986. This is approximately £2.50 per week which
compares with £2.10 per week - the lower rate in SB for the same year.

The State Housing Bank administers the main housing benefit along the lines that if
housing expenditure (not including energy expenditure in this case) is greater than
20 per cent of income the bank pays 70 per cent of the difference for those not in
work and 65 per cent of the difference of those in work.

INTERVIEWS WITH HOUSEHOLDS IN BRITAIN AND NORWAY

A sample of 20 low income pensioner households were interviewed in Newcastle and
Oslo. Fuel poverty, if it existed in Norway was considered most likely to be
evident in Oslo. This meant choosing a location for the interviews in Britain that
in some way matched. Newcastle was chosen because it was a city of similar size,
age, and with a climate which was not as different as it might have been. Existing
contacts with the local insulation project in Newcastle which does work in low
income households assisted in the selection of a sample.

The Newcastle Interviews Using a local insulation project as a source for the
sample might be considered to introduce a bias in that those who contact such a
scheme might at least be more aware of the help available to deal with fuel
poverty. An earlier study, (Hutton and others, 1985) however, of the clients of
such local insulation projects showed that there was little difference in terms of
income or household type or energy literacy between clients of the schemes and
comparison households interviewed in the neighbourhood which had had no contact
with these schemes. To reduce any residual bias, 15 households were selected from
the records of the scheme, half from those waiting to have work done and half from
those which had had some insulation work done. The remaining five interviews were
to be approached directly in the same neighbourhoods as those selected via the
scheme. Addresses were chosen randomly from the schemes records, ensuring that
they covered different areas of the city, types of housing and housing tenure.

Overall interviews were achieved with less than half of the addresses selected.
The call-in interviews were equally difficult to achieve, so that it required
around 40 attempts to gain five interviews. The survey of the comparison group in
the study of local energy projects had not led us to expect such difficulty but
there had been some recent warnings in the press to elderly people upon letting
people into their homes. In spite of these difficulties a sample was achieved of
20 interviews all of which were in receipt of the national insurance retirement

pension. All but two were on housing benefit and ten were dependant on
supplementary benefit. Most (14) were local authority tenants of which six were in
flats, four in a semi-detached houses, three in terraced houses and one in a
maisonette. Of the other six households two lived in private rented flats, three
were owner-occupiers in detached and semi-detached houses and one was in a housing
association flat.

The interviews took place in the week beginning March 9th and were completed on
March 18th. The outdoor temperatures during this time varied from $1^{o}C$ to $8^{o}C$ for
the later interviews.

The interviews and the information gathered   Interview schedules were drawn up
based on the questionnaires used in the 1983 study of the clients of local
insulation projects and as far as possible the same questions were asked in Oslo
and in Newcastle. Some of the questions about the use of gas for heating, for
example, were inappropriate in Oslo.

From the information gathered, the households were grouped as follows: (1) having
difficulty maintaining a warm home - nine households; (2) clearly managing to be
comfortable at home at a cost they can afford - six households; (3) and those who
do not clearly fall into either of these categories.

Evidence for difficulty in maintaining a warm home was based mainly on the
following pieces of information: how satisfied people were with the warmth in their
homes; how difficult they found it to pay for the fuel they consumed; desire to
heat more; being in arrears; how good their health; the temperature in the room of
the interview; and the interviewers' subjective impression. Temperatures were read
in the room of the interview in all but two households and for some households in
the hall or bedroom also. These temperature readings can only be used as part of a
broader indicator of the warmth in a dwelling.

The main question for this paper is to ascertain whether these groupings are a
result of policies for energy, income, or housing or some combination of all three.
The objective is to discover what separates the fuel poor from the comfortably warm
group and what could change the situation. This section discusses the household
interviews in the context of energy, income and housing.

Heating and energy   Newcastle City Council is one of the most enlightened in the
sphere of energy policy in Britain so the households interviewed are to that extent
unrepresentative of low income households in the rest of the UK. The council has
made considerable progress, for example, in modernising the heating systems in
council property so that of the 14 local authority tenants in the sample 11 had
some sort of central heating - eight had gas central heating, two had district
heating and one electric storage heaters. None in our sample had the notorious
electric under-floor or ceiling heating installed in the 60's.

The following table (Table 3) outlines the heating systems used by the fuel poverty
and 'warm' group. Lack of space prevents the presentation of the complete data.

TABLE 3 Heating in Newcastle

Interview number	Interview Temp. in living room °C	Central heating	Other main form of heating	No. of rooms in daily use not heated	Bedroom unheated	Satisfaction with warmth
Fuel poverty group 1.	14 pm	/	–	1	/	No
5.	15 am	/	Gas fire – most of day	1	/	Yes but draughty
9.	20 pm	–	One storage heater and electric fire in L.R.	3	/	No
10.	17.8 pm	–	Gas fire in L.R.	3	/	No – always cold
13.	20 pm	/	–	2	/	Fair
14.	12.5 am	/	Gas fire – on all day	0	/	Not entirely
16.	16.4 pm	/	Gas fire in L.R. – most of day	3	/	Yes but draughty
17.	14.5 am	–	Coal fire in L.R.	4	/	Very unsatisfied
21.	16.9 pm	–	–	0	N/A	No
Warm home group 2.	16 pm	/	Gas fire in L.R.	0		Yes
3.	22 pm	/	–	0		Yes
7.	21.9 pm	/	(storage)	1		Yes
11.	18 am (estimate)	–	Gas fires throughout home – on all day	1		Yes very
15.	21.8 am	/	–	0		Yes
18.	21 pm	/	–	0		Not entirely

A clear distinction between the warm group and the fuel poor group is that the warm group heated all the rooms they used daily. Gas central heating was the main heating system used except in one case where there were gas fires throughout the house. In contrast the fuel poor group had large parts of their accommodation which was in daily use unheated, and this was true whether they had central heating or not although those with central heating had fewer rooms unheated. Thus these households which have the means to heat their homes to a satisfactory level do not do so perhaps because of expense or attitude to heating. Three households in the fuel poor group did not have central heating and tended only to heat one room in their home. Respondents 9 and 17 had no other form of heating except for that in the living room and respondent 10 did have a portable calorgas heater which she used once a week to heat the bathroom before the nurse came to help her have a bath. She did not regularly use this otherwise and was determined that she would not heat her bedroom from choice. The main cause of fuel poverty for these three households was an inadequate heating system.

The other striking difference between the warm and the fuel poor group is that a substantial number in the fuel poor group did not heat their bedrooms at all. This is in spite of having the technical potential to do so. Of those with central heating among the fuel poor group, respondents 1 and 16 stated that they did not want to sleep in heated bedrooms, but respondent 13, living in a housing association flat with central heating had run up a large bill and was paying off arrears with a slotmeter. She only heated the living room and kitchen but would have liked to heat more.

Income   Examination of the incomes and expenditures of these households may indicate how far use of heating systems and ability to heat homes adequately is constrained by income or high fuel costs. In the following table (Table 4) the equivalent income after the deduction of fuel and housing costs is compared with worry about expenses and fuel costs.

TABLE 4   Income and fuel expenditure in Newcastle

Interview number	In receipt of SB	Equivalent income after fuel and housing costs* £	Fuel cost p.w.	Fuel exp. as ppn. of income less housing costs %	Central heating	Worry about expense	Would like to heat more
Fuel poverty group							
1.	SB	19.24	11.40	37	/	/	/
5.	SB	29.55	14.00	32	/	/	/
9.	–	30.18	5.74	16	No	–	/
10.	SB	34.94	6.50	16	No	–	
13.	SB	26.19	12.00	31	/	–	/
14.	SB	30.70	10.00	25	/	–	/
16.		22.13	8.59	28	/	–	
17.		39.50	6.20	9	No	/	/
21.	SB	33.53	5.18	13	/	–	–
Warm homes group							
2.	–	30.94	8.60	22	/	A little	–
3.	SB	27.10	12.00	31	/	A little	/
7.	–	31.21	7.50	13	/(storage)	–	–
11.	SB	34.18**	8.59	14	–	–	–
15.	SB	42.35	9.20	12	/	–	–
18.	SB	32.78	10.20	24	/	–	–

*   Equivalent income standardises for the number of people in the household

**   Respondent 11 did not have the bill for the winter quarter – this figure reflects the quarter up to October. It is therefore not counted in the average calculations

The table shows that income after deduction of housing and fuel costs was much the same across the groups. The proportion of households dependent on SB was also the same for the warm and fuel poor groups but none of the unplaced group were in receipt of SB. The notional fuel allowance plus lower rate addition which can be considered as the allowance within SB for fuel was £11.00 during the period of this consumption. Three of the households in the fuel poor group were spending more than this and only one of the comfortable group. All of these households had central heating.

Among the fuel poor group, the three lowest incomes were observed of £19.24, £22.13 and £26.19 and although the numbers are small it is worth saying that the average income was around £4 lower for the fuel poor group. A worry about expense was expressed by three households and for those with central heating the wish to heat more should perhaps be interpreted as a feeling of being limited by their low income. Fuel expenditure as a proportion of income less housing costs was computed and although paying a high proportion of income on fuel was not restricted to the fuel poor group it was more prevalent. It also seemed as though there might be an association between very high expenditure as a proportion of income and worry about expense.

It is noticeable that those with central heating among the fuel poor group generally have higher fuel expenditure than those who heat only one room. Expenditure does not seem to be clearly related to comfort. Some households with central heating in the fuel poor group have high expenditure on fuel – as high or higher than households in the warm group – but the temperatures were 14°C, 15°C and

12.5°C (for respondents 1, 5 and 14) respectively compared with 22°C and 21°C (for respondents 3 and 18).

Several households felt that they were only just able to afford what they were spending, respondent 1 was anxious not to use her central heating too much because of the expense. She said she could only afford to 'heat and eat' and could not afford any luxuries. Two other also said they found it difficult to afford the extras. One (14) said she could only manage extras like Christmas presents with her daughter's help and the other (17) saying they no longer bought their grandchildren Christmas presents.

Ten of the households interviewed paid their fuel bills quarterly, and only two used a slotmeter. Those in the warm group seemed possibly to be more likely to use a budget payment scheme.

Most of the households budgeted for their fuel and managed to afford their heating bills although this might be a precarious situation. One household (14) was burgled a few weeks before the interview and her pension was stolen; her lack of financial backup meant that she is paying back arrears weekly. Two households had difficulty with the first winter bill they received after their central heating has been installed. One resulted in arrears which are still being paid back one year later using a slotmeter and both cut-back their use of central heating in response.

With those in the fuel poor group who did not have central heating, income and cost did not seem to be the limiting factors but rather the lack of a good heating system.

Housing and insulation We looked at housing and there seemed to be little or no correlation between type and tenure of housing and fuel poverty.

Similarly, we looked at insulation - all households complained of at least some draughts and there seemed to be no difference in the use of draughtproofing and insulation between the groups. Several homes, particularly those in the 'fuel poverty' group however used home measures to prevent draughts - householders stuffed up cracks, keyholes and air vents with cardboard or newspaper, or 'sausage' draught excluders were pushed up against draughty doors.

Age, health and household type An analysis of the health age and household type of respondents in the three groups shows some fairly clear distinction. Those in the fuel poverty group were more likely to be older, widowed and living alone and have less good health.

Overview The interviews in Newcastle have highlighted some causes or factors associated with households having difficulty in maintaining a warm home: lack of an adequate heating system in a few cases; worry about expense particularly for those with central heating; only heating some rooms and particularly not heating bedrooms; and the greater likelihood of such households being older, widowed, living alone and in poor health.

A variety of polices are required to alleviate these symptoms. Some households need an efficient heating system installed. The numbers may seem lower in Newcastle than if the study had been done in some other large towns. Other households need advice and information particularly about the risks associated with unheated bedrooms and secondly when a central heating system is installed about how to avoid running up high bills. Doubt remains that the income available to the older widow, is adequate.

Interviews in Oslo   The full transcripts of the Oslo interviews have not arrived yet so cannot be analysed in the same way as those in Newcastle. Tøre Braend has sent a brief summary, however, and a fuller exposition will be available for the conference.

The interviews undertaken by four students at the Council for Environmental Studies took place on March 3rd and 4th. The outside temperatures were 'very cold' - well below zero. Ten households were selected from clients of the local Municipal Health and Social Services Unit and ten directly by the students. The students did not have difficulty in achieving the sample.

Housing   Some internal temperatures were quite low, 14°C. Most of the buildings in the neighbourhood are three to four stories high, with apartments of varying size. They often have several interior yards, one after the other in a row with passages from the street to the innermost yard. The departments facing the streets are usually the biggest, and the ones in the back are the smallest. Many apartments have only one room and a kitchen, with no bathrooms or toilets, and either a privy in the staircase or in the backyard. Many of the privies in the staircase have been renovated to flushing toilets, and some have installed showers in their kitchens. The east end of Oslo contains a concentration of some of the poorest and most low quality housing in the whole of Norway. This makes it not very representative of Norway as a whole, but very suitable for a pilot survey with the object to identify possible groups suffering from energy poverty.

Heating   Most of the respondents heated their homes with electric resistance heaters, plus in some cases with oil and wood. A larger survey in roughly the same area of Oslo confirms this finding.

Little difficulty was reported in paying the electricity bill or other costs of heating in some cases because the social worker or a relative dealt with it. On the other hand one respondent complained that when the quarterly bill for electricity was paid she could hardly afford to buy food that month. Although respondents were likely to have little left after paying for essentials and certainly not enough to lead the good life by the younger generation's yardstick, comments such as 'one shouldn't complain' may indicate different expectations.

The students found the definition of fuel poverty difficult to interpret when interviewing one respondent who said she could not afford to heat her home adequately because she was saving for a trip abroad. Two respondents were saving to go abroad. Tøre Braend suggests that pensions might be more generous in Norway.

Income   No clear connection was found between income and respondents use of energy or problems with keeping warm. There did however, seen to be clear connections between the rent and type of housing and peoples' point of view. They were much more likely to complain of fuel costs if living in recently renovated higher rent apartments although the renovations had given them warmer more comfortable homes. It is not known at this stage whether the total costs of rent and heating were higher overall.

Energy literacy and insulation   The respondents knew little of energy saving measures and indeed were of the opinion that it was the landlords' responsibility. If they did know they did not know where to go for help with installation and were any way reluctant to ask for fear of being a bother.

CONCLUSIONS

This pilot study, has yielded some interesting information on the ability of low income households to keep warm in different countries. From the postal enquiry

160

Britain is not alone in its concern for fuel poverty and although fuel poverty may not be as widespread in other countries lack of concern may also mean that the plight of many households is unknown. The lack of information on temperatures in homes in any country is striking. In terms of debt and disconnection Britain did not seem to be particularly bad compared with other countries but variation in seasonal mortality is higher.

The comparison between policies in Norway and Britain has not given any very obvious insights as to what policies could be transferred from Norway to the United Kingdom to combat fuel poverty. If anything, the project has made Norwegians more aware of the problem. Some lessons can be learned however, from the Norwegians about

(a) the comparison of investment in supply versus reduced demand
(b) The rôle of the utilities in energy efficiency
(c) More stringent building regulations
(d) Targeting of comprehensive insulation measures on elderly and older   housing.

From the household interviews similar problems emerge of lack of information and advice and worry about expense when the heating system or the apartment is upgraded. It is not clear at this stage whether the households in Oslo were better or less well able to afford warmth at home.

REFERENCES

Alderson, M.R. Season and Mortality, Health Trends, Vol. 17, 1985.
Berthoud, R. Fuel Debts and Hardship, Policy Studies Institute, 1981.
Bradshaw, J. and Hutton, S. Social Security Policy and Fuel Costs in Mannier, E., Gaskell, G., Ester, P., Joerges, B., Lapottine, B., Midden, C., Paiseux, L. Consumer Behavior and Energy Policy. An International Perspective Praeger, 1986.
Brown, I. The promotion of energy conservation in low income households - US experience. Lessons from America No. 6. Association for the Conservation of Energy, Neighbourhood Energy Action, 1986.
Collins, K.J. Low indoor temperatures and morbidity in the elderly, Age and Ageing, 1986, 15:212-220.
Deasy, L. and Montero, K. Fuel Poverty in Victoria, Energy Action Group, 1983.
Department of Industry Technology and Resources. Energy: an essential service for all Victorians. A Review of Energy Assistance Programs and Customer Services, 1986.
House of Commons, Energy Efficiency Office, Eighth Report. Session 1984-85, HC87, HMSO, London.
Hunt, D.R.G. and Gidman, M.I. A national field survey of house temperatures. Building and Environment, Vol. 17, No. 2, 1982.
Hutton, S., Gaskell, G., Pike, R., Bradshaw, J. and Corden, A. Energy Efficiency in Low Income Households: an Evaluation of Local Insulation Projects, HMSO, 1985.
Hutton, et al., op. cit.
Keatinge, W.R. Seasonal mortality among elderly people with unrestricted home heating: British Medical Journal. 20 September, 1986, 298 732-733.
Mant, D.C. and Gray, J.A.M. Health and Building Regulation, Building Research Establishment, 1986.
The Treasury Economic Progress Report No. 189, March-April 1987, HMSO.
Warkov, S. and Ferree, G.D. Energy Assistance in Connecticut: A survey of elderly, moderate/low income and hardship households. Institute for Social Inquiry, 1984.
Wicks, M. Old and Cold, Hypothermia and Social Policy, Heinemann, 1978.

He does point out that THE NATION saves total capital all the time (fewer power plants) & sometimes even the home owner saves capital in addition to electricity savings. Economies all screwed up in subsidies etc

The Benefits of Using Thermal Insulation in Buildings for Energy Conservation

Said M. A. Ibrahim

Mechanical Engineering Department, Faculty of Engineering, Al-Azhar University, Nasr City, Cairo, Egypt

Why the hell does anyone need this?

ABSTRACT

The present study deals with economic and engineering benefits of using thermal insulation in buildings as an effective measure for the conservation of air conditioning energy. The study is applied to typical buildings in a hot humid climate. It is cleary shown that substantial amounts of energy and consequently money can be saved on both individual and national levels. It is hoped that the present results would encourage the implementation of policies that make the use of thermal insulation in the walls and roofs of buildings compulsory. These measures will save the country large sums of money which may be invested in other economic sectors.

KEYWORDS

Thermal insulation; Building; Energy conservation; Cooling load; Customer index; Payback period; National index; Economic; Savings.

INTRODUCTION

Energy demand is increasing continuously in both developed and developing countries. Therefore, one can see in sight an energy crisis facing the world, and in essence, it is a fossil fuel crisis. Some alternative energy sources appear promising, but most of these new technologies are beset with difficulties of one kind or another. Two renewable energy sources may prove to be the ultimate sources for man, namely solar energy and nuclear fusion. Solar energy could be feasible for applications such as water and space heating and crop drying but still unfeasible for large scale electricity generation or in the fields of air conditioning and refrigeration. It may take a considerably long time before achieving these goals. The realization of fusion power is still confronted with combersome engineering and technical problems before it becomes economically and technically feasible. Other alternative technologies, e.g., wind power, geothermal energy, tidal energy, OTEC, photosynthesis... etc may be useful for either small scale or limited applications.

The two main energy sources available at this time are fossil fuels and nuclear energy. Most developing countries do not have yet the capability of using nuclear energy, and most of them depend entirely on fossil fuels and mainly on oil to generate their electricity needs.

The problem here is that both fossil and nuclear fuels are depletable resources, and the former, (mainly oil and gas), will be exhausted within the next few decades.

The rising cost of energy has a serious impact on economic development, and it is clear that the quality of life of the people all over the world depend on the availability now of large amounts of low cost energy in useful form. Therefore, new energy technologies and renewable resources should be developed and used in addition to energy conservation and the use of high efficiency systems.

As supplies dwindle and costs soar, it becomes obvious that energy conservation is no longer optional. We must develop new and better ways to stretch our energy money. For economic as well as philosophical reasons, it is imperative that we move boldly to reduce waste. Energy conservation policies should be adopted and implemented by governments, since such measures cannot be left for the choice of individuals. This should be looked at as part of the national security of the country concerned.

In 1980, the United States strategy was to be self suffcient in energy by implementing two measures : (1) increasing their energy production from oil, gas, coal, nuclear, solar... etc., and (2) energy conservation polcies (1975). The consumption of the USA, in 1980, was 12×10^6 bbl of oil equivalent/day and the goal of self energy sufficiency was to increase the national energy production to become 7.3×10^6 bbl of oil equivalent/day and to save 4.7×10^6 bbl of oil equivalent/day through energy conservation (1975). it is seen that the USA could have saved up to 40% of its total energy use via energy conservation . Such savings are essentially waste which is unjustifiable as long as it does not affect the standard of living of citizins.

Energy conservation is an essential feature of building design therefore, the thermal performance of buildings and their services have become of prime importance to all concerned, for it is in this area where significant savings in energy and consequently money can be saved. Amongest several factors that should be taken into consideration by the building designer, for energy conservation, the use of thermal insulation is one of the most effective measures.

The present study is an effort to indicate to individuals as well as energy policy makers the fact that the use of thermal insulation in buildings is economically feasible and will result in huge savings for the nation. As for individuals, very little money will be invested with comparatively short payback periods which will mean large savings on the long term. The present research is applied to buildings and data in the United Arab Emirates (UAE), which is situated in the Arabian Gulf.

The weather there is rather hot and humid. Although the people of the UAE are wealthy (with a per capita income of around $ 14000/year), however, it is important to implement energy conservation measures because :(1) UAE consumes about 70% of its electricity in domestic air conditioning which is essentially needed for almost 6 to 8 monthes of the year, (2) the resources of oil and gas constitute almost the entire revenue of the country and these are depletable resources in addition to the serious economic effects due to international oil price fluctuations, and (3) electricity is heavily subsidised by the government by almost 88% . These are enough important reasons for the country to look seriously without delay to implement energy conservation measures.

Three indices were calculated : (1) customer index, (2) government index, and (3) national index. These show the clear feasibility of using thermal insulation in buildings to conserve energy. The present conclusions should be applicable to other countries. The numerical values will be different but similar conclusions should be obtained.

METHODOLOGY

The aim of this study is to demonstrate the economic feasibility of using thermal insulation in the walls and roofs of buildings in the UAE. A common typical accomodation was chosen to perform this study which is a two-storey villa. The present calculations are based on actual weather data for the UAE. The methodology adopted is summarized as follows:

(1) The annual cooling load is calculated. This has been done for several insulation levels in the external walls and roof of the villa and for different window and balcony glass constructions.

(2) A complete economic analysis is then carried out for different insulation levels, in the walls and roof, and for different glass constructions . In this analysis, actual data relevant to the UAE are used throughout. The economic study was performed to give the savings for the customer, government, and the nation. For the customer, the net initial investment and payback periods were estimated. For the government and the nation, the initial capital and annual savings were calculated.

COOLING LOAD CALCULATIONS

The summer air conditioning cooling load was claçulated for the two-storey villa,of light colour, with a floor area of 250 m^2, The total area of solid walls and glass are 245 m^2 and 100 m^2, respectively. The location has an altitude of 25° N.

The cooling load has been calculated for the months of April, May, June, July, August and September. The Carrier Handbook Standards for Air Conditioning was used in these calculations (1965). The outside design temperature and relative humidity were taken as 46° C and 90%, respectively. The corresponding inside design conditions were 25° C and 50%, respectively. The method adopted was to clac<ulate the cooling load for the hottest day of the particular month and then multiply by the number of cooling hours for that month. The hottest day for each month was determined from previous weather records and the number of cooling hours was obtained from degree day profiles in which the base temperature is 23 °C (Ibrahim, 1987 a). The cooling loads were calculated for these levels of insulation in the walls and roof : 0, 30, 40, 50, 60, 70, and 100 mm. No insulation was used in the floors. The Present calculations show that heat gain through the floor is very small. Therefore, it is economically unfeasible to use insulation in the floor.

The insulation material used in the present study is an extruded polystyrene foam. Figure 1 gives details of insulation application to the roof and walls of the building.

The cooling loads were claculated, for the above given insulation levels, for these glass arrangements : single clear glass (6 mm), single tinted medium colour glass (6 mm), double glazing clear glass, and double glazing tinted glass. Figure 2 shows the variation of the total cooling load of the villa with thermal insulation thickness in the walls and roof, for the four glass arrangements. The use of tinted glass reduces the shading coefficient and the use of double glazing reduces the value of the coefficient of transmission . It is clearly shown that the cooling load decreases as tinted and double glazing glass are used .

Figure 3 gives variation of the heat transmission through walls, roof, and floor for different insulation levels. This figure serves also to determine

164

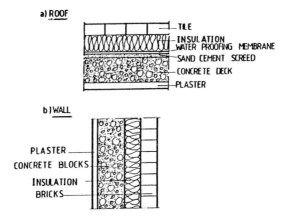

Fig. 1. Details of insulation application to roof and wall.

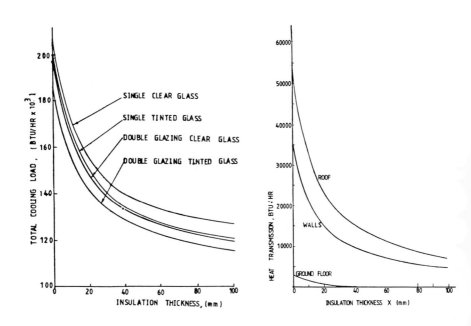

FIG. 2 TOTAL COOLING LOAD VS INSULATION THICKNESS
FOR DIFFERENT GLASS ARRANGEMENT.

FIG.3 EFFECT OF INSULATION THICKNESS ON HEAT TRANSMISSION

e insulation thickness which should be applied to the roof and walls, since
ne cannot increase the insulation thickness indifinitly in order to reduce
ne cooling load. It is seen that the large drop in heat gain occures at th-
cknesses of up to 70 mm for the roof and 50 mm for the walls.

ne external sensible cooling load for constant wall insulation thicknesses
ere calculated and are shown in Table 1 . These are sample results for the
4th of August at 4 p.m (i.e. for the hotteset day and peak load time for
ugust). The calculations were made for the four glass arrangements. The
all insulation thicknesses used in the calculations are 0, 30, 40, and 50mm.
he external sensible cooling load in Table 1 is that results from solar ga-
n through glass plus the heat gains through walls, roof, and floor, i.e.
ensible loads from outside air ; internal sources are not included (since
hese are constant). The results for only the two cases of single clear gl-
ss and tinted double glazing glass are shown graphically in Fig.4 (a, b).
hown in this figure are external sensible cooling loads against roof insu-
ation thickness for constant wall thickness and external sensible cooling
oads for different wall insulation thickness and constant roof insulation
hickness. It is shown clearly the effect of applying different insulation
evels in the roof and walls on the cooling load. The cooling load for diffrent combinations of roof and wall insulation levels may be determined.

ECONOMIC ANALYSIS

he incentive to conserve energy comes directly from the saving in operating
osts and sometimes indirectly through reduced size plant, reduced reliance
pon mechanical services, reduced availability of energy in the future, red-
ced maintenance, and improved morale.

n order to asses the true economic value of any investment in conservation
easures, the value of the benefits during the life of the building and the
lant must exceed the value of the additional investment, otherwise there
ill be no incentive to invest. Regretfully, such incetives are of no value
hen capital is provided by one organization or department, and revenue is
rovided by another. Common examples are government departments and rented
ccomodation. This is one reson why some form of legislation to conserve enrgy is desirable. There are many methods of economic analysis of investment
ecision. Economists, accountants, company directors, designers and lay peo-
le all appear to have their preferences . Whilst some techniques may be ade-
uate for certain straight forward investment decisions , some have flaws wh-
ch make them inappropriate for making decisions about energy conservation
ptions. All the techniques require the prediction of the value of the gross
nergy; hence the need for the prediction of energy consumption. The value
f any other merits may be incorporated if they are considered appropriate.

. Payback Period

ne of the simplest methods of investment decision making is to calculate
he period for which the savings will have to accumulate in order to equal
he extra investment. If this period is longer than the life of the invest-
ent, obviously there is no incentive to invest. In other words, the pay-
ack period is defined as the number of years required for the net cash flow
o equal zero. As this technique is usually apply to short life investments,
nflation or deflation is usually ignored. In the present work, the payback
eriod, n, was calculated from (Smith, 1981) :-

$$= \frac{\log [R/(R - iP)]}{\log (1-i)} \qquad (1)$$

Table 1. External sensible cooling load (Btu/hr) for various
levels of wall and roof insulation. 24th August at
4 p.m.

	wall insula- tion(mm)	Roof insulation thickness (mm)						
		0	30	40	50	60	70	100
Single clear glass	0	146419	110261	107158	104985	103278	102037	99399
	30	123411	87253	84150	81977	802770	79029	763911
	40	121521	85363	82260	80087	78380	77139	74501
	50	120051	83893	80790	78617	76910	75669	73031
Single tinted Glass	0	141164	105006	101903	99730	98023	96782	94144
	30	118156	81998	78895	76772	75015	73774	71136
	40	116266	80108	77055	74832	73125	71884	69246
	50	114796	78638	75535	73362	71635	70414	67776
Double clear glass	0	140319	104161	101058	98885	97178	95937	93299
	30	117311	81153	78050	75877	74170	72929	70291
	40	115421	79263	76160	73987	72280	71039	68401
	50	113951	77793	74690	72517	70810	69569	66931
Double tinted glass	0	135067	98909	95806	93633	91926	90685	88047
	30	112059	75901	72798	70625	68918	67677	65039
	40	110169	74011	70908	68735	67028	65787	63149
	50	108699	72541	69438	67265	65558	64317	61679

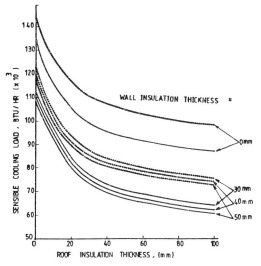

Fig. 4a. Sensible cooling load for different roof and wall
insulation thicknesses for double glazing tinted
glass ————— and single clear glass ‑‑‑‑‑‑ .

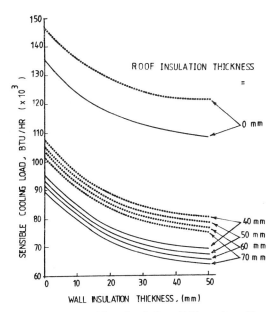

Fig. 4b. Sensible cooling load for different wall and roof
insulation thicknesses for double glazing tinted
glass ————— and single clear glass ‑‑‑‑‑‑‑‑ .

Where R = annual savings, P= initial investment, and i = interest rate.

a. Analysis

The economic analysis was carried out for the villa for several combinations of wall and roof insulation levels. The wall insulation thicknesses considered were, 0. 30, 40 and 50 mm and those for the roof were 40, 50, 60 and 70 mm. The analysis was performed by fixing the wall insulation thickness and varying that of the roof. This was repeated for the four glass arrangements.

In the present analysis, three economic indices were evaluated : customer index, government index, and national index. Each of these indices was divided into initial capital and annual savings.

The customer index consists of :

(1)Savings due to reduction in electricity consumption.

(2)Savings due to reduction in maintenance cost.

(3)Interest on annual savings.

(4)Savings from reduced installed air conditioning (A/C) capacity.

(5)Capital invested in energy conservation measures.

(6)Net capital invested which is the capital invested in energy conservation measures minus that saved due to reduced A/C capacity.

(7)Payback period.

The government index includes :

(1)Savings due to reduction in electricity generation (subsidies)·

(2)Interest on annual savings.

(3)Capital saving due to reduction in installed electric power.

The national index is composed of :

(1)Annual savings which is the sum of customer and government savings.

(2)Capital savings which is the sum of customer and governmentsl savings.

The following data are used in the analysis : (1 $ = 3.7 Dirhams (DH), 1 DH = 100 fils) :

. Cost of electricity to customer   7.5 fils/kwh
. Cost of maintenance   100 DH/ton refrigeration
. Cost of installing A/C equipment   2000 DH/ton refrigeration
. Cost of tinted glass over clear glass
  (6 mm) $30 \ DH/m^2$
. Cost of installing insulation $160 \ \$/m^3$
. Double glazing costs about four times
  as much as single glass
. Cost of electricity to government   73.02 fils/kwh
. Cost of installed electric power to
  government   3000 DH/kw
  Interest rate   10%

The key step in the calculations of the economic analysis is to determine how much refrigeration capacity was saved due to the application of thermal insula tion of different levels in the roof and walls of the building under consideration. This was done by comparing the refrigeration load for any insulation level, and given glass arrangement, with the reference cooling load. The reference load is that for the case of no insulation in the roof and walls and clear single glass.

RESULTS

The calculatios of the economic analysis of the four glass arranagements are given in Tables 2-5. The results indicate clearly that the use of thermal insulation materials in buildings is economically feasible for the customer government, and the whole nation.

The results for the customer show the following :

(1) Very little sums of money are involved in installing the required energy conservation measures.

(2) Short payback periods in comparison with the life time of the building and A/C equipment are involved. In many cases the payback period did not exceed even one year. In most cases, n, ranges from 1 to 3 years. In a few cases, (very high insulation levels and double glazing tinted glass), n exceeds the three and four years by a few months. Short payback periods means large saving in the long term.

(3) In some cases, the customer pays nothing for his thermal insulation, and even adds money to his own pocket.

(4) The results show that it could be more economical to use high insulation levels, when double glazing or tinted glass are used, rather than using lower insulation thicknesses. This is so because the saved A/C capacity for higher levels of insulation is greater than that for lower levels. This means smaller initial capital investment.

(5) The use of double glazing has advantages over single glass even though more money is invested. These advantages are not only to reduce the solar gain but also to reduce the noise level and to prevent dust and sand penetration.

(6) It is desirable to use high insulation levels whenever possible even if this means a somewhat longer payback period. This is so in order to keep the mechanical strength of the insulation material at better values. This means a longer life time for these materials. This is still economically feasible since the differences in the payback periods are not large.

(7) In the present analysis, the cost of installing one ton regrigeration and its annual maintenance were considered of moderate values. The customer will save more money if these values are higher.

The following conclusions are drawn for the government and the nation :

(1) The government and the nation never invest any money in implementing the energy conservation measures ; they always save.

(2) Large sums of savings are involved. These are mainly due to reduction in installed electric power and due to savings in the electricity production cost.

(3) The given large savings are for one building only. The savings will be in terms of millions of Dirhams on a whole national scale.

CLOSURE REMARKS AND CONCLUSIONS

The building designer should take into consideration the proper orientation with respect to the sun. Exterior shading devices, such as sun hoods and trees, can be used to control solar radiation. Proper joint design and adequate weather stripping can help minimize heat gains and losses caused by

Table 2. Economic analysis for the case of single clear glass.

Insulation Thickness x (mm)	Savings in load (Ton Refrigeration)	Customer Savings				Customer Investment (DH)		Payback period for customer (YR)	Government Savings		National Savings	
		Annual (DH / YR)			Capital (DH)	Initial	Net		Annual (DH/YR)	Capital (DH)	Annual (DH/YR)	Capital (DH)
		Electricity Consumption	Maintenance + Interest	Total	A/C Equipment	Insulation & Glass	Initial Capital		Electricity cost + Interest	Installed power	Customer + Government	Customer + Government
x_wall = 0												
x_roof = 40	3.27	1552.19	504.92	2067.11	6540	5900	None	Credit	16640.90	17250	18708.01	23790
50	3.45	1637.63	543.26	2180.89	6900	7375	475	0.233	18238.98	18195	20419.67	25095
60	3.59	1704.08	565.31	2269.39	7180	8850	1670	0.810	18978.63	18930	21248.02	26110
70	3.70	1756.30	582.63	2338.93	7400	10325	2925	1.415	19379.33	19515	21718.26	26915
x_wall = 30												
x_roof = 40	5.19	2463.56	817.26	3280.82	10380	10325.0	None	Credit	27438.19	27375	30719.01	37755
50	5.37	2569.00	845.60	3394.60	10740	11800.0	1060.0	0.336	28389.27	28320	31783.87	39060
60	5.51	2615.46	867.65	3483.11	11020	13275.0	2255.0	0.709	29128.99	29055	32612.10	40075
70	5.62	2667.67	884.97	3552.64	11240	14602.5	3362.5	1.165	29385.11	29640	32937.75	40880
x_wall = 40												
x_roof = 40	5.35	2539.51	842.45	3381.96	10700	11682	982	0.312	28169.25	27270	31551.21	37970
50	5.53	2624.95	870.80	3495.75	11060	13157	2097	0.655	29234.99	29160	32730.74	40220
60	5.67	2691.41	892.84	3584.25	11340	14632	3292	1.021	29976.21	29910	33560.46	41250
70	5.77	2738.87	908.59	3647.46	11540	16107	4567	1.417	30504.59	30635	34152.05	41975
x_wall = 50												
x_roof = 40	5.47	2596.47	861.35	3457.82	10940	13127.5	2187.5	0.692	28917.64	28845	32375.46	39785
50	5.65	2681.91	889.69	3571.60	11300	14602.5	3302.5	1.028	29870.54	29805	33442.14	41105
60	5.79	2748.37	911.74	3660.11	11580	16077.5	4497.5	1.389	30610.27	30540	34270.38	42120
70	5.90	2800.56	929.60	3735.60	11800	17552.5	5752.5	1.771	31190.57	31110	34926.17	42910

Table 3. Economic analysis for the case of single tinted glass.

Insulation Thickness x (mm)	Savings in load (Ton Refrigeration)	Customer Savings				Customer Investment (DH)		Payback period for customer (YR)	Government Savings		National Savings	
		Annual (DH/YR)			Capital (DH)	Initial	Net		Annual (DH/YR)	Capital (DH)	Annual (DH/YR)	Capital (DH)
		Electricity Consumption	Maintenance + Interest	Total	A/C Equipment	Insulation & Glass	Initial Capital		Electricity cost + Interest	Installed Power	Customer + Government	Customer + Government
x_wall = 0												
x_roof = 40	3.71	1761.04	584.20	2345.24	7420	10439.30	3019.30	1.460	19612.76	19560	21958.00	26980
50	3.89	1846.49	612.55	2459.04	7780	11914.30	4134.30	1.950	20565.66	20520	23024.70	28300
60	4.03	1912.94	634.59	2547.53	8060	13389.30	5329.30	2.486	21305.49	21255	23853.02	29315
70	4.14	1965.15	651.92	2617.07	8260	14864.30	6584.30	3.070	21887.51	21840	24504.58	30120
x_wall = 30												
x_roof = 40	5.63	2672.42	886.54	3558.96	11260	14864.30	3604.30	1.131	29763.04	29605	33322.00	40945
50	5.81	2757.86	914.89	3672.75	11620	16339.30	4719.30	1.457	30715.88	30645	34388.63	42265
60	5.95	2824.32	936.93	3761.25	11900	17814.30	5914.30	1.812	31455.67	31380	35216.92	43280
70	6.05	2871.78	952.68	3824.46	12100	19141.80	7341.80	2.156	31984.04	31905	35808.50	44005
x_wall = 40												
x_roof = 40	5.78	2743.62	910.16	3653.78	11560	16221.3	4661.3	1.446	30556.52	30480	34210.30	42040
50	5.97	2833.81	940.08	3773.89	11940	17696.3	5756.3	1.753	31563.16	31500	35337.05	43440
60	6.11	2900.26	962.13	3862.39	12220	19171.3	6951.3	2.101	32301.07	32220	36163.46	44440
70	6.21	2947.73	977.87	3925.60	12420	20646.3	8226.3	2.491	32829.44	32745	36755.04	45165
x_wall = 50												
x_roof = 40	5.91	2805.33	930.62	3735.96	11820	17666.8	5846.8	1.803	31244.32	31170	34980.28	42990
50	6.09	2890.77	959.00	3850.00	12180	19141.8	6961.8	2.113	32195.39	32115	36045.39	45030
60	6.23	2957.23	981.02	3938.25	12460	20616.8	8156.8	2.458	32935.12	32850	36873.37	45310
70	6.33	3004.49	996.75	4001.24	12660	22091.8	9431.8	2.850	33465.32	33390	37466.56	46050

Table 4. Economic analysis for the case of double glazing clear glass.

Insulation Thickness x (mm)	Savings in load (Ton Refrigeration)	Customer Savings				Customer Investment (DH)		Payback period for customer (YR)	Government Savings		National Savings	
		Annual (DH/YR)			Capital (DH)	Initial	Net		Annual (DH/YR)	Capital (DH)	Annual (DH/YR)	Capital (DH)
		Electricity Consumption	Maintenance + Interest	Total	A/C Equipment	Insulation & Glass	Initial Capital		Electricity cost + Interest	Installed Power	Customer + Government	Customer + Government
$\frac{x_{wall} = 0}{x_{roof}} = 40$	3.78	1794.27	595.23	2389.50	7560	13465.5	5905.5	3.010	19983.54	19935	22373.04	27495
50	3.96	1879.71	623.57	2503.28	7920	14940.5	7020.5	3.486	20934.61	20880	23437.89	28800
60	4.10	1946.17	645.62	2591.79	8200	16415.5	8215.5	4.040	21676.16	21630	24267.95	29830
70	4.21	1998.38	662.94	2661.32	8420	17890.5	9470.5	4.659	22256.46	22200	24917.78	30620
$\frac{x_{wall} = 30}{x_{roof}} = 40$	5.70	2705.65	897.57	3275.65	11400	17890.5	6490.5	2.104	30133.82	30060	33737.04	41460
50	5.88	2791.09	925.91	3379.09	11760	19365.5	7605.5	2.425	31155.14	31005	34872.14	42765
60	6.02	2857.04	947.95	3459.54	12040	20840.5	8800.5	2.786	31826.44	31755	35631.88	43795
70	6.21	2905.01	969.70	3868.71	12240	22168.0	9928.0	3.141	32354.82	32280	36223.53	44520
$\frac{x_{wall} = 40}{x_{roof}} = 40$	5.85	2776.85	921.19	3698.04	11700	19247.2	7547.5	2.418	30927.29	30855	34625.33	42555
50	6.04	2867.04	951.10	3818.14	12080	20722.5	8642.5	2.718	31932.12	31860	35750.26	43940
60	6.18	2933.49	973.15	3906.64	12360	22197.5	9837.5	3.073	32671.84	32595	36578.48	44955
70	6.28	2980.96	988.90	3969.86	12560	23672.5	11112.5	3.479	33200.22	33120	37170.08	45680
$\frac{x_{wall} = 50}{x_{roof}} = 40$	5.98	2838.56	941.66	3780.22	11960	20693.0	8733.0	2.783	31618.72	31575	35398.94	43535
50	6.16	2924.00	970.00	3894.00	12320	22168.0	9848.0	3.088	32566.17	32490	36460.17	44810
60	6.30	2990.45	992.05	3982.50	12600	23643.0	11043.0	3.440	33305.89	33225	37208.39	45825
70	6.40	3037.92	1007.79	4045.71	12600	25118.0	12318.0	3.845	33834.27	33750	37879.98	46550

Table 5. Economic analysis for the case of double glazing tinted glass.

Insulation Thickness x (mm)	Savings in load (Ton Refrigeration)	Customer Savings				Customer Investment (DH)		Payback period for customer (YR)	Government Savings		National Savings	
		Annual (DH/YR)			Capital (DH)	Initial	Net		Annual (DH/YR)	Capital (DH)	Annual (DH/YR)	Capital (DH)
		Electricity Consumption	Maintenance + Interest	Total	A/C Equipment	Insulation & Glass	Initial Capital		Electricity cost + Interest	Installed Power	Customer + Government	Customer + Government
$\frac{x_{wall} = 0}{x_{roof}} = 40$	4.22	2003.13	664.51	2667.64	8440	18004.8	9564.8	3.703	22310.21	22260	24977.85	30700
50	4.40	2088.57	692.86	2781.43	8800	19479.8	10679.8	5.130	23261.29	23205	26042.72	32005
60	4.54	2155.02	714.90	2869.92	9080	20954.8	11874.8	5.657	24001.01	23940	26870.93	33020
70	4.64	2202.49	730.65	2933.14	9290	22429.8	13149.8	6.300	24529.39	24465	27462.53	33745
$\frac{x_{wall} = 30}{x_{roof}} = 40$	6.14	2914.50	966.85	3881.35	12280	22429.8	10149.8	3.211	32460.49	32385	36341.84	44665
50	6.32	2999.95	995.20	3995.15	12640	23904.8	11264.8	3.509	33411.57	33330	37495.21	45970
60	6.46	3066.40	1017.24	4083.64	12920	25379.8	12459.8	3.856	34151.29	34065	38234.93	46985
70	6.56	3113.84	1032.98	4146.82	13120	26707.3	13587.3	4.205	34679.67	34590	38826.49	47710
$\frac{x_{wall} = 40}{x_{roof}} = 40$	6.29	2985.71	990.47	3976.18	12580	23786.8	11206.8	3.507	33253.97	33180	37230.15	45760
50	6.47	3071.15	1018.82	4089.97	12940	25261.8	12321.8	3.797	34205.04	34125	38295.01	47065
60	6.62	3142.45	1042.45	4184.90	13240	26736.8	13496.8	4.124	34998.52	34920	39184.42	48160
70	6.72	3189.82	1058.18	4248.00	13440	28211.8	14771.8	4.526	35526.89	35445	39774.89	48985
$\frac{x_{wall} = 50}{x_{roof}} = 40$	6.42	3047.41	1010.94	4058.35	12840	25232.3	12392.3	3.859	33939.94	33855	37998.29	46695
50	6.60	3132.86	1039.29	4172.15	13200	26707.3	13507.3	4.144	34982.84	34815	39064.99	48015
60	6.74	3199.31	1061.33	4260.64	13480	28182.3	14702.3	4.483	35632.57	35550	39893.21	49030
70	6.84	3246.78	1077.08	4323.86	13480	29657.3	15977.3	4.886	36160.94	36075	40484.80	49755

172

infiltration of outside air. The use of glass in new buildings should be planned carefully. Passive options and passive - active options to reduce energy
consumption should be considered. Care should be taken to make sure that a
building's heating, cooling, lighting, and electrical systems are all designed and ⸝perated in such a way as to conserve energy. In planning for energy
conservation on national basis, other energy consuming sectors should be included in addition to the building sector. These sectors are : electricity
generation, transportation, and production. These constitute the four major
consuming sectors in a nation.

It is important for the citizin to change and improve their energy use habbits
Energy shoul⁴ be used in the proper and most economical way when it is needed.
For example, air conditioning equipment and lights, and any appliances should
be switched off if not in use ; A/C thermostates should be set for a little
higher temperatures; fix broken windows ; use lightcoulered roofing and wall
materials to reduce solar gain in buildings which use A/C ; turn off unneed-
ed elevator after normal business hours ; replace single glazing with double
glazing where appropriate ; use tinted glass instead of clear one whenever
possible ; and fit windows with solar screens to reduce cooling loads.

The present results indicate clearly the economic feasibility of using thermal insulation in buildings. Millions of Dirhams can be saved if such energy
conservation measure is implemented. The saved money can be directed to the
industrial and agricultural sectors or indeed to any other economic sector.
This may reduce the heavy reliance on oil and gas revenues which has serious
social, economic, and political impacts on the country.

The load calculations showed that the heat gain through the walls and roof
can be cut down by up to 60% to 70% for not a too high insulation levels.
These results were verified experimentally (Ibrahim, 1987a, 1987b).

In the light of the present results, one may conclude that it could be economically feasible for the country if the total cost of the present energy
conservation measures be payed for by the government. In any event, the government should subsidise the cost of such energy conservation measures which
will be, no dought, much cheaper than subsidising the electricity generation
cost . It seems that the heavy governmental subsidies of energy, encourages
citizens to use energy without conservation or rationalization. This matter
should be given careful thought to decide on an alternative policy. The necessary legislations should be devised to force the use of thermal insulation
in buildings. This of course needs detailed studies to set the required codes
and speciafications.

REFERENCES

Carrier Air conditioning Company (1965). Handbook of Air Conditioning
system Design, Mc Graw Hill, New York.

Congressional Quarterly Inc. (1975). Continuing Energy Crisis in America,
Washington, D.C. 20037, USA.

Ibrahim, Said M. A. (1987a). Experimental study on the use of thermal insulation in buildings to conserve air conditioning energy. Int.J. Energy
Research, 11, No. 1,1.

Ibrahim, Said M. A. (1987b). The thermal behavior of thermally insulated
and uninsulated buildings. Energy, the International Journal. Accepted for
publication in Vol.12,No.1.

Energy Efficiency—a Leading Role for an Energy Supplier

Robert J. Jones

British Gas plc, National Westminster House, 326 High Holborn
London WC1V 7PT

INTRODUCTION

When the call for papers for this conference went out, I was struck by the statement that the event, "is aimed at encouraging innovation and new approaches to energy efficiency". In describing British Gas energy efficiency activity, this paper will show that the organisation has brought innovation and new approaches to the subject, not only in the expected technological areas but also in the areas of marketing and business opportunity. For those whose interest is energy efficiency this paper aims to give them knowledge and understanding of those whose business is energy supply.

Last year British Gas underwent a major change in status, moving from what is termed "the Public Sector" to what is quite generally understood as "the Private Sector". For the purposes of this study, I am suggesting that this change is put aside and that the place of energy efficiency in the business represents a continuum of usual practices and attitudes. In fact, the expression, "Business as usual" has been used extensively within the organisation in both planning and operational discussions concerning energy efficiency. This is not entirely surprising, given the characteristics of a nationalised industry in the UK and the characteristics of a controlled, privately-owned monopoly in most of the rest of the world. Perhaps a few words of explanation may be helpful to visitors from overseas.

In general terms, a UK nationalised industry is not at all like the sort of body that other nations might recognise as "State Enterprises". We in the UK would categorise such enterprises in other nations as being as close to civil service culture as our own UK nationalised industries are to the market culture. For example, the nationalised British Gas was a market-conscious entrepreneurial organisation that was aggressive in the market-place and that turned-in profits. Those very profits, of course, were judged more than satisfactory by the millions who responded to our recent flotation. It was in this climate that energy efficiency was exploited as a marketing opportunity. As the Company intends to stay lively and profitable, it is likely that such attitudes will prevail in the future. What follows, therefore, is not a history of a recently dead

173

body but an account of some of the activities of a very lively organisation that happens to have had a change of ownership.

1. ENERGY SAVED

Before discussing British Gas energy efficiency marketing projects, it would be useful to dispose of the question, "Has it saved energy?"; if only to satisfy the curiosity of those whose business it is to save energy. Research by British Gas has shown that the energy efficiency of many gas customers has increased whilst, at the same time, comfort levels have improved for domestic customers and quality and output have improved for industrialists. What is particularly interesting about these conclusions is that gas users, compared to other fuel users, have far less economic reason to save energy; the fuel is not only competitively priced, but can be used more flexibly, conveniently and effectively than other fuels or energy sources. From this I conclude that our energy efficiency activities have been effective in saving energy for individuals and in improving the nation's energy efficiency. Research in the domestic sector, for example, shows that gas-users invest more in energy-saving measures than non-gas users: 90% of gas central heating users have loft insulation against only 81% for electric central heating users. There is a similar pattern for other energy saving measures. To put this into context, there are 10.4 million users of gas central heating in the UK, representing some 50% of all domestic energy users. In addition, there has been a reduction of 13% since 1979 in average individual household consumptions by gas central heating customers. Two influences probably account for this reduction; firstly, the effect of customer education and "Thermsaver" marketing and, secondly, the "technical fixes" as a result of British Gas R&D investment. (Both of these are discussed later in their marketing context.)

In the non-domestic sector (industrial and tertiary), it is less easy to assess results with an overall average figure. Quite apart from the irrelevance of the word "average" to the wide ranges of size, activity and energy-intensiveness represented by these markets, the restructuring of industry through "de-industrialisation" has involved the biggest energy users. Over the period of study gas has also made significant gains in market-share; so that it is difficult to talk about "savings" in global terms for gas. Even the very act of changeover from less controllable energy sources has increased process efficiencies over and above specific "Save It" measures. To illustrate the magnitude of the change, we should look at market-share by fuel in these sectors between 1977 and 1986. In 1977 gas had 25% of the end-use industrial and commercial energy market. By 1986 this had grown to 36%.

Because global efficiency figures are difficult to prepare and, probably, misleading however prepared, it is instructive to look at some typical industrial processes that have benefited from British Gas R&D efforts. The gas recuperative burner has produced savings of up to 40%. The regenerative burner, which has figured in four UK Department of Energy Demonstration Projects, can achieve savings of up to 60%. Over 250 had been sold into the process industries up to last year. Ceramic tube-heaters have not only improved working conditions, by reducing the amount of heat involved in processes, but commonly show fuel savings of over 50%. One aluminium bale-out furnace fitted with such a heater saved 80% on running costs. The conclusion that I suggest should be drawn from this is that British Gas energy efficiency marketing has contributed to improved energy efficiency in the UK, has reduced costs for domestic consumers and contributed to the prosperity - perhaps the survival - of many industrial and commercial enterprises.

2. ORGANISING FOR ENERGY EFFICIENCY IN INDUSTRIAL AND COMMERCIAL MARKETING

In the aftermath of the "oil crisies" British Gas was the first of the UK fuel-supply industries to establish an Energy Conservation Department, to co-ordinate energy conservation activities nationally and between organisational functions. It is a part of the Marketing Division but also has to co-ordinate the Company's own-use energy management. It is probably this single-minded approach that has enabled the industry to dominate the field and also to give strong support for successive governments' energy saving campaigns and to use that support to produce marketing advantage.

Just because formal co-ordination of British Gas policy started around 1978, it would be misleading to conclude that nothing was being done before that. It was, perhaps, a factor in the creation of a national and corporate approach that there was already so much activity that needed to be consolidated and that so many energy efficiency initiatives had become well-established over a long period. For example, the British Gas School of Fuel Management had been set-up (We believe as a "World First") as long ago as 1975, based on an establishment that had been teaching energy efficiency to our own staff since 1967. The School is adjacent to one of the Industry's R&D facilities, the Midlands Research Station, whose scientists and engineers can be called upon to teach on energy efficiency courses at the School, to augment the work of the permanent staff. To some extent this is similar to research-undergraduate-teaching relationships of a university. As far as the nation is concerned, the best qualified and knowledgeable people are on hand to support the teaching: As far as British Gas marketing interests are concerned, existing and potential industrial and commercial gas-users are being made aware of the very latest technical advances in gas utilisation. I would like to offer the situation of the School as a case-study to illustrate how an energy efficiency need can become a useful part of a marketing strategy.

3. ENERGY TEACHING CASE STUDY - THE BRITISH GAS SCHOOL OF FUEL MANAGEMENT

The British Gas School of Fuel Management is just one example of an energy efficiency activity being used for vital marketing support. Although the School is operated by the Company's Education and Training Department and is concerned so much with technology and engineering development, four members of its Management Board are drawn from the HQ Marketing Division. Its courses are developed according to the market for such training, are priced in a market-related manner and form part of the overall strategy of the TCS (Technical Consultancy Service) marketing approach.

Though stressing the marketing utility of the energy efficiency activities of the School, I would also draw the Conference's attention to the fact that it is also used to support the Government's own energy energy efficiency efforts. For example, the facilities, including development staff, were loaned to the UK Department of Energy in 1978 to start their National Energy Management Courses and a British Gas manager was a member of the Management Committee. As the Department of Energy's staff became experienced in the task, they were able to take-over the management and marketing of the courses; then it was a British Gas initiative that enabled the Department to combine with the British Institute of Management to run the courses, which have prospered and are now known as the "Energy Managers Workshops".

The School consists of staff and facilities much as one would expect at a modern technical training establishment. There are conventional classrooms but also specialist laboratories for group and individual tutoring. There are also well-equipped workshop-classrooms, with a wide range of gas-burning equipment to enable hands-on training to take place. There is a growing computer capability and a demonstration collection of insulation materials that is probably unique, at least in the UK. There are no residential facilities but the School's Manager already had a network of agreements with local hotels and guesthouses when the School was used only for our own staff. When the School was reorganised to include residential courses for outsiders, these contacts were expanded and have proved satisfactory and economic.

Since those early days the market for energy efficiency teaching has changed. Apart from the Government's own "Workshops", many other bodies have introduced courses for energy management and, at the same time, the customers' requirements have changed. In response, the School has withdrawn from courses for higher management and concentrated on material for supervisory and technician levels. Customers are also less willing to release staff for even week-long residential courses; so the School has responded with shorter, modular courses, sometimes away from the School and sometimes on the customer's own premises. As part of this trend some regional energy management centres have been established, staffed and managed locally but working as part of the original School of Fuel Management concept. Recent examples are the Centre for Energy Management at Fulham in British Gas, North Thames and the Energy Management Centre at Bishops Stortford in British Gas, Eastern; the former opened in 1982 and the latter in 1986. To give a flavour of the current topics and levels of teaching, the following are the titles of short courses now offered by the School of Fuel Management and the regional centres, under the generic title, "Elements of Fuel Efficiency":
  1. Space Heating Controls
  2. Combustion and its Control
  3. The Recovery of Heat
  4. Building Insulation
  5. Process Control
  6. Investment Strategy in Energy Management

The last of these, "Investment Strategy in Energy Management", is interesting as a case-study of an innovative energy efficiency initiative, which was exploited as a commercial opportunity whilst being used to support a Government energy efficiency project.

4. THE "IMAGE" PROJECT CASE-STUDY - AN ENERGY MANAGEMENT GAME

As a case study, the "IMAGE" Project is typical of the British Gas approach to energy efficiency. Internal technical resources became part of the energy efficiency programme, with subsequent benefit to the business, to the customer and to the national government's energy efficiency drive.

To understand this case study it is necessary to have some knowledge of the UK Energy Management Groups, which the UK Department of Energy started in 1978. The idea was to encourage like-minded people, with energy management interests and problems, to meet on a regular but voluntary basis to exchange ideas - much the same as many other business interest groups; such as, accountants, transport managers, personnel managers etc. Companies encourage their own people to attend, to assist with organisation and to hold office. The Department of Energy assists with a modest budget and with some administrative support, through its regional representatives, the Regional Energy Efficiency Officers.

At present there are nearly 80 groups, divided into ten regions and with a membership of over 10,000. They meet, usually at members' business premises, to hear guest speakers and to visit successful energy management projects. (A session at the British Gas School of Fuel Management is a feature of many of their programmes and local British Gas engineers often give lectures and demonstrations.) Each region also nominates a representative to the National Energy Management Advisory Committee (NEMAC), which meets regularly with Department of Energy officials and ministers to represent the views of experienced energy managers. The movement is, therefore, an important aspect of the government's energy efficiency programme and an influential source of expert opinion for government policy-makers.

As it is past history, it is probably no secret that around 1982 the Energy Management Groups were experiencing something of a crisis of identity, stagnant membership and falling numbers at local meetings. In consultation with the NEMAC Chairman and with Department of Energy officials, British Gas produced the "IMAGE" project as a revitalisation exercise. A computer-based mathematical model had been produced at the Midlands Research Station, which could be used in a management game mode to produce a stimulating and instructive competition for individual or group players. It became known as "IMAGE" (Investment and Management Game in Energy) and was offered to NEMAC as an exclusive activity for the Energy Managers Groups.

British Gas has had a long tradition of harnessing the competitive spirit for its own marketing purposes and had considerable experience of doing this as an innovation in energy efficiency activities. Two major initiatives would illustrate this: The "GEM" Awards and the Design for Energy Management Competition, both highly successful marketing efforts.

The "GEM" (Gas Energy Management) Awards were started in 1975 and were the first national-scale application, by British Gas, of the competitive spirit to promote energy efficiency. At the time of "the First Oil Crisis" British Gas had reorganised and redirected its industrial and commercial technical sales force to provide energy efficiency consultancy for its customers. The British Gas Technical Consultancy Service (TCS) was established. To stimulate enthusiasm, it was decided to make a national award, jointly to the TCS team and to the customer who had produced the most praiseworthy energy saving scheme over the previous year. There were no cash prizes, merely prestige and national recognition through a very high-quality, high-profile approach. The awards prospered and are clearly recognised in the UK as the "Oscars" of energy efficiency.

There are now regional heats and awards that are themselves important events in local business calendars. In 1986/87 the 327 entrants to the competition saved between them 43 million therms of energy but also produced 73 million therms of new gas business.

In 1978 British Gas introduced another national energy efficiency competition, this time to encourage architects to design energy-efficient buildings. In association with the Royal Institute of British Architects, an ideas competition was launched for students of architecture. Like the GEM Awards initiative, the competition is based on high prestige and high profile but, in addition, as we were targeting students, cash prizes were included! The first prize for the winning student is currently worth £1,500. Such has been the success of the competition that, from 1986, it has been upgraded and repositioned as a postgraduate event and supplemented with local undergraduate awards for energy efficiency for each of the UK's 37 Schools of Architecture. As with the GEM Awards, these competitions have shown that the competitive spirit can be harnessed for the cause of energy efficiency and, of course, to promote "gas mindedness" among fuel decision-makers. With such traditions and experiences behind us the 1982 problems of the UK Energy Management Groups seemed a good case for treatment by a British Gas competition.

The computer model that had been developed by some members of staff of the British Gas Midlands Research Station could be run on a microcomputer to simulate the energy use of a small fabrication factory. It was possible to test a number of energy management strategies and to simulate the outcome. Developed as an energy investment appraisal program and used as a teaching tool, it was immediately obvious that it had great potential for a wide range of users. However, the most novel suggestion was to use it as the basis for an energy management game. Having been approached by NEMAC for support in revitalising the Energy Management Groups, British Gas Marketing Division took the brainchild of the R&D people and proposed the "British Gas/NEMAC National Energy Managers Competition". It was widely advertised and was run over a five-month period for three-man teams. The prizes included personal microcomputers for each team member. Eighty-two teams took-up the challenge. The game and its progress has been explained elsewhere and was a great success. The fictitious factory had an annual energy consumption of 970,000 therms for process work and 190,000 for space-heating. The winning team cut the energy account by £530,000. A very high-level of interest was generated over a period of about a year and a further event was organised for 1984/85. Letters were received from participants concerning action taken, in their own factories, as a result of playing the game. Many enquiries had been received from teams wanting to take part but who were not members of an Energy Managers Group. In every case they were told to join an Energy Management Group to get access to the game. Papers were presented at three international conferences, giving useful PR advantages for both the UK government and for British Gas. Subsequently, the game has been marketed by the School of Fuel Management as part of an energy investment course, also as a distance-learning game and as a stand-alone computer package, the "Gas Energy Management Industrial Game".

This has been a very brief account of a complex marketing activity. It was certainly innovative and illustrates how an imaginative marketing organisation can develop a new approach to energy efficiency, not only for its own ends but also for the benefit of its customers and in support of government help for industry. Similar help for domestic customers has presented similar challenges.

5. ORGANISING FOR ENERGY EFFICIENCY IN DOMESTIC GAS MARKETING

41% of total end-use energy in the UK is consumed by domestic users. Unlike the industrial and commercial sector, the domestic sector has a large number of decision-makers, millions of them, each responsible for only a small fraction of the overall energy use - and each with a savings potential of barely a £100 worth of energy out of a home-heating bill that is less than is spent on petrol for the family car; perhaps far less financially significant than the food and veterinary bills for the family dog! Innovation and new approaches are certainly needed in this area to encourage energy efficiency and it is an area in which the dominant fuel supplier has a leading role.

The British Gas approach has been to make a two-pronged attack; first by adopting the "technical fix" and, second, by giving advice and selling energy efficiency services to customers. It is a pity that energy efficiency commentators become divided over the "People-in" or 'People-out" question[2]. This has been discussed elsewhere and the most profitable conclusion seems to be that neither approach is sufficient on its own. There is a view that we merely have to design and install energy-using equipment that makes it difficult for users to be wasteful; for example, by having low-energy structures, by fitting permanently-closed windows, by fitting sealed controls etc. That is the "People-Out" view. But there is also a need to educate users to act energy-efficiently. This is the "People-in" view, that harnesses intelligence and goodwill so that people become involved with their energy-using systems to use them economically and efficiently. Back in 1977 the UK Advisory Council for Energy Conservation (ACEC) stated:

> "..... users will manage their energy demands responsibly if they know what is expected of them and if these expectations are reasonable and seen to be useful ....."[3]

Whilst bolder people than ourselves continue to disagree on this subject, the British Gas approach does at least have the virtue of prudence; we target both technology and people.

For the "People-out" protagonists, British Gas has had a longstanding policy of ensuring that only the most energy-efficient appliances, compatible with market economics, are available to gas users. The British Gas Watson House Laboratories are mostly concerned with research and development of energy-efficient domestic gas appliances. Gas central heating systems are marketed to domestic customers under rigorous energy efficiency standards; either as "Gaswarm", in new homes, or as "Heatsaver", for renovation and rehabilitation projects. The "Gaswarm" concept has been especially successful in the UK new homes market. Over 70,000 Gaswarm homes have been built since the scheme started in 1982. This compares favourably with only 75,000 Electric "Medallion" homes, though both schemes show that it is possible to influence builders and developers by marketing low-energy as a selling advantage. It was British Gas that pioneered the low-capacity gas boilers of the seventies and the same marketing and R&D effort has recently brought super-efficiency condensing boilers to the UK market. In summary, it is very difficult for a householder in the UK to buy an inefficient gas appliance or gas-burning system, so long as he deals with British Gas. To that extent we have adopted a "People-out" approach. But that is only one side of our two-pronged attack. Involving people and employing classic retail marketing, are addressed on the other side. The "Thermsaver" initiative, a brand-name case study of innovation, is typical of the many energy efficiency activities that are targeted at British Gas domestic customers.

6. THE "THERMSAVER" CONCEPT CASE-STUDY

Since the mid-seventies most regions of British Gas were providing energy efficiency advice to their domestic customers, either through leaflets, posters, media-advertising, showroom displays or a combination of all or some of these activities. Some were selling energy saving materials and services, either directly or through sub-contractors. Some were using the business, more or less successfully, to support gas-appliance sales; for example, one region was offering free roof-insulation with all new gas central-heating boilers.

Whilst there was no uniformity of approach, a general enthusiasm to support - or to be seen to support - the government's "Save It" campaign of that era seemed to keep the various approaches going in the market-place. However, none of the initiatives was a commercial success, in terms of items sold, but they were certainly earning the Industry useful PR points at a time when businesses that made their livings from selling energy felt themselves to be in an ambiguous position, as the world began to beat its breast for past energy profligacy. In fact, British Gas did not find much ambiguity in the situation. It had a long tradition of selling energy-efficient systems and appliances; in the immediate post-war years, merely for survival, because of price, and, after that, because it clearly made business sense to persevere in that vein, even when we were on a winning-streak with price, control, cleanliness etc. Adding energy conservation to the sales pitch merely reinforced the message to the customer. However, there was still a need to establish a degree of uniformity because energy saving activities were, at that time, assuming a high-profile. Whatever one region did immediately prompted the question from observers, Why not the other eleven? That is a recipe for an open-ended commitment. Also, there was a need to get some sort of marketing science behind something that was patently dependant on a lot of local enthusiasm.

In the Summer of 1979 a pilot scheme was launched in what was then Eastern Gas - now British Gas, Eastern. There were two strategic objectives, a research objective and an operational objective:

"- To illustrate that British Gas actively supports the responsible and wise use of gas in the domestic market.

- To ensure that energy conservation measures, if taken, are taken expertly and do not inhibit gas and gas-appliance marketing.

- To establish the market potential for package sales of insulation and controls.

- To provide showrooms with additional customer traffic."

A package of energy conserving goods and services was prepared under the selling theme, "Don't Waste Your Energy". Three major showrooms were equipped with Energy Advice Units, with eye-catching displays, racks of energy conservation packs for sale and a technical adviser in attendance. Strong local newspaper advertising was implemented and even a few weekend out-of-town energy conservation cash-and-carry fairs were organised in service depots that were not operational on Saturdays. There were strong tie-ups with local insulation and double-glazing contractors. Shortly after, a similar pilot exercise was launched by Segas (British Gas, South Eastern), with a different display style but including a free home energy audit.

Neither exercise was particularly successful in terms of products sold. Even the Segas free home energy audit attracted few customers. We felt that that was particularly significant and passed on the information to government and others. We were not surprised when a later government-sponsored, fee-paying home energy audit pilot exercise (HEAT Limited) failed as a commercial venture. It remains the British Gas view that home energy audits that are not linked to remedial packages are not a merchandisable commodity. As far as the sale of products was concerned, the failure to develop volume business was something of an irritation to an organisation that was used to continuing success with its retailing activities. We suspected that the concept of a package of services and products was not being grasped, either by the general public or by our own sales people. A spin-off from the Segas exercise produced the name "THERMSAVER", to describe the generic nature of the various products and services on offer. At a national "post-mortem" on both exercises it was decided to use the name as the brand-name of British Gas domestic energy efficiency services. It was adopted by all regions and in 1981 national advertising began. Both public and staff quickly grasped the name and concept. The establishment of a brand-name made it easy to talk and think about energy efficiency activities as a sales package. Politically it produced a much stronger overall PR impression and lent itself to use as showing that British Gas does support what we termed, "the resposnsible and wise use of gas".

As a stand-alone retailing activity "Thermsaver" was not much more successful than the earlier pilot exercises. It has proved to be very difficult to sustain a viable retailing business in energy efficiency services for dwellings. However, research has shown that most customers do go away and do something positive to save energy after visiting a "Thermsaver" unit or being exposed to a "Thermsaver" leaflet or advertisement; even if they do not actually buy from the unit. When associated with a larger purchase; such as a new or replacement heating system, energy conservation is sold more successfully - Our "Gaswarm" and "Heatsaver" schemes illustrate that. In other cases, there is probably a need for strong external forces - such as the sort of government assistance seen in other countries - to generate public interest and action. However, the value of branding the service remains. The 'Thermsaver" name has kept the concept alive in theory, even if it is sluggish in practice. This has been politically invaluable, especially when the Industry, as a nationalised body, had to respond to the government's Energy Efficiency Year in 1986. A re-launch of "Thermsaver" was a painless, low-cost way of making a high-profile response.

"Thermsaver" is still with us. As an innovation, it has given product status to a diverse range of energy conservation services and its chief value to the Company has been to give our staff something tangible to work on. The pilot exercises have shown that there is no great market opportunity in this area and that a low-energy society is unlikely to arise if we are to rely solely on "market forces" to motivate the domestic customer.

My final case study of innovation and new approaches to energy efficiency concerns activity related to a section of society whose members are quite unable to react to "market forces". I refer to the socially disadvantaged, particularly those who are termed poignantly, but succinctly, "The Old and Cold". Here, I suggest, an energy supplier has a leading role to play and has much to offer, combining public duty with business responsibility.

7. A COMMUNITY SERVICES CASE-STUDY - NEA

For many years there has been a strong lobby in the UK for special weather-proofing services for those who could not afford the commercial rate for the work. This lobby has been inclined to suggest that such charitable services should be provided by the fuel industries. As gas accounts for about 60% of end-use energy used in the home, British Gas seemed to be the natural provider. That always struck the Industry as being just about as logical as expecting the oil industry to pay for pensioners' road vehicles to be tuned and serviced. However, we also feel that there are strong and practical social reasons why we should not be directly involved in tackling this particular problem. We do not think that we should try to dabble in social service work, for which our staff are neither trained nor organised. Instead, British Gas has chosen to help this disadvantaged sector of society by giving strong support to Neighbourhood Energy Action (NEA). The organisation was pioneered in the City of Newcastle-upon-Tyne, to promote energy efficiency initiatives for the socially disadvantaged and to combat "fuel poverty". Our activities in this area illustrate, I suggest, yet another leading role for an energy supplier in energy efficiency - not to attempt to do what we are not suited for but to encourage and enable those who have the expertise to do the job properly. In an introductory note to the 1986 Conference of NEA, the Chairman of British Gas, Sir Denis Rooke, put it like this:

> "NEA makes a unique and priceless contribution to improving the quality of life of many people and deserves the support of those of us in the energy business who are probably not as skilled as the NEA specialists in operating in this sensitive area" (My italics)

As mentioned earlier, attempts to establish market-led home energy audit services among ordinary people in the UK have failed. Even the government-backed one in 1985, Heat Ltd., progressed no further than a disappointing pilot scheme. However, just because those who can afford to help themselves choose not to do so, it is not necessary to leave the socially disadvantaged, who cannot help themselves, to the mercy of "market economics". On the other hand, if the socially disadvantaged can be kept out of the clutches of a bureaucratic machine whilst being helped out of "fuel poverty", so much the better. It is also a fact of commercial life that field operations are best controlled by local management, if they are to be carried out effectively and economically.

NEA has evolved an ideal working system; locally managed, avoiding bureaucracy, tapping many sources of finance and resources and obtaining some help from manufacturers who actually benefit from selling the materials that are used. Because of this direct economic benefit for the materials manufacturers, we in British Gas feel that it is they who should bear the lion's share of providing cash help, if government funding is inadequate. For our part, we are pleased to step in with resources, services and some cash for specific purposes where our own interests are involved and even where we can take some commercial benefit whilst helping NEA. We have sponsored promotional events, a national conference, a training video, local publicity and so forth. In addition, we have loaned training facilities for Project Leaders and other members of local NEA groups, for both national and local programmes. Along with other industries we loan meeting and catering facilities for NEA Council meetings and have made premises and reception facilities available for press conferences and media events, including entertaining a foreign delegation. These sorts of requirements, commonplace and accepted by a large business organisation, could be a crippling drain on NEA resources, especially in terms of hard cash to pay for them. In this way we are able to give far more in value than if we made cash grants to buy services that we ourselves can provide. I would suggest that this is a very real way in which industry can help an organisation such as NEA. The donor is able to exercise a certain level of control whilst enabling the recipient's officials to become closely associated with and understand the workings of the large organisations that they have to relate to.

In a sensitive area such as NEA operates in, understanding is needed from all sides. There are so many areas of potential conflict between a fuel supplier and those combating "fuel poverty" that I think our method of working with NEA has much to recommend it. In some areas NEA people have visited the Arrears Sections of Customer Accounting departments of British Gas Regions. In the Autumn of 1986 the NEA Director was the guest of our National Customer Accounting Committee at one of its meetings. Both sides benefit from these encounters. British Gas is a business, not a charity. NEA is a charity that has succeeded because it is business-like. I like to think that its close relations with British Gas have helped it to be so, chiefly because they are not relations based on a cash handout between "Lady-Bountiful" and a needy recipient but are the relations that arise from a business association. We have developed this model of operating with other organisations and I think others may follow where we have led but especially in this sensitive area of helping the socially disadvantaged to benefit from improved energy efficiency.

8. CONCLUSION

These brief case studies have been chosen to illustrate the wide range of British Gas energy efficiency activity across all energy markets. Each is, I suggest, innovatory and has relevance to other fuel suppliers and businesses related to energy efficiency. They also show that a leading role in energy efficiency can have businesses advantages to a fuel supplier and that such business interest is wholly compatible with support for a nation's drive towards becoming a low-energy society.

References:

184

1. "Microcomputer-Based Management Game". R.J. Jones and J. Masters. Association Techniques pour les Economies d'Energie. Paris. November 1984.

2. "Effective Line-Management for Energy Conservation". R.J. Jones. Design Industries Association/RIBA. London. May 1982.

3. "Energy Paper Number 25". Advisory Council on Energy Conservation. HMSO. 1977.

The Economic Evaluation of Energy Saving Investments

Walter L. Lom, Consultant

Hotham Park House, Bognor Regis PO21 1HW

Abstract.

Energy saving investment, while inherently desirable and in the interest of any national economy, should be carefully selected in order to produce optimum results. When deciding whether and in what form to invest a forecast will have to be made of the future cost of the form of energy which is being replaced. Furthermore present and future interest rates will play an important part. Finally the expected life of the equipment or modification in question will have to be forecast.

There are a number of alternative investment pzarameters which can be applied and, unfortunately, they do not always yield the same result: thus we may find that on the strength of pay-out time one form of energy saving is justified and that e.g. a DCF calculation leads to a different conclusion. A genuine dilemna arises if investment is justified by one method and continued "wastage" is preferable according to another!

The writer, in his function of technical advisor to the European Investment Bank, has had to comment on the desirability of very substantial investments, and in order to be consistent a basic methodology had to be developed and applied to all requests for loans in this domain. The method and parameters which were applied throughout are described in the paper.

185

Text.

The purpose of energy saving investments, whether they aim to reduce losses in existing equipment, improve production processes so as to reduce specific energy consumption, utilise sources of energy which normally go to waste or simply generate new energy which in the past had not been available, is, quite obviously, to produce long term economies. In order to be justified at all the investment in question must result in financial savings; in order to be selected from a number of possible alternatives the main criterion will be minimum expenditure for maximum results.

However, since the savings in question will occur in the future any recomendation for energy saving investment will implicitly include long term forecasts in regard to at least two variables, the cost of energy and the cost of money. If either of these forecasts turn out to be wrong the recomendation to invest at all or the choice of investment could often be erroneous.

In addition to these two critical parameters which must be studied and forecast to the best of our ability we also must decide on the particular method used to compare the various pos- sible alternatives. A number of financial/economic parameters exist which are used by accountants and economists to assess and put in economic order alternative investment strategies. They include:
        —the classical return on investment concept
        —the pay back time
        —the present value
        —the discounted cash flow return
        —the net present value
        —the discounted final value
                and several other parameters.

Unfortunately we have found on occasions that, not altogether surprisingly, different results are obtained according to which of the above is used to deciode whether to invest, or which par- ticular form of energy saving should be chosen.

It is not proposed to discuss in detail the pros and cons of the different parameters. Suffice it to say that the direct rate of return on capital, due to any net savings, is the most basic standard of comparison, and that comparing the alternatives or the basis of the discount rate which reduces all in- and out- flows to zero, i.e. the discounted cash flow return (DCFR), some- times called interral rate of return (IRR), is probably the most common method.

Itis also closely linked with the concept of net present value i.e. the residual discounted value which results if a given IRR is applied (e.g. a threshold value) to a series of annual cash flows. The IRR has, apart from its universality, the advantage of not requiring any assumptions regarding the future cost of money Admittedly the same applies to the classical rate-of-return-on capital method, but using the latter we have to make further as sumptions regarding depreciation and also — somewhat misleadingl — disregard the time value of money.

However, even if we eliminate the first source of ambiguity, the chosen financial/economic parameter, and decide to apply through- out IRR's in combination with net present value, there remain two further degrees of uncertainty. Future savings will necessarily depend on future energy prices. And further assumptions will have to be made in regard to the relative prices of the different forms of energy.

One of the more obvious problems which arise at this stage is, therefore, whether in fact different forms of primary energy can maintain, in the long run, different unit prices. A second ques- tion which needs an answer is whether relative prices of dif- ferent energies are likely to remain the same.

Wile, no doubt, different answers to these two questions will be forthcoming from different experts the balance of opinion at the European Investment Bank seemed to suggest that differences be- tween the prices of different forms of energy are here to stay. Although energy value at the burner tip does only depend very slightly on fuel type, the effects of logistics, convenience of operation, and simply the fact that certain fuels will price themselves out of certain markets will ensure that overall sub- stantial differences remain.

As to the price relativities these are so much a function of market forces that it would be a brave man who could assert that today's price ratios will continue into the future.

A particular difficulty of forecasting energy price levels is the fact that they are generally quoted in U.S. dollars; and while changes in dollar parity are not normally reflected in world price levels, and I am not prepared to say whether this is due to the lack of sophistication of energy economists or the weight of U.S. domestic energy prices, there is little doubt that in the longer term changes in dollar parity will have to affect the world's pricing system. The very much simplified price forecast below is thus a combination of expected price changes due to economic factors, on the one hand, and curency effects, on the other.

	1986-1990	1991 and beyond
Gasoil	$ 200/ton	$ 240/ton
Fuel Oil (3%S)	$ 120/ton	$ 146/ton
Fuel Oil (1%S)	$ 135/ton	$ 156/ton
Coal	$ 40/ton	$ 40/ton
Crude Oil (34'API)	$ 15/barrel	$ 22/barrel

There is little doubt that these forecasts are tied to particular moments in time and will have to be revised fairly frequently.

Having decided on the above level of energy prices for the two periods, which are meant to cover the economic life of most of the energy producing asnd energy saving investments submitted for valuation, we next had to specify threshold values of IRR for different types of projects. And while these particular levels, clearly, will not be automatically acceptable to other assessors, they were, in our case, to some extent linked with the social and political aims of the European Investment Bank.

The latter, it will be remembered, is one of the European Institutions created under the Treaty of Rome, and its main tasks are
to reduce, by means of judicious investment, regional discrepan-
cies in GNP in the EEC, to contribute to the economic integration
and growth of the Community and, in certain instances, to help
the restructuring of sectors of the economy which find themselves
in particular difficulties.

In the fulfilment of these tasks primary banking criteria, such
as financial viability of the preomoter and his financial standing and the profitability of the investment under review,
clearly, had to be borne in mind; but for particularly desirable
operations, especially those which corresponded to several of the
above mentioned criteria, a lower level of returns would be
tolerated. The following table summarises IRR threshold values
for a range of investment types linked to energy saving and
energy generation in the EEC.

It also indicates the type of fuel saved and the assumed life of
the investment as well as the annual operating cost, expressed
for simplicity as a percentage of the investment:

	DCFR Threshold	Fuel Replaced	Opcost % Invest	Life Years
Buildings:				
Energy saving equipment	5%	Gasoil	4%	1?
Heat Insulation	5%	Gasoil	−	30
Industry:				
Energy saving processing	5%	Fuel Oil	4%	15
Improved insulation	5%	Fuel Oil	−	15
Infrastructure:				
Gas Distribution Systems	8%	Electricity,	6%	30
(in special develpt areas	5%)	Gas oil,LPG		
Hydropower	5%	Fuel Oil,Coal	2%	30
Waste Incineration,Biogas	5%	Fuel Oil	−	15
Cogeneration of Heat and Electricity	5%	Fuel Oil	−	25

Some of these assumptions call for further comment. The lowest
acceptable rate of return of 5% e.g. implies that the project has
certain not quantifiable benefits such as environmental improvements, safety of supply, security etc. Returns on gas distribution investments e.g.can be lower initially in areas where
reticulated supply is considered to be beneficial to overall economic growth. It will also be noted that we do not attribut
operating costs to waste incineration, biogas and cogeneration of
heat and electricity. This is again a simplification since we
have assumed that opcosts of the systems which have been replaced
or modified by the new plant, e.g. waste disposal by dumping, are
of the same order of magnitude.

The conclusions from the above set of assumptions are best
expressed in terms of maximum permissible investment per unit of
energy saved or generatedper year, and it should perhaps be un-
derlined that there is no basic difference in regard to investment considerations between the two. These values are shown i

the table below first in terms of U.S.dollars per ton of oil
equivalent and then, at today's conversion rates, as permissible
investments in local currency for a number of European countries
in which energy saving investments are common, and for the U.K.
which is probably somewhat behind the others in that respect –
certainly as far as requests for EIB loans for energy saving
projects are concerned.

Currency	Ratio U.S.$/toe	Italy M Lire	France K FF	Denmark K DKr	U.K. £
Buildings:					
Energy saving equpt	1570	2.8	14.8	10.7	1023
Heat Insulation	3330	6.0	19.9	22.6	2170
Industry:					
Energy saving process	1060	1.9	6.3	7.2	691
Improved Insulation	1060	1.9	6.3	7.2	691
Infrastructure:					
Gas distribution	930	1.7		n.a.	380
Hydro Power	1770	3.2	10.6		n.a.
Waste Inciner.,Biogas	1500	2.7	9.0	10.2	977
Cogeneration	1500	2.7	9.0	10.2	977

It should perhaps once more be emphasised that the above figures
are entirely based on certain assumptions and, therefore, should
not be used in isolation, except perhaps as a rough guide to the
order of magnitude of investment which could be accepted by
either a company board or a financing institution. The main purpose of this paper is, clearly, not to list maximum permissible
investment figures for various types of enrgy saving investment,
but to provide a suitable methodology for such calculations and
an indication as to how a major bank in the EEC would approach a
demand for loan finance in the field of energy saving or energy
generation.

Process Integration—an Innovative Approach to Energy Strategy

E. K. Macdonald and I. C. Kemp

Atkins–EPI, Process Engineering Research Centre, B.488.T6,
Harwell Laboratory, Didcot, Oxfordshire OX11 0RA, United Kingdom

Process integration or "pinch technology" was recognised as a major advance as soon as it appeared. For the first time a process engineer could find out exactly how much energy his plant should use – a rigorously derived yet practical target rather than an impossible theoretical target or a simple experience value. The techniques have developed to encompass combined heat and power (CHP) systems, heat pump placement, and separation processes in addition to heat exchanger network (HEN) analysis. In its present form, pinch technology is a major tool in the formation of an energy strategy for a wide range of industries. This paper highlights the strengths and weaknesses of its use, and shows how it can give rise to innovative solutions to energy efficiency problems.

Process integration; energy efficiency; evaporation; pinch technology innovation.

INTRODUCTION

Whether process integration itself is an innovation is open to question. After all, heat recovery is not a new idea, and for many years process designers on complex chemical plants have been designing and optimising heat exchanger networks (HENs) to reduce energy consumption. However, the simplicity of the underlying concepts – achievable energy targets and the concept of the pinch – has led to a much wider appreciation of the benefits of systematic energy saving campaigns and an awareness of the need to employ innovative changes in process design and operating practices. In the past such changes may have been considered but rejected because of real or perceived problems. Once a process integration analysis identifies process change as a crucial step in the energy efficiency campaign, it is clearly supported by technical and economic evidence which helps to overcome the fear of change. Thus process integration helps to stimulate innovation in process design and, by its very nature, forms the central part of an overall energy strategy designed to improve profitability and competitiveness.

THE PROCESS INTEGRATION METHOD

The key to understanding the use of process integration lies in identifying where energy is available within a process and where it is needed. This is done with the aid of composite curves (Fig. 1) (Hohmann, 1971; Linnhoff, 1979; Linnhoff, Boland and others, 1982).

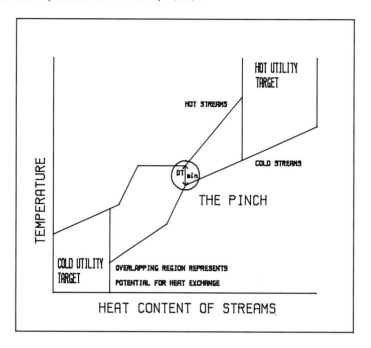

Figure 1
Hot and Cold Composite Curves

These graphs show the combined heating needs of the cold streams and the combined cooling needs of hot streams at given temperatures throughout the process. Using these graphs or a simple calculation it is possible to derive a rigorous but practical target for the energy requirement, and also to identify the critical point, known as the pinch, which helps the designer make decisions on how energy should be used within the process.

The energy targets themselves depend on how close the two composite curves are allowed to approach. At some point in the middle of the process, the temperature difference between the hot and cold streams is a minimum. The choice of this DT_{min} fixed the relative positions of the hot and cold composite curves and hence the targets. If we know (or can estimate) heat transfer coefficicents for the streams, and the cost for heat exchangers and energy, we can base the choice of DT_{min} on a sound economic judgement which balances energy and capital costs, rather than an experience value which relies heavily on guesswork and personal judgement.

Knowing DT_{min} and the pinch location, we can apply simple design rules to ensure that the plant is correctly designed to meet the calculated energy targets. These arise from the division of the process at the pinch into a system which needs heat above the pinch, and one which has too much heat below the pinch. The shortage above the pinch is met by supplying hot utility; the surplus below the pinch is removed by cold utility. In energy terms, the two systems are independent, and any flow of energy from hot to cold across the pinch is inefficient. This simple insight leads to the pinch design rules (Linnhoff and Hindmarsh, 1983).

1. Do not transfer heat across the pinch.
2. Do not use hot utility below the pinch.
3. Do not use cold utility above the pinch.

These rules then guide the design task, which is to identify and eliminate pinch violations in a retrofit study, and avoid them in a new plant design.

PROCESS CHANGE

We often wonder whether a process change, such as a change in distillation column operating pressure, or a new flash vessel, will have repercussions elsewhere in the process although the unit operation itself will be improved. Process integration is an ideal tool, not only for assessing the effects of change, but also for stimulating innovation in process development. An engineer experienced in the use of the techniques can interpret the composite curves and the derived grand composite curve as to what changes will lead to a reduction in energy use, and whether they are practical. An innovative approach, coupled with an awareness of realistic changes, can lead to significant savings.

Consider the following example. Figure 2 shows the grand composite curve of a chemical process (the grand composite curve is derived from the composite curves to show the overall heat flow through the process: the pinch is the point where the curve touches the temperature axis and indicates zero heat flow). The main features of the graph are the two horizontal lines which represent the evaporation and condensation of water from a solids crystallization stage in the process. The enclosed area represents potential heat recovery from the vapours to other process streams that require heating below the pinch. Above the pinch all the heating is done by steam. The energy target for the process is 1000 units. Figure 3 shows the same process, but with a four-effect evaporator superimposed on the same graph (Kemp, 1986; Macdonald and Kemp, 1986). The evaporator acts as a pre-concentrator for the feed to the crystallisation process, and its efficiency is such that one unit of steam evaporates four units of water. However, the operating temperatures of the evaporator are such that heat supplied to the first effect is above the process pinch, whereas the vapour from the fourth effect is condensed below the pinch. The evaporator therefore transfers heat across the process pinch, and in doing so increases the energy use of the system as a whole to 1107.8 units. Perhaps the obvious way to reduce energy consumption is an evaporator is to increase the number of effects, and from the grand composite curve it is possible to envisage an additional effect between the existing third and fourth effects. This could save perhaps 20% of the evaporator duty - a good saving in the stand alone context, but it only represents 2% of the total process energy use.

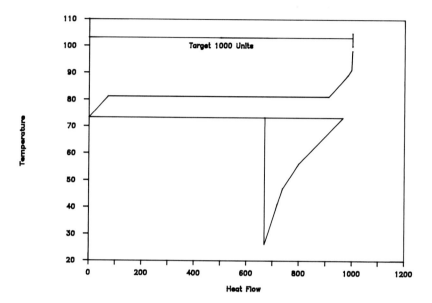

Figure 2
Process Grand Composite Curve

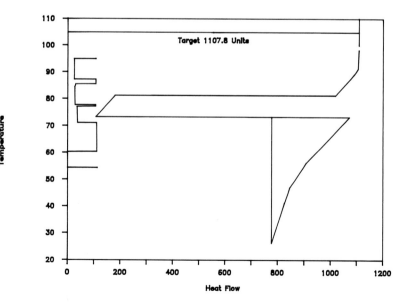

Figure 3
Process and Evaporator Grand Composite Curves

The way forward in this case is to consider a single effect evaporator. At first glance, the idea sounds ridiculous, as the evaporative heat duty would rise by a factor of about 4. However, if the evaporator can operate entirely above the operating temperature of the crystallization process, the vapour from the evaporator could be used to drive part of this process, and the loss of heat across the pinch could be eliminated. The result of this analysis is shown in Fig. 4. The combined process now needs only 1016.2 units of energy, a saving of 8.3% compared with the original process. In this case the traditional approach of optimising unit operations in isolation led to a better overall solution being missed. Such an overview is essential if full benefit is to be gained from process integration analysis.

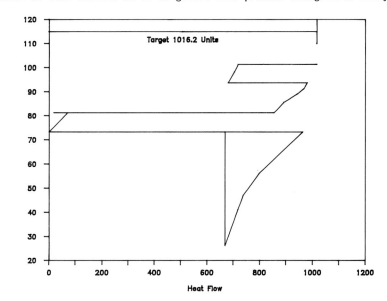

Figure 4
Modified Process Grand Composite Curve

CASE STUDIES

Harwell's Energy and Process Integration Service (EPI) has performed a number of studies in a wide range of process industries. The following case studies illustrate a cross-section of the innovative process modifications which led to significant savings.

Procter and Gamble (Clayton, 1985)

An area of significant cost was the use of liquid nitrogen to cool the air used in a pneumatic conveyor for detergent granules in the summer months. However, detailed examination of the cooling needs using the grand composite curve showed that such a low temperature coolant was both unnecessary and costly. An alternative refrigeration process was proposed (and is now operating successfully) which had much lower operating costs and allowed tighter product quality control. This illustrates an important aspect of process integration: selection of utility levels is made much easier using the grand composite curve.

Courtaulds Fibres (Clayton, 1987)

Significant energy using processes included multi-effect evaporators to recover valuable chemicals from the textile spinning process. It was thought that the only way to improve the evaporation efficiency was to add more effects or to consider vapour recompression cycles. The EPI analysis showed that about 10% savings could be achieved by improved heat recovery within the evaporators themselves - a much simpler solution than the original ideas. In addition, the analysis showed ways in which the spinning process itself might be modified to reduce the energy consumption of the plant.

Scottish Grain Distillers

An innovative approach to the analysis of the grain cooking and mashing process led to the identification of an alternative, more efficient process configuration. This involved the specification of an optimised flash system to increase the heat recovery from the hot cooked grain slurry as it is discharged from the cooker to the mash tun. The potential savings from the proposed modification are 36% of current operation.

LIMITATIONS

Of course any technique such as this must have its limitations. An inexperienced user can make major errors simply in the selection of process data and definition of study boundaries at the start of a study, resulting in analysis of the wrong problem. And it must be remembered that pinch technology is only an analytical tool, not a magic wand to provide instant solutions. Projects must still be designed, engineered and installed before savings can be made. The over-riding influence on what is possible is therefore the economic position. Sometimes heat recovery is difficult because of fouling, corrosion or contamination risks. In other cases the value of energy savings is so outweighed by product values that it is not economic to devote resources to energy projects other than simple housekeeping measures. Process integration cannot guarantee a better return than other schemes competing for limited capital resources. What it does guarantee is that the selected projects will make the best use of capital investment allocated to energy efficiency.

On the other hand, just because a study does not reveal any startling new ideas we should not assume that it was a pointless exercise. The benefits can still be seen in the formulation of an overall energy strategy that puts projects, previously developed in isolation, into a site context, and ranks them in order of economic benefit and practical necessity. This provides a clear statement of the way forward to a more efficient process. In addition, a real cost can be attributed to traditional and technical process constraints so that a value judgement can be made of the benefits of taking action to remove the constraints. Finally, even in mature industries the value of a process integration study at the design stage cannot be overstated. The potential savings in capital cost and lifetime operating costs will certainly justify any additional time spent on the process design.

CONCLUSIONS

Process integration is a classic example of an apparently simple idea put to very effective use. From the original applications within ICI to the worldwide use now in evidence, its use has shown the way forward to many technically aware companies, through identifying innovative solutions to complex energy recovery problems. The benefits of a clearly defined objective in the rigorous energy target are seen in the incentive for the engineer to achieve the target, and confidence of management in making investment decisions based on sound technical and economic justifications. Such an approach helps to clear the way for the innovative approach to energy efficiency, overcoming technical and traditional constraints. The end result is an overall energy strategy to boost efficiency, competitiveness and profitability.

REFERENCES

Clayton, R.W. (1985). Cost reductions on a site producing soap and detergents identified by a Process Integration study at Procter and Gamble Ltd. Energy Efficiency Office, R&D Demonstration Report No. RD/8. Original report by EPI, Harwell.
Clayton, R.W. (1987). Cost reductions on a man-made fibre plant identified by a Process Integration study at Courtaulds Fibres Ltd. Energy Efficiency Office, R&D Demonstrate Report No. RD/21/28. Original report by EPI, Harwell.
Kemp, I.C. (1986). Process Integration applied to separation systems. Paper presented to Institution of Chemical Engineers Annual Research Meeting, Bradford, April 1986.
Hohmann, E.C. (1971). Optimum Networks for Heat Exchangers. Ph.D. Thesis, University of Southern California.
Linnhoff, B., D. Boland, D.W. Townsend, G.F. Hewitt, B.E.A. Thomas, A.R. Guy and R.H. Marsland. (1982). A User Guide on Process Integration and the Efficient Use of Energy. Chapter 2.
Linnhoff, B. and E. Hindmarsh. (1983). The pinch design method of heat exchanger networks. Chem. Engng. Sci., 38(5), pp. 745-763.
Macdonald, E.K. and I.C. Kemp. (1986). Process Integration gives new insight on evaporators. Process Eng. (GB). November 1986, pp. 25-27.

Managing to Ensure that Industrial Energy Conservation Measures are Effective over the Longer Term

J. T. McMullan* and D. J. Raymond**

* Centre for Energy Research, University of Ulster, Coleraine, Northern Ireland
**Harland and Wolff Plc, Queen's Island, Belfast, Northern Ireland

ABSTRACT

This paper sets out to indicate the structural changes which are needed at the management level of both small and large companies to ensure that appropriate energy saving measures are adopted and sustained. The suggested changes bring into question the need for an independent energy manager with the responsibilities and limitations that such people normally have at present. Instead, they emphasise the importance of making energy control and budgetting the day-to-day responsibility of production managers. Possible structural changes will be suggested and some of the necessary supporting technical measures will be indicated, though no specific energy saving measures will be discussed in detail.

INTRODUCTION

Over a period since the first oil price shock there have been significant developments in the practical techniques for energy conservation and the rational use of energy. There has also been a propaganda war with almost continuous exhortation to save energy in all sectors of the economy, including industry. This publicity has been punctuated by occasional high-profile government sponsered campaigns with catchphrases such as "Save It" and "Monergy". Given such activity over a period of very high fuel prices, one might be tempted to ask why the message has not got through and why it is still necessary to continue the pressure - especially in the industrial sector which is supposed to be very cost conscious.

The answer lies with human inertia. The message has usually got through when it was presented, and while it is relatively easy to take local energy saving measures which will be effective for days, weeks or sometimes even months, it is very difficult to maintain the impetus and to ensure that the

savings which can be realised in the early stages are
maintained or even enhanced.

As with any other resource, the effective management of energy
in a large organisation can be achieved most successfully
through the establishment of a common direction for all of the
interacting functions in the organisation. It will usually
require changes in operating habits and behavioural patterns,
the modification of existing equipment to be more energy
efficient, the introduction of advanced technologies, or the
recognition and acceptance of opportunities to produce more
energy efficient products.

To ensure their continued effectiveness, the changes introduced
must be permanent and must not depend on individual enthusiasms
for their survival. The changes must therefore be such that
their reversal or removal requires positive and obvious
initiatives to be taken.

It is a well recognised phenonemon that it is relatively easy
to make energy savings over the short term, but it is extremely
difficult to make them permanent. This results from the
effects of boredom, inertia, laziness, difficulties of
accounting and pressure of other events such as this week's
crisis. The first consequence of all of this is that, if
energy conservation measures or steps to promote the rational
use of energy are to be successful in the long term, the
wholehearted cooperation of the work force must be obtained.
Thus, the usual range of publicity campaigns, employee
suggestion schemes, etc. must be created to generate and
sustain the necessary momentum.

The second consequence, however, is much more far reaching,
more difficult to achieve, and largely unrecognised in the
schemes generally in use today. Basically, energy management
and budgetting must become part of the day-to-day job of all
line and production managers. Until this is the case, each
day's production crisis will always take precedence over the
more diffuse problem of reducing energy consumption.

This paper will discuss the introduction of energy conservation
methods in a large shipyard on a large site (about 1 sq. mile),
with very large and widely separated buildings and with much of
the work being carried on outdoors or in extremely large,
partially heated, sheds. Part of the discussion will consider
the particular problems of dealing with such a site, the
measures which were introduced, their effectiveness and the
lessons learned. The rest of the discussion will consider the
managerial, administrative and technical measures which must be
introduced if the measures are to remain successful over the
longer term.

APPROACH

The simple objective of the programme is the reduction of
energy use. In view of this, easily achieved savings should
not be held up awaiting the design and application of an
overall energy management programme. Clearly there will be
different lines of effort which may not progress at the same
rate nor necessarily have identical short to medium term
priorities and objectives.

At this point it is necessary to devise a structure for the
programme. This can be begun by identifying positive action
areas and arranging them in order of importance or impact.

There should be a positive decision to introduce effective
energy management throughout the organisation. It must come
from the most senior level as it is unlikely that effective
changes could be introduced without the continuing support of
the most senior executives. Their degree of commitment depends
on the implications of energy costs to the organisation.

Information on the energy costs is most easily obtained from
supplier invoices. Values will not be presented here as this
is a widely known exercise and no particularly novel data
appeared.

The important factor here, however, is that the impetus for the
programme came from board level and the proposed changes were
actively pressed through.

The programme must be able to create an effective change in the
way in which the Company deals with energy use, and the changes
will fall into one of three general categories as shown in
Figure 1.

FIGURE 1. Types of Change Needed in Energy Programme

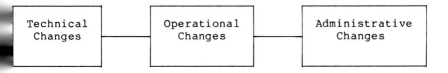

The structure of the programme must reflect these divisions,
but be able to react with flexibility. Areas of opportunity
for change can be identified for each section of the company,
but changes will only occur as the result of decisions taken on
the basis of the comprehensive information which becomes

available on the circumstances in any area. In a large
organisation such as Harland and Wolff it would not normally be
possible to produce a detailed analysis at an early stage of
the programme formulation as the information required would
involve inputs for a wide range of functions not yet involved
in or committed to the programme. Because of this lack of
detail it may not always be clear to departmental heads how
their sections can be involved in energy management. They have
an image of energy management as simply a need for them and
their staff to use less energy in their daily work, but they do
not see how their objectives may be modified to make more
complete contributions.

Presentation of the information which is available will,
however, encourage thought about the subject, which can be
followed up by more specific involvement at a later stage. It
is a matter of judgement as to how widely information should be
disseminated at this stage. It is reasonable to suppose that
it may be used to compile an edited overview for wider
consumption and be backed up with localised examples to which
individuals within the organisation can easily relate and
thereby begin to be involved in conservation and change.

Compressed air is an example which can be used to great effect
in a shipyard. Shipyards tend to make extensive use of
compressed air and most of the workforce either use it
directly, or work in areas where it is widely used by others.
Indeed, one of the most characteristic sounds of a shipyard at
rest (eg during the lunch break) is the hiss of compressed air
leaking from pipes and fittings. It is useful, therefore, to
present a simple calculation which relates to pneumatic tools
and other equipment and which shows how the waste of compressed
air can affect jobs.

Let us assume that there are leaks around valves and couplings
along a 550 kpa (80 psi) hose line which together add up to an
equivalent of a 6.4 mm (1/4 ") diameter hole. The leakage loss
will be 2.4 m^3min^{-1} (85 cfm) which will require 14.15 kW of
compressor power. If electricity costs £0.053/kWh then the
cost of this leak for a 40 hr. week will be £30.00, which
represents a total of £1500 per year (50 weeks).

This is a very straightforward calculation which gives a clear
indication of the cost of a very small leakage.

For publicity on the shop floor, a simplified version of the
facts can be shown - possibly in the form of posters as
illustrated in Fig 2 and 3. These can draw very clear
connections between energy waste and jobs.

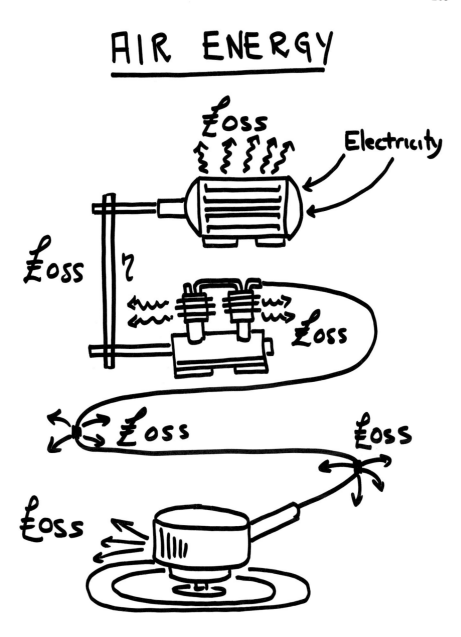

FIGURE 2.   Poster

204

FIGURE 3.   Poster

It is interesting to note that it can be extremely difficult to convince shop floor workers that opening and closing a large door merely to drive trucks in and out can be extremely expensive. For example, it came as a revelation to one group of foremen that to open a particular 21.5 m x 21 m door for access purposes six times per day during the heating season with the actual internal shed temperature of only 11 ^{o}C could cost over £20,000/year.

In order to identify possibilities for change, a full energy audit is required. Each area of activity must be analysed, its effect on the overall energy profile identified and positive changes implemented.

An energy audit was carried out for Harland and Wolff, and identified the usual range of opportunities. These included, improvements in the compressed air supply system, air recirculation in the very tall assembly sheds to reduce the temperature at the ceilings, (which reduces both heat loss and the number of times the crane drivers down tools, or go to sleep) possiblities for co-generation, fuel substitution, boiler control, changes in lubricating oil usage, changes in production methods particularly in the steel production facility and in the paint shops, improved equipment and building maintenance, modified lighting systems, power and load factor improvements etc. Additionally, an unsuccessful attempt to renegotiate the input metering of electrical power with the supplier to avoid incurring peak demand changes at different parts of the site at different times.

Extensive effort was also put into examining both the possibilities for design changes and for changes in the production process. These were extremely effective and have led to improvements such as that in the assembly of the main engine fuel oil booster pump. Here assembly time has been cut from 320 hours to 99 and there have been attendant direct energy cost reductions.

Another identified area was procurement policy - particularly in dealing with monopoly or effective-monopoly suppliers, and it was found that cooperation between large users could produce a level of cooperation and control in the consumer side of market which is similar to that on the supply side, can lead to significant reductions in fuel costs.

Lubricating oil was an other profitable area of investigation, and changes in policy regarding recycling, preventive maintenance, testing and waste disposal here led to savings of over 60%. The possibilities for burning over 50,000 litres of waste oil per year are now being investigated.

It is interesting to note that during this exercise, a number of possibilities for improving the energy efficiency of ships themselves were identified and are now being carried forward into the company's new designs. Some of these arose out of the paint shop investigations and involve the use of anti-fouling and self-polishing paints and better initial paint finishes.

All of this is now routine, however and the main complications
arose not from the procedures themselves or from the
identification of possibilities, but from the physical size of
the site and the necessary degree of segregation of activities
such as steel production and primary assembly, painting,
fabrication, etc. Much more interesting problems arise when
implementation is considered.

IMPLEMENTATION

The programme was established and carried out by the Company
Utilities Manager who was given responsibility for energy usage
on the site. He was given strong support by the Chairman and
Managing Director, and was provided with open access to senior
executives of the Company. He also had a supporting group with
the necessary technical expertise to ensure the eventual
adoption and implementation of the programme.

From the outset it was realised that the role of such a group
was not to implement changes directly in any of the identified
areas but rather to work with area staff in identifying changes
and to smooth the path throughout the rest of the organisation.
The group had to see energy management as important and as a
vehicle for both job satisfaction and career development. It
was recognised that the group's efforts would otherwise lack
the necessary commitment for success. It was appreciated that
not all members of the group should be permanent but that
people with special skills should be brought in for special
problems and that work should be allocated in blocks with
clearly defined objectives. Continuity should be maintained by
permanent group members whose knowledge and ability would span
the fields of

> Energy Technology
> Production Technology
> Administrative Techniques
> Maintenance and Services
> Personnel

The role of the group was primarily to be that of a change
agent. However, it was also found that, because of specialist
knowledge, overall responsibility for some aspects of the
programme were better vested in the group, and in this case the
role was clearly managerial, i.e., in procurement, personnel,
service or operation.

The eventual structure was roughly as indicated in Fig 4.

Given this organisational structure, the close involvement of
the production departments and the high level of company
commitment, it was possible to make significant changes to the
energy efficiency of the plant.

FIGURE 4.   Energy Change Management Structure Adopted

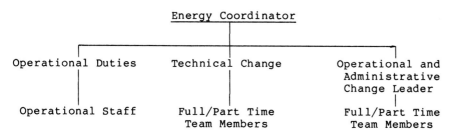

The practical efforts were backed up by an extensive education and training programme which was intended to provide all employees from apprentice to senior management with a better appreciation of energy matters in general, and with the skills necessary for improving matters in their own areas.

The programme was also followed up by a monitoring programme to assess the effectiveness of the various measures which had been adopted, and to watch for any regression towards the old patterns of energy use. This required changes to be made in the system of metering so that information could be available at a more local level, and as a consequence a better picture of plant energy use slowly became available.

The picture that began to emerge was very much the classic one that each new measure had an instant effect which was measureable and usually quantifiable. Over a period, however, unless it was a permanent process change, the improvement began to disappear and had to be restimulated. This could be traced to three main factors:

-    an inbuilt preference for doing things the way they were done before,
-    a measure of familiarity breeding contempt after some period of operation of a new system so that short cuts tended to be taken if possible,
-    the over-riding priority in production departments to meet deadlines at whatever cost - particularly if the cost is being carried by some other part of the organisation.

The first two, while different in their operation, are both related to a desire to take the easier route if at all possible. The third reflects the day-to-day priorities of manufacturing industry. It is worth noting, however, that the pressures of the third may lead to acceptance of the previous two.

How can all three be avoided?

There is only one effective way. Energy management must be
made a part of the responsibility of the line or production
managers. They must be given an energy budget in exactly the
same way as they are given budgets for the other aspects of
their operation, and they must have to account for it in
exactly the same manner.

That is, the energy use of every sector of the plant must be
part of the data produced at the Monday morning managers'
meeting. Each manager must be held responsible for
over-expenditure in the energy area just as he already is for
wages or metal.

This changes the emphasis of the discussion in the only way
possible if measures initiated by the present energy manager
are to remain in place. It also carries other implications.
If the energy usage of each department is to be provided weekly
on a routine basis, then facilities have to be provided for
adequate metering of all energy inputs to all departments.
(Obviously, it might be necessary to combine some activities
for logistical reasons.) This in turn means that the
administrative and accounting systems will have to be modified
to cope with the additional information flows. The structure
for ensuring the rational use of energy therefore becomes as
shown in Figure 5 which shows a double system involving both
the technical and administrative arms of the company.

FIGURE 5. Two Sides of a Company Energy Conservation Programme

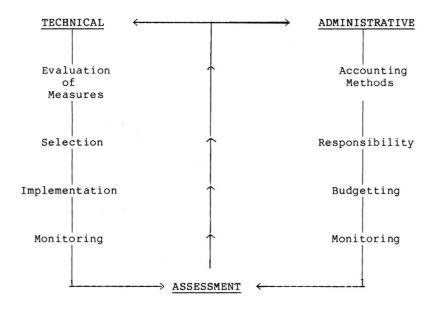

Interestingly, this change also modifies the role of the energy manager himself. No longer is he in the position of trying to convince production departments that changes should be made, or even that questions should be asked. Instead, he will be invited in by the appropriate departments to advise on how the proposed ten per cent reduction can best be achieved. Indeed, the energy manager can now be regarded by the production departments as a desirable asset rather than as a necessary nuisance.

The importance of these administrative and other changes was not realised until the energy rationalisation programme was well underway at Harland and Wolff, and it was gratifying to realise that the necessary structural elements had already been implicitly incorporated into the procedures. They are now being formalised as part of a company policy of ensuring that beneficial changes are permanent - or at least as permanent as they can be in a rapidly moving and competitive manufacturing environment.

The sting, however, is in the tail. The energy conservation and management programme introduced at Harland and Wolff has been so sucessful that there is no useful function for an Energy Manager to play. The person who played their role has been transferred to other responsibilities, and his function has been incorporated into the responsibilities of other managers - exactly as suggested above.

Microcomputer Based Design Concepts to Improve Solar Water Heaters

Y. Palierakis

Postgraduate School of Control Engineering, University of Bradford, U.K.

ABSTRACT

This paper presents an analysis of the limits in energy saving expected from solar water heaters when they operate under varying demand conditions and suggests new design concepts which improve the efficiency of solar heaters. Such options are realized by incorporating microcomputers to enable the solar heater to perform complicated operations while maintaining a user friendly environment.

KEYWORDS

Solar water heater; solar heater control; microcomputer control; split tank solar water heater.

INTRODUCTION

Solar energy is inherently non polluting, saves valuable fossil fuel and reduces the effect of fuel price increases. Solar energy is being used extensively for heating water in the domestic and the industrial environment (Brinkworth, 1972; Field, 1978; Halacy, 1973).

Much of the past research work in solar water heaters has focused in improving their performance by means of improving their physical design (Duffie, 1980; Halacy, 1973; Wieder, 1982;). It is thought that to improve the performance of a solar water heater means to make efficient the process of conversion of solar energy to heat stored in the water, while minimizing heat loss from the system . Extensive research has taken place in designing efficient solar collectors because they primarily determine the absorption efficiency of the heater and they incur the majority of heat losses from the heater.

Designers of solar heaters (Field, 1978) justify the economic feasibility of a solar water heating installation, by computing the value of the energy stored

212

by the system per annum. This is equal to the cost of conventional fuel
required to provide the same amount of energy. Subtracting the value of solar
energy from the total cost of fuel that would be necessary to heat the volume
of water consumed per annum if no solar energy was used, gives the expected
annual saving from the solar system. However such computation is only
appropriate if the demand for heated water is assumed to be either equal or
greater than the capacity of the solar heater.

In practical applications the demand for heated water (volume and temperature
varies daily, and often it is found that the volume of water is much smaller
than the capacity of the solar heater. It is then that recalculation of the
economic parameters shows that installing a solar heater does not always
guaranty savings.

This paper discusses design options for solar water heaters which offer a
positive rate of saving even when they operate under varying demand
conditions. Such options are realized by incorporating microcomputers to
enable the solar heater to perform complicated operations while maintaining a
user friendly environment.

LIMITS TO ENERGY SAVING

It is known (Szokolay, 1977; Wieder, 1982) that the operating efficiency of
solar heaters decreases with temperature (Fig. 1). Thus, the use of solar
heaters operating in the low temperature range is advocated. Consequently,
solar heaters with large storage tanks are used in order to extend the time
during which the heater operates in the low temperature range and increase
their potential for energy storage.

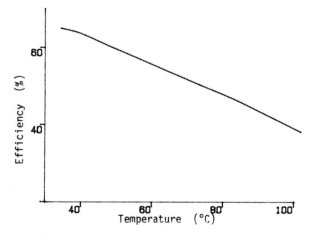

Fig. 1. Collector's efficiency curve.

However from the consumer's point of view efficiency is determined by the
ability of the solar heater to provide the required volume of heated water a
less cost than a conventional water heater would do. Therefore, the consumer

definition of efficiency depends on the demand for water (volume and
temperature), the capacity of the solar heater and the amount of solar energy
absorbed, and it is defined by the ratio,

$$n= \frac{CostN - [CostE - CostS]}{CostN} \times 100\%$$

For a known demand of volume and temperature of water, CostN represents the
cost of fuel that would satisfy this demand if no solar energy was used.
CostS is the value of fuel equivalent of solar energy absorbed by the system,
and CostE is the expense in fuel if the full volume of water in the solar
heater was heated to the required temperature without the contribution of
solar energy.

It is apparent from the definition, that the consumer's efficiency ratio takes
into account the real volume of water in demand, unlike the designer's
definition which assumes that the demand is greater than the water capacity of
the solar heater. When the volume of demanded water is greater or equal to
the capacity of the solar heater, CostN is greater or equal to CostE and the
efficiency ratio is either positive or zero. Because CostS appears in the
numerator, maximization of the efficiency n is achieved through maximization
of the absorption efficiency of the solar system.

However, when the demand for heated water becomes less than the capacity of
the solar heater, CostE is greater than CostN. It is then possible that the
situation may arise when CostS < (CostE-CostN) which would translate to a
negative efficiency ratio. Negative efficiency ratio means that using the
solar water heater is more expensive than heating the required water by
conventional fuel.

It is therefore essential to re-examine the economic parameters related to
solar water heater installations by taking into account variations in water
demand. The obvious improvement would be offered by heaters that allow
adjustment of the parameter CostE with demand, so that its value remains
always less than CostN. Such adjustment would ensure that the numerator of
the consumer's efficiency ratio is always greater than zero and improvements
in absorbing efficiency are reflected by an increase in the value of the
consumer's efficiency ratio.

THE SPLIT TANK APPROACH

Solar water heaters comprise three main units; the collector, the storage tank
and the heat exchanger, Fig. 2. Solar heaters differ in the structure of
their collector, the efficiency of their heat insulation and their size.
Collection of solar energy is based on the high absorption of radiant energy
by black surfaces and the "greenhouse effect".

A solar collector is essentially a liquid-cooled flat surface. The front face
of the collector is covered by one or two sheets of glass and is exposed to
the sun, whilst its rear face is well insulated. The liquid in the collector
is heated and then used to heat the water in the storage tank. Flow of the
liquid from the collector into the storage tank takes place either under the
thermosyphon effect or is forced by a pump. The liquid in the collector may

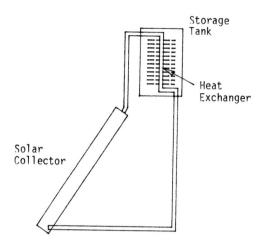

Fig. 2. Typical solar water heater

flow directly into the storage tank if it is water, alternatively it may flow through a close circuit incorporating a heat exchanger in which case it may be a mixture of antifreeze and water.

Currently available solar water heaters employ the above principles and come with collectors and storage tanks of predefined dimensions. These are determined during the specification period on the basis of the local levels insolation, expected maximum demand and cost (Field, 1978). Such systems are economically justified if the expected demand for heated water is greater or equal to their capacity, but as shown in the foregoing they may not be justified when their storage capacity exceeds their demand.

The split tank approach offers a substantial improvement in the efficiency of the solar heater. In the "split tank system" the storage volume consists of an array of small tanks rather than a single capacity tank. The maximum storage capacity of the heater is determined by the sum of the volumes of al tanks. As the storage capacity consists of many small tanks, the user may reduce the effective storage capacity of the heater by switching off a number of tanks in the array. If the tanks in the array have different volumes, the user has greater flexibility in determining the effective storage capacity depending on the combination of tanks he selects. This approach effectively allows the user to determine the value of CostE in the consumer's definition of efficiency, so that it is never more than CostN. Thus the consumers efficiency ratio always takes positive values and the heater generates savin at every level of demand.

An additional advantage offered by the split tank approach arises from the fact that a circulating pump is used. Thus the storage tanks may be placed indoors which reduces the size of the outdoor installation and makes the sol heater architecturally more attractive.

Large commercial installations (ie hotels, camping sites, hospitals) may implement the split tank approach easier because they typically consist of a large array of storage tanks and collectors. Further, the implementation of the split tank approach is of greater benefit in such an environment because the demand for water normally varies in a predictable way.

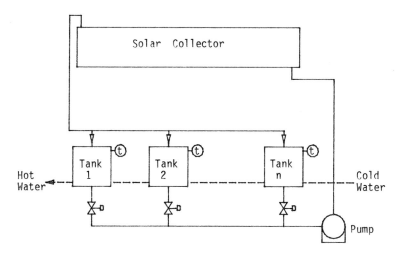

Fig. 3. The split tank solar water heater.

Figure 3 illustrates the schematic lay out of a split tank solar heater. A large solar collector is used to heat the water in an array of storage tanks of equal volume. Because the circulation of the collector liquid is forced by a pump the water tanks may be placed at any level in respect to the collector unit (ie indoors). The water tanks are connected in series to the water supply. The tank from which the hot water supply is drawn is tank 1, then tank 2 and so on. The maximum water capacity of the system is given by the sum of the volumes of all tanks. The minimum quantity of water that it is possible to heat on solar energy at any time is given by the capacity of tank 1. The outlet from the collector is connected in parallel to the heat exchangers in the water tanks. The flow through the heat exchangers in each tank is allowed or inhibited by the solenoid valve connected at the outlet of the element. By restricting the flow of hot fluid through a different combination of tanks, the user effectively determines the volume of water to be heated on solar energy.

MICROCOMPUTER IMPLEMENTATION

The manual control of a split tank solar heater is a fairly complicated operation for the average user which makes the introduction of such systems unattractive. However, the use of a microcomputer to control the heater, substantially simplifies the user's task and makes possible the introduction of split tank solar water heaters, (Palierakis, 1986).

The microcomputer normally maintains the operation of the heater at some default level unless the user determines a different set of parameters. Input

216

parameters supplied by the user are; volume of water, temperature and the time by which the water is required.

On the basis of this data the microcomputer enables a combination of tanks to determine the correct volume of water. It then checks if the time available is sufficient to heat the volume of water required by using only the auxiliary heating element. If the time is sufficient, the input parameters are accepted otherwise a warning message is issued and the request is rejected. The microcomputer in regular intervals compares the remaining time with the time necessary to raise the temperature of the water at the required level by auxiliary heating. When the two times are equal the auxiliary heater is switched on.

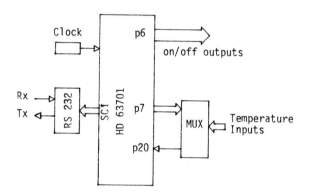

Fig. 4. Single chip microcomputer controller.

The use of microcomputers broadens the flexibility of the heater while it maintains a user-friendly environment. Further to controlling the effective water capacity of the split tank heater the microcomputer offers a number of additional functions such as the following. Readily available information to the user about the effective water capacity of the heater and its current temperature. Temperature projections which are computed on the basis of the past rate of water temperature increase due to solar insolation. The user requests a volume of water to be heated within a set time and is informed in advance if his request will be realized. If the user determines the volume of water only, the heater enables the correct combination of tanks and runs exclusively on solar energy.

With the advent of single-chip microcomputers today it is possible to package all control and monitoring functions for the split tank solar water heater into a vary small board. The component requirement is minimal while the only accountable power consumption would arise from the operation of the circulating pump.

Figure 4 shows the block diagram of the required microcomputer system. Input signals to the system come from the sensors monitoring the temperature of the liquid in the storage tank and the collector. Output comprise the on/off signals that operate the circulating pumps and the solenoid valves. Interaction of the user with the heater takes place either through a terminal as was in the case of the prototype system, or via a dedicated i/o panel. Finally the software routines are permanently stored in the rom which is integral to the single-chip microcomputer.

CONCLUSION

The split tank solar water heater is a realistic alternative to conventional solar heaters and it is best realized by using microcomputer based monitoring and control techniques.

The split tank solar water heater can vary its effective storage capacity, thus it offers a positive rate of saving even when operating under varying demand conditions. It also offers comprehensive information to the user, it is easy to operate, and potentially it is architecturally more attractive.

REFERENCES

Brinkworth B.J., (1972), Solar energy for man, Wiley, N.Y.
Duffie J.A., (1980), Solar Engineering and Thermal Processes, Wiley, N.Y.
Field R.L., (1978), Design manual for solar heating of Buildings and domestic hot water, Solpub Co., 2d ed.
Halacy D.S., (1973), The coming age of solar energy, Harper and Row, N.Y.
Palierakis Y., (1986), Microcomputer-based monitoring and information system for solar heaters, J. Microcomputer Apl., Vol.9.
Szokolay S. V., (1977), Solar Energy and Building, 2nd ed., New York, John Wiley.
Wieder S., (1982), An Introduction to Solar Energy for Scientists and Engineers, John Wiley, New York.

*Here is an idea not much
talked about applied to
industrial setting.*

Industrial Case Study of a Latent Heat Storage

J. Persson and B. Karlsson

*Department of Mech. Eng., Div. of Energy Systems, Linköping Inst. of Tech.,
S-581 83 Linköping, Sweden*

ABSTRACT

Small and medium sized industries have a great potential for short term
energy storage. Energy audits conducted in Sweden show possibilities of
saving 5 TWh annually. One way of taking care of surplus heat from industrial processes is by using a latent heat storage.

A heat recovery installation at a corrugated board mill is discussed. A
latent heat storage is used combined with an heat exchanger. In designing
the storage experimental results and numerical simulation are used. The net
present value method is used together with simulations to maximize profitability of the storage installation.

Profitability is reported for different combinations of energy prices and
price of latent heat storages. The complete installation show pay off
periods ranging from 1.3 to 3.3 years. Pay off periods for the latent heat
storage vary between 3.4 and 9.7 years.

KEYWORDS

Industry; heat recovery; latent heat storage; experiments; simulation ;
economics.

INTRODUCTION

The main reason for energy conservation measures is to reduce the need of
primary energy. This can be done in various ways e.g. heat exchanging, heat
pumping and by using a heat storage. Heat exchanging is applied when the
supplied surplus energy from the source process is available simultaneous
with the energy demand, the sink process. This requires an adequate temperature level of the source, if this is not fulfilled heat pumping is required. However simultaneousness is not always at hand, which implies using an
energy storage in order to overcome the time lag between source and sink.

Provided that this energy storage can be achieved at reasonable cost the
potential for short term energy storage is substantial. Energy audits
conducted in Sweden, (Söderström, Johansson, 1983) show a potential for
short term energy storage of 5 TWh per annum. This figure is calculated for
small and medium sized industries with a total energy use of 37.4 TWh per
annum. The interaction between two processes can be shown as in Fig. 1.

219

220

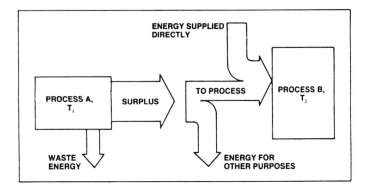

Fig. 1. A source-sink pair and the attached nomenclature
(Söderström, Johansson, 1983).

Energy storages, in this case heat storages, can be achieved in different
ways. There are sensible heat storages, e.g. a water tank, chemical heat
storages, using energy stored in chemical bindings, and latent heat storages. The latter uses a materials ability to absorb respectively emit heat
at a phase change process. Advantages with phase change material (PCM) heat
storage is, high amount of energy stored per unit volyme, access to heat at
constant temperature and simple (passive) heat recovery solutions.

In a full scale experiment (Solmar, 1983) a latent heat storage was installed in a bakery. Reduction in heating cost of 50% was achieved. A conclusion was that a connection to the district heating system could have been
avoided.
This paper describes the application of a latent heat storage at a corrugated board mill. An air flow equipment is used to test different storages.
In design and sizing of the storage a mathematical model of the storage is
used. On site measured parameters of a surplus energy flow are recreated
under laboratory conditions.

CORRUGATED BOARD MILL

General

The corrugated board mill is one out of six plants in the Esselte Well
Corporation. The factory employs 50 people and the production is about 8000
tons of corrugated board annually. Corrugated board produced is mainly used
for wrapping. Energy use amounts 5 GWh annually (Wigö, 1984), roughly
divided on oil for steam and hot water production, 3.7 GWh, and electricity
for machinery and lightning, 1.3 GWh.

Paper is delivered to the factory in rolls which are mounted in the corrugated board machine. The paper is corrugated, glued and edgecut to produce
sheets of corrugated board for different wrappings. The sheets are run
through a punching machine and a printing press, before transported to the
delivery storage.

Possible heat recovery

The factory is in operation between 6 am and 10 pm, divided in two working
shifts. Waste material is evacuated through a pneumatic tube system where
material and air is separated by a cyclone. The evacuation system is in
operation during both working shifts, that is, there is a need for heating
inlet air replacing the evacuated air. The corrugated board machine, emitt-
ing airborne surplus energy, is in operation only in the morning shift, 6
am to 2 pm. This indicates that surplus energy can used by heat exchanging
combined with heat storage during the morning shift. Figure 2 show schematically how this can be done.

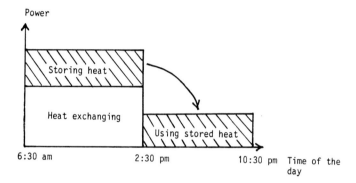

Fig. 2. Schematical use of surplus energy from the corrugated
board machine.

METHOD

Measurements in the corrugated board mill

Collection of data is done by a computerised measurement system (Karlsson
et al 1986). The system gives on site graphical presentation of collected
data. Measurements has been made in 15 minute intervals, generating average
values of temperatures and power levels. Temperatures of inlet and outlet
airflows has been measured as well as consumption of electricity. Operation
time of oil boilers has also been recorded in order to get an overall
picture of the energy use of the factory. Transmission of data has been
made by telephone to the Linköping Institute of Technology. Measured airflows are considered to be constant.

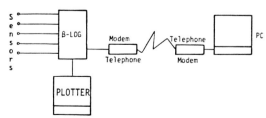

Fig. 3. Measurement system used for energy audits.

222

Experimental equipment

An air flow equipment is used for performance testing of energy storage units. Test programs are controlled by a computer allowing set values to be defined according to a test procedure. Recordings of surplus heat temperatures from an industry can serve as set values for a test. Hereby possible installments of industrial energy storages can be studied in the laboratory.

Simulation model

In numerical simulation of a PCM storage it is an advantage to use a simple model. Numerical simulation aims to calculate the the temperatures of the dynamic process of charging and discharging the storage. Knowledge of inlet and outlet temperature ables us to calculate power levels and amount of energy stored and discharged. A lumped heat capacity model is used (Persson, Solmar, 1987).

Simulation program

The program used for simulating the PCM storage is Simnon (Åström ,1982). Simnon is a special programming language for simulating dynamical systems. Systems may be described as ordinary differential equations, as difference equations or as combinations of such equations. The program is a general purpose program, that is, there is no energy system component library. The language has an interactive implementation which makes it easy to work with. Simnon allow optimization, introduction of experimental data and parameter fitting.

RESULTS

Performance testing of PCM storages

Four types of storages are tested. Encapsulation is in two cases flat rectangular capsules, one with corrugated sheet metal between the capsules for increasing the heat transfer area. One encapsulation is spherical, another cylindrical. In Fig. 4 the different encapsulations are shown.

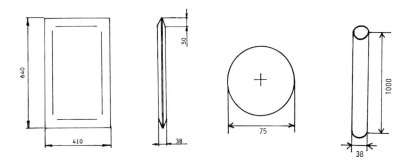

Fig. 4. Tested encapsulations of PCM material.

The PCM storages are tested in the air flow equipment. Different charge and discharge lapses are used. A comparision of the storages is made for a test run were at first the PCM is tempered to 12 °C. Charging of the storages is made for 12 hours with a temperature of 40 °C, followed by discharging for 12 hours with temperature 12 °C. Two different airflows, approximately 2000 m³/h and 4000 m³/h are used. Table 1 shows the conditions for the tests.

Table 1. Conditions for performance test of PCM storages.

type of storage	PCM	trans temp [°C]	latent heat [kWh]	volume [m³]	mass [kg]
flat rect.	CaCl+6H₂O	29.5	41.6	1.2	1000
flat rect. corr.	CaCl+6H₂O	29.5	31.1	1.2	750
spherical	CaCl+6H₂O	27	34.6	0.9	730
cylindrical	sodium sulf.	31	30.1	0.9	645

Mass is including capsule. To determine the latent heat of the PCM contained in flat rectangular plates a calorimetrical test is used (Friser, 1987). The heat of fusion is found to be 46.1 Wh/kg (166 kJ/kg) and the temperature of phase transition 29.5 °C. These values differ from values given by PCM manufacturer, (Dow Chemical, 1981). Possible reason is that the PCM was delivered dry and water was added by non experts, resulting in incorrect proportion of ingredients. Heat of fusion of the others are, spherical capsule, q_{tr} = 190 kJ/kg, cylindrical capsule, q_{tr} = 270 kJ/litre (1 litre = 1.6 kg including capsule). The results of performance test are shown in Fig. 5 and 6.

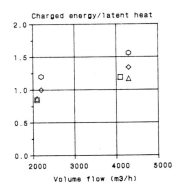

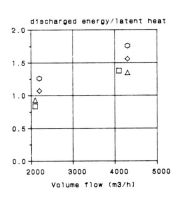

◻ : Flat plate, ◇ : Flat plate with corr. sheet metal,
○ : Spherical, △ : Cylindrical

Fig. 5. Stored and discharged energy divided by latent heat of respective storage.

224

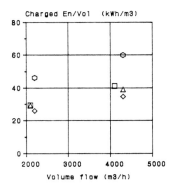

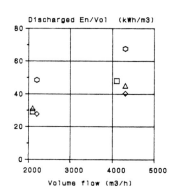

Fig. 6. Stored and discharged energy divided by volume of respective storage.

Measurements on the corrugated board machine

Measurements has been carried out in two periods, april 1986 and february 1987, the latter to verify temperature level of surplus energy flow during low outdoor temperatures. Total supply air flow of the factory is about 50000 m³/h and total outlet air flow is about 40-65000 m³/h. Outlet air flow varies depending on whether the corrugated board machine is in operation or not. Surplus energy air flow from the corrugated board machine amounts 28000 m³/h. Temperature level as of Fig. 7.

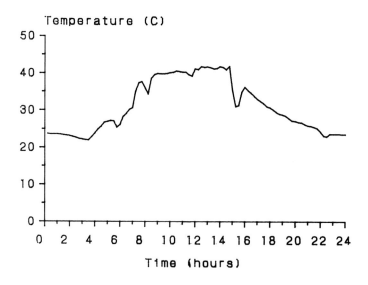

Fig. 7. Temperature level of outlet airflow from corrugated board machine, measured on the 18th of February, 1987.

Latent heat storage installation

General Air is supplied to the factory in a number of places. One major inlet air flow amounts 33000 m^3/h. This inlet flow is suitable to be connected to the surplus energy flow from the corrugated board machine. This installation is favourable due to a reasonable amount of tube installations. A latent heat storage combined with a heat exchanger can be installed as in Fig. 8. The heat exchanger will always meet the demand on temperature of inlet air if the temperature drop over the PCM storage is less than 10 °C (Persson, 1986).

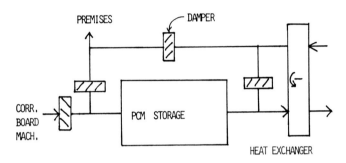

Fig. 8. Installation of PCM storage and heat exchanger.

Technical and economical performance Recreating the surplus energy flow combined with a energy storage is done in laboratory and by mathematical simulation. Discussed here are results from the flat rectangular capsules. This is due to the fact that the simulation model is verified only for this geometry. However results from the performance test of spherical and cylindrical encapsulation are used to show how these encapsulations may work at the corrugated board mill.

In finding out possible performance of a storage there are three stages of execution;

1. Tests are carried out in the laboratory with the PCM storage in question. This test is used in the simulations to identify important parameters. Parameter identification is done to achive good accordance between experimental results and mathematical simulations. Parameter identification can be done by using optimization.

2. Having achieved reasonable accordance as described above, the size of storage is scaled to make use of actual surplus energy flow. Variations in e.g. geometry and size are tested to find a storage of good performance. Economical performance is also examined.

3. Apply found enhancements to a physical storage in laboratory and verify the simulations.

Discussed here are point one and two above. Fig. 9 shows results after identifying parameters on an storage with flat rectangular capsules.

226

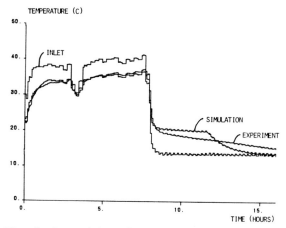

TEMPERATURE (C)

INLET

SIMULATION

EXPERIMENT

TIME (HOURS)

Fig. 9. Comparision of experimental and simulated results of a PCM storage. Inlet temperature to storage varies according to measured values from the corrugated board mill.

Scaling to actual surplus energy flow is done combined with increased convective heat transfer coefficient by introducing enlargement of heat transfer surface. Here it is assumed that it is possible to introduce surface enlargement doubling the convective heat transfer coefficient compared to the results in Fig. 9.

Numerous simulations are performed. In the simulations a temperature level from the corrugated board machine as of Fig. 7 is used. The result is presented as actual charged energy divided by size of storage (latent heat).

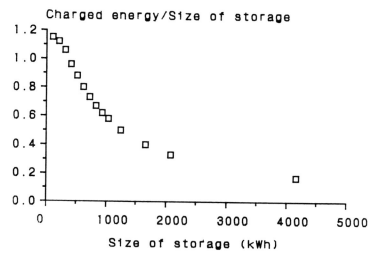

Charged energy/Size of storage

Size of storage (kWh)

Fig. 10. Performance of latent heat storage at the corrugated board mill as a function of size of storage.

From Fig. 10 it is understood that a small PCM storage will be fully
charged and discharged. However, the amount of energy saved on each cycle
will be correspondingly small. On the other hand, using a large storage
will save more energy and thus more money. Therefore sizing the storage is
done to maximize the profitability for a given life time of the investment
and a given interest rate. Here an economic lifetime of ten years and an
interest rate of 5% (above the inflation rate) is used. To compare the cost
with the revenue of the storage installation the net present value method
is used (Söderström, 1986). Economical performance of the complete installation, Fig. 8, will be discussed later.

Annual revenue due to the PCM storage is calculated by using the outdoor
temperature duration (VVS handboken, 1974). Installation cost is divided on
cost for machinery and cost for encapsulated PCM. Cost for machinery is
estimated to 104 kSEK (Tube plåt, 1986). Fan is not included in this figure
as it is possible to use an existing fan. Two figures are used for the cost
for encapsulated PCM, 300 SEK/kWh and 100 SEK/kWh, the latter representing
a possible cost at large scale production. The energy savings are worth
either 0.3 or 0.5 SEK/kWh. Two values are used to show the importance of
the energy price. In Fig. 11, revenue and total investment cost as a
function of storage size are shown.

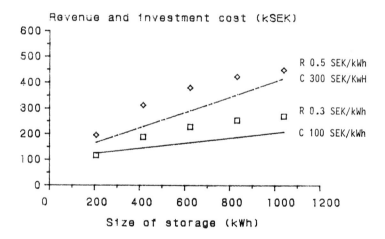

Fig. 11. Revenue of PCM storage installation for energy prices 0.3
SEK/kWh and 0.5 SEK/kWh. Installation costs for PCM prices 100
SEK/kWh and 300 SEK/kWh.

Profitability is calculated as revenue minus total installation cost for a
given size of storage. Figure 11 gives four possible ways to calculate this
figure. The results are shown in Fig. 12.

228

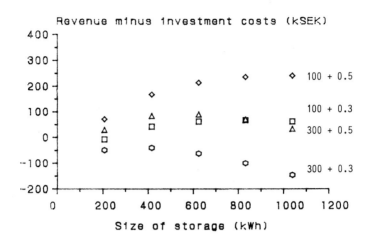

Fig. 12. Profitability as a function of storage size.

The size of storage to be chosen, is were the curves are at their highest
points. As observed in Fig. 12 the highest points of the curves represents
different sizes of storages. In the following discussions a storage size of
624 kWh is used, representing an intermediate value on the scale of Fig.
12. The result from simulating an installation of a storage of this size is
shown is Fig. 13. Inlet temperature as of Fig. 8.

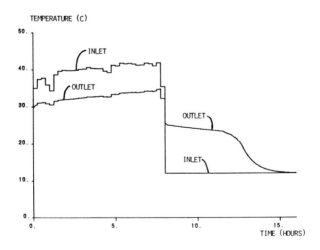

Fig. 13. Simulation of an installation of a PCM storage, size 624
kWh latent heat.

229

The results show that charged and discharged energy amounts to about 500 kWh. In table 2 an overall picture of installation costs and annual revenue is given. Annual revenue due to the heat exchanger is calculated on basis of the outdoor temperature duration. Inlet air needs to be heated to 20 °C. Price is given by manufacturer (PM-luft AB, 1987). Table 3 shows corresponding pay off periods for various combinations of energy prices and price of PCM storages.

TABLE 2. Installation costs and annual revenue for different configurations of heat recovery equipment and energy prices, in kSEK.

Installation	Cost of installation		Annual revenue	
	100 SEK/kWh	300 SEK/kWh	0.3 SEK/kWh	0.5 SEK/kWh
PCM storage	167	291	30	49
Heat exchanger	139	139	82	137
PCM st + Heat ex	247	372	112	186

TABLE 3. Pay off periods for different configurations of heat recovery equipment, on basis of combinations of energy and storage prices..

Installation	100 + 0.5 [SEK/kWh]	300 + 0.5 [SEK/kWh]	100 + 0.3 [SEK/kWh]	300 + 0.3 [SEK/kWh]
PCM storage	3.4	5.9	5.6	9.7
Heat exchanger	1.0	1.0	1.7	1.7
PCM st + Heat ex	1.3	2.0	2.2	3.3

Benefits from heat recovery installations may also include peak load reductions in the use of electricity. At the corrugated board mill a power charge of 200 SEK/kW is applied. Possible reductions in peak load are 100 kW for the PCM storage and 250 kW for the heat exchanger (Persson, 1986). Correspondingly the pay off periods are lowered. Pay off periods including benefits from peak load reduction are calculated to; PCM storage, ranging from 2.8 to 7.4 years, heat exchanger, 0.8 and 1.1 year.

CONCLUSIONS

Numerical simulation is shown to be a useful tool in designing latent heat storages. Experiments are neccesary to verify the model of a storage. When this is achieved simulations show the performance of various storage designs.

The net present value method combined with several simulations serve as a powerful tool in sizing the storage for maximum profitability. The effect of diffrent prices of storage and energy to the size of the storage is shown.

The installation of a heat exchanger show short pay off periods, less than two years. The latent heat storage show longer pay off periods. However higher energy prices, compared to the Swedish level of about 0.3 SEK/kWh,

230

together with large scale production may show the latent heat storage competitable.

REFERENCES

Dow Chemical (1981), Developmental Product Technical Data Sheet for XFS-43076, TES-81. The Dow Chemical Company, Midland, Mich..

Friser, M., (1987). Tests were carried out by M. Friser at Chalmers Inst. of Tech., Gothenburg.

Karlsson, B.O., Björk, C.O, Karlsson, B.G. (1986). Microcomputer Based System for Efficient use of Electricity in Industry, Journal of Microcomputer Applications 9, 279-288, Academic Press Inc, London.

Persson, L.J., and A. Solmar, (1987). Design of a Latent Heat Storage using Numerical Simulation and Experimental Results, a Industrial Case Study, Proc of the 5th Int. Conf. on Numerical Methods for Thermal Problems, Montreal .To be published.

Persson, L.J, (1986). Inledande mätningar vid Esselte Well AB, Vikingstad (Measurements at the Esselte Well Corp. , Vikingstad), In Swedish, LiTH-IKP-R-449. Linköping Inst. of Tech., Dept. of Mech. Eng., Div. of Energy Systems, Linköping.

Solmar, A. (1983). Energy Storage in a Bakery - Full Scale Experience, Proc. of 2nd BHRA Fluid Engineering International Conferance on Energy storage: Energy Storage for Energy Management. Published by BHRA Fluid Engineering, Cranfield, Bedford.

Söderström, M., and T. Johansson (1983). Use of Energy Storage in the Industry, Proc. of 2nd BHRA Fluid Engineering International Conferance on Energy storage: Energy Storage for Energy Management. Published by BHRA Fluid Engineering, Cranfield, Bedford.

Söderström, M., (1986), Evaluation Method of Industrial Thermal Storages, Linköping Studies in Science and Technology, Dissertations, No. 146, Linköping Inst. of Tech., Dept. of Mech. Eng., Div. of Energy Systems, Linköping.

Tube plåt, (1986), Installation were discussed with C. Andersson at Tube Plåt Corp., a heat recovery installation company, Linköping.

VVS handboken, (1974), Reference book for building material, constants, climate etc., Förlag AB VVS, Stockholm.

Wigö, H. (1984). Energikartläggning av ett Wellpappföretag: Esselte Well i Vikingstad (Energy Audit of a Corrugated Board Mill: The Esselete Well Corp. in Vikingstad), In Swedish, LiTH-IKP-EX-466, Linköping Inst. of Tech., Dept. of Mech. Eng., Div. of Energy Systems, Linköping.

Åström, K.J, (1982). A Simnon Tutorial, Dept. of Automatic Control, Lund Inst. of Tech., Lund.

The Development of New Energy-Saving Technologies—the Role of the European Commission

D. A. Reay

David Reay & Associates, PO Box 25, Whitley Bay, Tyne & Wear NE26 1QT

Get # firm Govt !

ABSTRACT

The fields of energy efficiency and process innovation are ones in which participants can receive substantial funding from a variety of sources. This paper presents information on schemes managed by the European Commission. Some guidance to potential applicants is given, based in part on the Author's work for the Non Nuclear Energy R&D Programme administered from Brussels. This, together with the Energy Demonstration Programme and BRITE, are described.

KEYWORDS

Energy Efficiency; Innovation; Industry; European Commission; Sources of Funding.

INTRODUCTION

The fields of energy efficiency and process innovation are ones in which participants in industry, research laboratories and educational establishments can receive substantial funding from a variety of sources. The UK government, principally via the Dept. of Energy and the Dept. of Trade & Industry, is able to support industry in a number of ways, either as a manufacturer of energy-efficient or energy-conserving products, or as a user of energy, and some aspects of this will be discussed by other speakers at this Conference.

In this paper, it is intended to concentrate on the programmes organised and funded by the European Commission which are relevant to the themes of the Conference.

The European Commission, operating principally from Brussels, has many programmes supporting innovation. These are directed both at energy saving and at improving industrial competitiveness - often similar or complementary routes can be used to meet these aims. The BRITE and EURAM programmes, together

231

with EUREKA (although not funded by the Commission), the Non Nuclear energy R&D and the Energy Demonstration Programmes, are among the major initiatives taken by the EC.

In this paper it is intended to present data on some of the schemes which could benefit companies concerned with process energy efficiency and innovation. The Author will also give advice on proposal preparation, and list some examples of projects which have been recently funded, based in part on his experience as a successful contractor and as a consultant to the Commission non-nuclear energy R&D programme.

2.   THE EUROPEAN COMMISSION ENERGY PROGRAMMES

There are several programmes managed by the EC which have relevance to energy use in processes. In some, notably BRITE, EURAM and EUREKA the energy interest tends to be peripheral, but nevertheless of some interest. However two schemes which to some extent parallel those operated by the UK Department of Energy are solely related to energy. These are:

    (i)   The Non-Nuclear Energy R&D Programme - managed
            by DGXII

    (ii)   The Energy Conservation Demonstration Programme
            - managed by DGXVII

(DG's are the Directorate Generals responsible for administering different areas of Community policy and legislation-roughly equivalent to a UK government Department).

2.1   Non Nuclear Energy R&D Programme

This programme, which involves calls for proposals on a variety of topics on a large scale every 3 to 4 years, with occasional intermittent more restricted calls, covers several topics. These include renewable energy sources, the better exploration and exploitation of liquid, gaseous and solid fuels and energy conservation R&D. This section concentrates on this last subprogramme. However, the full range of topics covered in the programme now underway are listed in Table 1, and a budget of about 175 million ECU was allocated to the programme.

The current (3rd) programme commenced in 1985, and is scheduled for completion in 1988. A 4th programme is currently being planned. There are 3 principal objectives of the energy conservation subprogramme:

    (i)   "The promotion of energy conservation in the
            Member States by means of new and improved
            energy technologies, processes and products.

    (ii)   The development of new and improved energy
            technologies which will additionally lead
            to a reduction in pollution.

    (iii)   The encouragement of industrial involvement
            in the above by financial incentives, publicity
            and example."

TABLE 1
EC Non-Nuclear Energy Research & Development
Subprogrammes of 3rd (1985-1988) Programme

A)   Development of Renewable Energy Sources

       1)   Solar Energy
       2)   Energy from Biomass
       3)   Wind Energy
       4)   Geothermal Energy

B)   Rational Use of Energy

       5)   Energy Conservation
       6)   Utilization of Solid Fuels
       7)   Production and Utilization of New Energy Vectors
       8)   Optimization of the Production and Utilization
              of Hydrocarbons
       9)   Energy Systems Analysis and Modelling

TABLE 2
Subprogramme on Energy Conservation

The main objective of this subprogramme is to promote energy
conservation in all Member States by means of improved technologies, pocesses and industrial products which have the potential of leading to demonstration projects and ultimately to
commercial applications.

This subprogramme covers the following areas:

A.   Energy Conservation in Buildings

       Development of heat pumps, in particular, absorption
       types, new fluid combinations and components, study
       of dynamic behaviour of systems.

B.   Energy Conservation in Industry

       Development of advanced integrated heat recovery
       systems considering waste-heat sources and applications
       for the heat and power produced.

       -      Development of high performance heat exchangers
              for application in corrosive fluids and high
              temperature exhaust gases.

       -      Development of advanced heat pumps and heat
              transformers.

       -      Development of fluids and components.

       -      Improved combustion techniques (including
              fluidized beds, stationary internal combustion
              engines and turbines) for low calorific value
              gases, heavy fuel oil, wastes, etc.

- R&D work aiming at new energy saving processes and products in energy intensive industrial branches (e.g. petrochemistry, chemistry, metallurgy, purification and transport of gas, food processing, etc).

C. Energy Conservation in Transport

Basic long-term investigations on critical aspects of energy-sensitive components and techniques mainly for application in road transport.

- Improvement of engine and propulsion efficiency considering changes in fuel characteristics and pollution abatement.

- Development of material, components and manufacturing processes.

D. Energy Storage and Fuel Cells

The development of energy storage systems and fuel cells should be aimed at rational use of energy and be considered mainly with a view to their future application in transport and power generation. Their economic aspect should be taken into account.

- Development of new electrical energy storage systems and study of their application.

- Development of fuel cells and study of their application.

Approximately 100 projects are currently being supported, and funds of up to 26.5 MECU have been made available. (In May 1987 1 ECU = £0.69).

These projects are covered in the main by the categories listed in Table 2.

2.1.1 Proposal Procedures

A strict deadline exists for submissions, and the selection procedure is exhaustive, involving a number of assessment hurdles. Recently the EC has set out the evaluation procedure in some calls for proposals, and in the most recent call in this programme it was as follows:

- receipt, registration & acknowledgement
- distribution to experts appointed by the EC for assessment
- consideration by Management & Co-ordinating Advisory Committee
- Final conclusions by the EC
- Notification of decision to proposers
- Submission of more detailed administrative and financial form by proposers
- Contract Negotiations
- Signature of contract.

During implementation of the project, formal progress and final reports are normally specified. These are assessed by the experts. Project review meetings are held in conjunction with other contractors at least once during the course of a programme, and a full conference is likely to review the programme. These may be in Brussels or at a centre elsewhere in Europe.

The EC and/or its representatives (eg an expert) may visit the site of the work during its progress. This is likely to occur at least once during the course of the contract.

With regard to financing, up to 50% of the cost of the R&D project can be provided by the EC. In the case of Universities and Polytechnics, marginal cost funding is provided. In practice this pays the full salary of the researcher, (not his supervisor), and up to 40% overheads.

There is in some programmes a necessity to link with laboratories/companies in other Member States before funding for a project can be made available. This is not the case in this programme, but it is advisable to attempt to formulate an international project if possible. The EC may suggest that you co-ordinate activities with another contractor, as is happening with the expert system project later used to illustrate the scheme. The author suspects that the trend will be towards more projects involving several contractors, as in BRITE.

2.1.2 Advice for Potential Applicants

The documentation associated with the programmes of the EC tend to spell out in some considerable detail the topics which will be regarded as acceptable. For example, in the case of heat exchangers for heat recovery, high temperature corrosion resistance may be cited as a particular goal, but the development of units for operation below the acid dewpoint may not be mentioned. It is imperative that the proposer takes note of the specific requirements, as requests for support in the second category would be unlikely to pass the first screening. (Such proposals may be directed by the EC to other programmes where they may be more applicable). These requirements have been arrived at following detailed discussions involving the EC, its experts, and possibly subcontracted organisations whose role it may be to assess European capabilities in a specific area of technology.

Implicit in this is the desirability to "make your mark" with the Commission or one of its advisers at an early stage, preferably before a programme is announced. This can be done at seminars occasionally organised by bodies promoting European collaboration, via the D.En. who have representatives on the steering committee, via the experts, or directly with the Commission. It is most effective, at some stage, to include a visit to the offices of the appropriate personnel in Brussels.

The proposal form gives precise information on the data required from the proposer. Accurate costing is essential at this stage as while any cost escalation may receive sympathy, it is highly unlikely to extend as far as financial bailing-out!

It is worth noting that the technical activities, once agreed, form part of the contract. It is therefore useful to all concerned if the technical programme is spelt out in some detail and in a form such that "milestones" and specific tasks can be readily identified. A bar chart of course complements this.

Implementation of the results of the R&D is a requirement if the contractor is to retain rights and/or funding received from the Commission. A period of 3 years is normally allowed between termination of the R&D contract and commencement of exploitation. The Commission may thereafter offer rights on a non-exclusive licence basis to others, or demand reimbursment. Renegotiation is always of course a preferred third option, insofar as timescale for exploitation is concerned.

Finally, language difficulties in general do not occur when dealing with the Commission. Documentation is in English and all the personnel with whom you are likely to come into contact will speak English well.

2.1.3   Project Example - The Development of Expert Systems for Heat Exchangers

The subject of Expert Systems comes within the area of computing developments known as intelligent knowledge-based systems (IKBS). The broad area of IKBS has received attention within the EC ESPRIT programme, but no Community-wide initiative had been taken for the application of expert systems, and the opportunity to include projects in the non-nuclear energy R&D programme is of considerable importance.

The principal objective of this project is to develop expert systems linked to data bases to assist users in the specification of heat recovery equipment. More specifically, the system being developed at NEI-IRD with the assistance of Pergamon BPCC will relate to heat exchangers. Financial support from the EC is matched by inputs from NEI, ETSU (representing the UK Dept. of Energy) and Pergamon BPCC.

Note that in this particular project funding has been obtained from the EC and the EEO. Normally this will be unlikely to exceed 50% of the total project cost, but if the technology is of particular interest to ETSU, for example, support by way of a form of "buy-in" may lead to a small increase above the 50% limit.

A parallel project is being undertaken by Comprimo BV in the Netherlands, concentrating on a system for heat pumps in buildings. Close collaboration is taking place between Comprimo and IRD.

The project involves a number of stages leading to the testing of the expert system and preparation of packages. These include minimum hardware specification, software specification and development, rule bases for equipment selection and preparation of the data bases.

The project is administered by the Commission with the assistance of a Project Officer in Brussels and an expert. The expert assesses the progress of the work, makes recommendations as and when necessary, and attends some progress meetings.

Examples of some of the projects being funded in the UK and on mainland Europe are listed in Table 3.

TABLE 3
Examples of Projects Receiving Support in the
Energy Conservation Sub-Programme

-   Improving solid Lithium batteries to permit operation at room temperature (10 organisations in France, Italy and UK).

-   Development of an engine fuel management system for large stationary and marine Diesel engines (Lloyds Register of Shipping, plus 27 collaborators)

-   Basic research on molten carbonate fuel cells (6 laboratories).

-   Development of ceramic heat exchangers for heat recovery at up to 1300°C (Bertin, France)

-   Improving absorption cycle heat pumps for domestic, commercial and industrial applications. (Several companies and universities)

-   Development of advanced cyclone preheaters for cement kilns (Cardiff University).

2.2   Commission Energy Demonstration Programme

2.2.1   Programme Aim & Nature of Projects Supported

The aims of the demonstration projects supported by the European Commission are broadly similar to those in the UK. However the level of support available, and the governing rules differ. Also, the areas covered by the EC programme are wider-ranging, particularly in the fields of alternative energy and solid fuels processing. The topics included in the programme announced in 1986 are listed in Table 4.

It is worth noting that within each category certain projects may be ineligible for support. For example sports centres and hospitals were among those excluded from the "Energy savings in buildings" sector in 1986. Considerable help is given to the potential proposer in this way in the call put out by the Commission, in common with that of the R&D programme.

As in the UK, there is a requirement to ensure that the project demonstrates an innovative technique, process or product, or involves a new application of a known technique. Proposals may be submitted by users or producers of equipment. However as the onus to stimulate replication is to some extent on the party being funded, it is more likely that the supplier will take the lead in an EC project. In the UK the EEO takes on itself the task of promoting successful projects to encourage replication.

If the project is likely to be implemented without EC support, the criterion for funding related to technical and/or economic risk disappears, and the case for support would be difficult to justify.

2.2.2   Financial Support Available

Financial support may cover the whole project, or specific stages of it.   In the 1986 call, an upper limit of 40% of the eligible cost was put on support available from the EC. The eligible cost must include the cost of evaluating the results and all report preparation.

(Note however that in the case of the Commission-funded demonstration projects, no requirement has to date existed for the use of independent monitoring agents.   Reliance is put on the host site and/or equipment supplier to assess the energy savings resulting from the installation or modifications. You may well, of course be subjected to a financial audit by EC accountants)!

A degree of flexibility in funding is possible due to the policy of the UK government to consider "topping up" suitable projects.   It should be emphasised that this is never meant as a contribution to bale out an under-costed project, but is made where the project may have particular local interest, or could complement the UK demonstration strategy.

Retrospective funding is not possible, except in some cases related to new phases of projects already receiving financial support.

The total level of funding allocated to the 1986 scheme was about 90 MECU.   It was recognised that only in exceptional cases would the allocation to individual projects be likely to exceed 1.5 MECU, and most projects would be supported at a level well below 1 MECU.   (This, nevertheless is often greater than the average funding for projects within the UK scheme, although it should be noted that the nature and scope of the projects differ).

Repayment of Commission funding is no longer requested.   In previous programmes the Commission attempted to recoup part of its investment by negotiating repayment as either a levy on sales or as a proportion of the value of the energy saved per annum.

It should also be pointed out that the two regional studies reported at this Conference - those for the North West and North East of England - are funded in part by DGXVII.   Thus funding is not limited to Demonstration Projects.

3.   TECHNOLOGY PROGRAMMES - BRITE (Basic Research in
         Industrial Technology for Europe)

The BRITE programme, administered like the Non-Nuclear Energy R&D Programme by DGXII, is directed at improving the technological base of the more traditional industries in the

Community - hence it has a degree of relevance to this region. It does this by providing financial encouragement leading to the introduction of advanced technologies in these industries. As with most other R&D programmes supported by the Commission, the participants can include academic and research institutes, as well as industrial companies.

Over the period 1985-89, the budget allocation to BRITE is about 125 MECU, and the programme was implemented in two stages, the second call closing in May this year. Projects supported cover the following areas:-

-   Reliability, wear and deterioration

-   Laser technology and powder metallurgy

-   Joining techniques

-   New testing methods
-   CAD/CAM and methametical modelling

-   Polymers, composites or other new materials

-   Membrane science and technology

-   Catalysis and particle technology

-   Flexible materials

Brite projects are not intended to lead directly to marketable products, but clear industrial objectives should be seen - ie it is precompetitive R&D. It is particularly directed at encouraging trans-frontier collaboration, and projects which would normally be funded by national programmes are excluded.

The amount of funding is limited to no more than 50% of the cost of a single project, except in the most exceptional cases. The objective was encourage contractors to obtain the other 50% from industry, rather than from national government or other bodies. It was also emphasised that for research work carried out at Universities and Polytechnics, the Commission would only fund marginal costs, which remains attractive to many academics.

In the first series of submissions to the programme, a total of 559 proposals were received. Fifty of these were described as bad. A further 100 proposals were rejected at a fairly early stage for a number of reasons. In some cases it was suggested that the proposers should drop the subject because it was too near the marketplace. In others, the topic was considered too academic, or more appropriate to the Esprit or non-nuclear energy R&D programmes. In a small number of cases the proposers did not follow the rules for preparation for the proposal or submission to the Commission, or showed that they had sufficient support available from other sources. In conjunction with experts employed by the Commission, the list of proposals was reduced to 195. Of these, approximately 100 projects were selected for the first tranche, and the total funding available was 65 MECU. This left 60 MECU remaining for the second phase. 100 proposals were put on a reserved list, and of these nine were subsequently selected due to some of the original proposers withdrawing. The other 91 had no chance of being selected, and were not put forward for the second phase, as the Commission considered that the R&D proposed would by then be out of date.

A small prejudice has been shown in favour of projects involving smaller countries and smaller enterprises. Some extra help has been allocated to sectors which are not historically strong in the fields of research and development. In particular the textile industry was cited.

The average project cost as received by the Commission was 2 MECU. Because of the availability of funds, it was decided that some cost reduction would take place. As a result of this, some projects are only receiving funding for their first phases, while in large projects, a two year programme has been agreed and funded. A number of contractors have been persuaded to reduce their costs, typically up to levels of 30%.

The experience on contract negotiations has been mixed. It was pointed out that agreement on some of the Technical Annexes was difficult as the research programme was rarely sufficiently detailed in the proposal. This was considered particularly important in future proposals. Proposers need to quantify the likely results, improvements etc. Also the proposer must be prepared to answer queries on the management of complex international projects.

In the second tranche a number of rule and priority changes were implemented. These included the following. Firstly, as a general rule, projects should have a minimum of two industrial partners, each of these being in a different country. Secondly, the preference given to small and medium sized (SME) enterprises was much more explicit. (These enterprises are classified as those employing less than 500 people). The involvement of large companies in projects into which SMEs are brought was also encouraged. It was also made easier for companies outside the European Community, but still within Europe to be involved in projects. In particular Sweden, Norway, Switzerland and Austria could benefit from this.

In deciding the priorities for projects, it was emphasised that a knowledge of the areas already covered was necessary, as was a knowledge of areas where problems are now solved or the emphasis needs changing. There was a need to make it easier to see what project areas would be appropriate for Brite and what could go into other areas.

Bonus points were awarded to proposals which covered the following points:

1.   Those involving small firms.

2.   Those involving partners in smaller or new community countries.

3.   Those projects which were highly innovative.

4.   Proposals which were very well prepared.

Interestingly, the Department of Trade & Industry at a meeting to promote Brite, listed the following points which could lead to a successful proposal:

1.   Good quality.

2.   Comply with the conditions.

3.   Collaboration with other industrial partners in other countries.

4.   Show evidence of management capability.

5.   Projects must gain from a European dimension.

6.   Include small firms and small or new member states.

7.   Include all the required information in the proposal.

8.   Match the proposals to the size of the programme (not too small or too large).

4.   CONCLUSIONS

Three of the major European Commission programmes which offer support to organisations with an interest in innovation and energy efficiency are described. The experience of the author in his association with these programmes will, it is hoped, benefit the reader.

Some data on how to build low energy houses that work!

Comfort and Low Energy in Medium Priced Housing

P. Ruyssevelt,* J. Littler* and P. Clegg**

* Research in Building, Polytechnic of Central London, 35 Marylebone Road London NW1 5LS
** Feilden Clegg Design, 1 Canton Place, London Road, Bath BA1 6AA

ABSTRACT

n January 1985 construction began in Milton Keynes on a group of twelve ouses (Fig. 1) four of which were built to a superinsulated standard, three imes more stringent than the UK Building Regulations. All twelve houses vere sold by September 1985 for prices in the region of £35,000. An ntensive monitoring programme began in January 1986. Thermographic urveys have shown the standard of construction in these timber frame ouses to be very high. Experiments revealed the superinsulated houses to ave a specific heat loss rate of ~70W/K and a background air change rate f less than 0.2 per hour. The average heating bill for these houses for 986 was £10. Comfort conditions are achieved by virtue of the internal ains alone, for the majority of the heating season.

KEYWORDS

uperinsulated; timber frame houses; ventilation heat recovery; under floor sulation; low 'E' windows; monitoring.

INTRODUCTION

ne superinsulated houses project at Milton Keynes project was devised by e Research In Building (RIB) group at the Polytechnic of Central London CL) and the architects Feilden Clegg Design (FCD) of Bath. Partial nding for the demonstration project was obtained from the Commission of e European Communities, Directorate General XVII for Energy Saving.

e houses, four of which were built to the superinsulated standard, were cupied from June 1985 and eight of them have been intensively monitored nce 1st January 1986. The space heating costs, illustrated in Fig. 2, for e four superinsulated houses for 1986 were £7, £6, £20 and £6. These gures are very impressive, representing only 15% of the cost of heating e control group of houses and probably less than 5% of the cost of ating similar houses built over the last ten years. However, the results of

243

244

Fig. 1 A pair of semi-detached superinsulated houses.

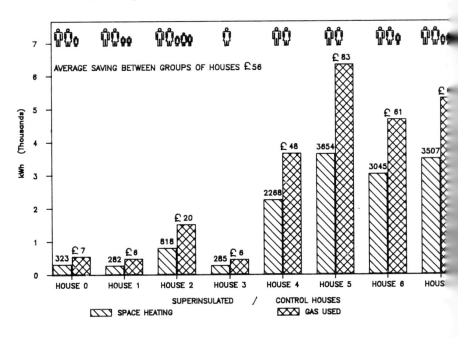

Fig. 2   Space Heating Costs for Superinsulated and
Control Houses.

the first years monitoring raise several questions about the design of the control houses, and the space and water heating systems installed in both groups of houses.

Throughout this paper the individual houses will be referred to as house 0 to house 7. Houses 0-3 are superinsulated, houses 4-7 are the control group, and house 0 is the main superinsulated demonstration house.

ARCHITECTURAL AND THERMAL DESIGN

The main aim of the architectural design process was to produce a with a very low level of background infiltration and a low heat loss rate. The result shown in Fig. 1 is a three bedroomed, semi-detached house, with a floor area of 75m², and an attached garage.

In parallel with the architectural design process the Research in Building group used the thermal simulation model SERI-RES to evaluate the energy savings attributable to different levels of airtightness and insulation. These savings were then compared to the extra capital expenditure on the basis of repaying a standard house mortgage over a twenty five year term.

Performance Predictions

The thermal specifications and cost-effectiveness calculations upon which the final designs for the superinsulated and control houses were based are shown in Tables 1 and 2.

TABLE 1 Thermal Specifications	Superinsulated	Control House
Constructional Element	U value W/m²K	U value W/m²K
Walls	0.24	0.46*
Roof	0.11	0.35
Ground floor	0.21	1.03£
Windows	1.44	5.00
Doors	1.56	3.40
Total fabric heat loss rate (UA) W/K	53	156
Predicted Mean air change rate (ACH)	0.35¹	0.75
Resulting heat loss rate	22	47
Combined heat loss rate W/K	75	203
Auxiliary energy requirements kWh (including boiler efficiency)	1470	11620
Predicted Annual heating energy costs	£19	£149
Predicted Annual saving	£130	

Building Regulations require 0.6, but 90mm timber frame gives 0.46W/m²K.
Estimated value, as SERI-RES models heat flow to a ground temperature.
Equivalent in heat loss to 0.15 background + 0.6 with 66% heat recovery.

The design and costing of the difference between the control and superinsulated houses performed in 1983/84 gave rise to an estimated additional cost of £1865. The designers believed that this figure could fall £1500 per house with replication on a larger scale. The additional extra mortgage repayment for £1500 was calculated to be £150/annum at 1984 rates.

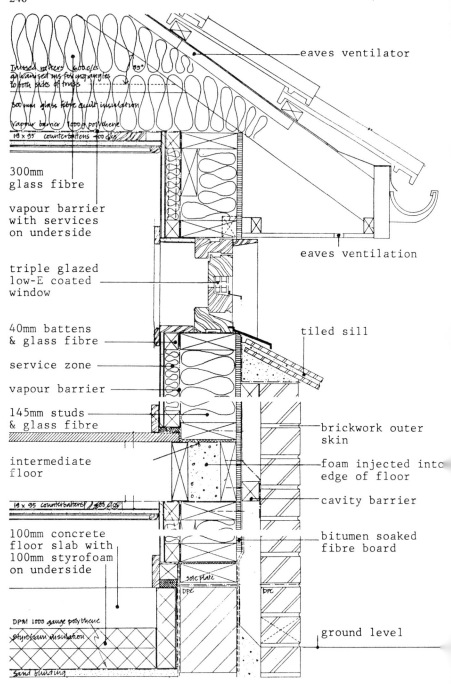

Trussed rafters 600c/s
galvanised ms fixing angles
to both sides of truss

300mm glass fibre quilt insulation

vapour barrier / 1000g polythene

18 x 95 counterbattens 400 c/s

eaves ventilator

300mm
glass fibre

vapour barrier
with services
on underside

triple glazed
low-E coated
window

40mm battens
& glass fibre

service zone

vapour barrier

145mm studs
& glass fibre

intermediate
floor

18 x 95 counterbattens 400 c/s

100mm concrete
floor slab with
100mm styrofoam
on underside

DPM 1000 gauge polythene

styrofoam insulation

sand blinding

eaves ventilation

tiled sill

brickwork outer
skin

foam injected into
edge of floor

cavity barrier

bitumen soaked
fibre board

sole plate

DPC

DPC

ground level

Fig. 3   Section through external wall of superinsulated
         house.

TABLE 2 Predicted Cash Flow per Year for a Superinsulated House						
Year	1	2	3	4	5	26
Extra loan repayments	£150	£150	£150	£150	£150	£0
Saving (5% inflation)	£130	£137	£143	£150	£158	£440
Within 4 years fuel savings will exceed extra outgoings ^						

The predicted cash flow analysis in Table 2 suggested that in the initial years the house owner would be paying a small premium for the benefits of the superinsulated house. However after the fourth year it was predicted that the savings would balance the outgoings.

CONSTRUCTING THE HOUSES

Construction Techniques

A timber frame system was chosen as the most convenient method of achieving the high level of wall insulation required for the superinsulated houses. The kits were supplied by a Milton Keynes based company, Finlandia. As the control houses would have to sell alongside the superinsulated houses a standard timber frame from the same manufacturers was chosen, giving a wall U value of 0.46W/m²K.

The kits arrived on site from Finland in the form of large panels completely finished from sheathing through to plasterboard, with doors and windows installed and glazed. Each house required only one day to erect from finished floor slab to felt and battened roof. All joints between panels were sealed on site with expanding UF foam. The construction drawing in Fig. 3 shows the factory installed vapour barrier which is located between the 145mm structural stud and the 40mm cross battening. This technique provides a protective zone in which electrical conduit and other services can run without puncturing the vapour barrier. Only the first floor ceiling vapour barrier is installed on site.

The ground floor concrete slab in the superinsulated houses is insulated on the underside with two layers of 50mm Styrofoam (extruded polystyrene) and a 50mm layer surrounds the edge of the slab.

Space and Water Heating Systems
The Cormorant space and water heating system, developed by Gledhill of Blackpool and British Gas Research and Development, was chosen for installation in all the houses. The systems for the control and superinsulated houses were matched as far as possible, so that all the savings achieved would be attributable to the extra insulation measures and not to any increased efficiency. This decision was later to be regretted.

The principle feature of the Cormorant system is the insertion of a large thermal store between the heat generating device and the heat/hot water distribution system. The thermal store is kept at a constant 70-80°C. The main advantage of the system is that the heat generator size can be significantly reduced as it does not have to cope with peak demands. Hot water is provided at mains pressure by passing the incoming supply through two heat exchangers connected in series in the thermal store.

Fig. 4 shows a schematic of the space and water heating systems as installed in the control and superinsulated houses. These installations differ from a standard Cormorant system in the use of a wall hung gas boiler rather than a circulator adjacent to the thermal store. The control houses are heated by a series of radiators throughout the house, whereas, the superinsulated houses are heated by a 2kW heat exchanger coil in the input air stream of the ventilation system.

Ventilation

To maintain a controlled air change rate in the superinsulated houses a mechanical ventilation system with heat recovery was installed (see Fig. 5). This system manufactured by BAHCO had been previously tested in trials carried out by the Electricity Council (1984), and found to perform very well. As reported by Ruyssevelt (1986), one of the main attractions of this particular system is the grouping of all the major components in one unit in the kitchen, making for easy access to carry out regular maintenance. Locating the unit in the kitchen means that the low level of noise it generates is completely masked by other domestic appliances. The duct distribution system is neatly integrated into the intermediate floor, avoiding numerous entries through the loft vapour barrier.

Unfortunately a package of measures designed to ensure adequate natural ventilation in the control group of houses were not carried out by the builders. In order to alleviate severe condensation problems, frame ventilators have subsequently been fitted to the windows of these houses.

Construction Costs

The predicted extra construction cost at the design stage was £1865 per house. The actual extra costs incurred are shown in Table 3. Overall the estimate was some £330 short of the actual cost. To put the figure of £2193 into context, it represents 6% of the sale price of the houses and 4% of their current market value. Looked at another way, it is less that the cost of an en-suite shower room, or one might consider that £2000 is probably equivalent to the first two years depreciation on a new family car!

TABLE 3 Actual Construction Costs			
Item	Cost £	Cost £	Balance
Timber frame kit & erection*	9712	8658	+1054
Heating & hot water system#	734	964	-230
Extra 200mm loft insulation	181	0	+181
Under floor insulation	331	0	+331
Ventilation system	841	0	+841
Window frame vents	0	70	-70
Bathroom extract	0	80	-80
Cooker extract	0	120	-120
Extra cost of superinsulation			£1907
Plus builders profit @ 15%			£2193

* Difference = +95mm of wall insulation and triple glazing with Kapafloat.
# Difference is 7 radiators to one heat exchanger.

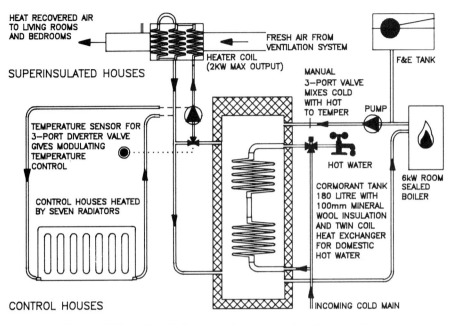

Fig. 4   Schematic of Cormorant space and water heating system.

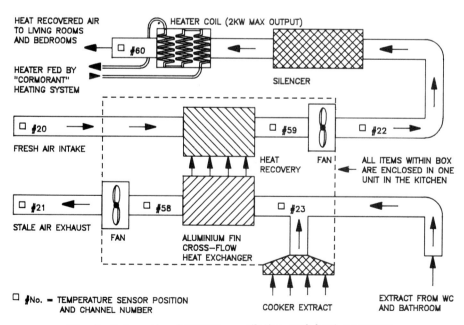

Fig. 5   Schematic of BAHCO ventilation and heat recovery system.

THE MONITORING INSTALLATION

The purpose of the demonstration project is to assess the performance of the houses and this can only be achieved by a carefully planned monitoring programme. PCL are using equipment supplied by the Energy Monitoring Company Ltd. to monitor; three room temperatures, total electricity and gas consumption, status of the heating and ventilation systems, heat output of the boiler, and heat input to domestic hot water.

The site values monitored are; wind speed and direction, external dry bulb temperature, solar radiation on the east and west vertical planes. One of the superinsulated house is more intensively monitored as the main demonstration house. A total of 168 channels of data are stored each hour.

EXPERIMENTATION

As well as continuously monitoring 168 channels, a number of one-time experiments have been carried out to analyse in detail the performance of the houses.

Thermographic Surveys

Two thermographic surveys were carried out. The first shortly after completion of the first two superinsulated houses and the second after all the houses were occupied (Dudek & Valentine, 1986). Two superinsulated and four control houses were studied. In general the thermographic study showed the houses to be considerably better constructed than the standard UK timber frame house. The external walls were found to have no missing or misplaced insulation. Joints between the panels were shown to very tight with few cold bridges, excepting areas close to the floor slab. The loft was shown to be the weak area in all the houses. Typically the insulation had been incorrectly installed at the eaves, allowing cold bridging to occur.

Co-heating Experiments to Determine the Specific Heat Loss Rate

To date co-heating experiments have been performed in three superinsulated houses and one control house. Table 4 gives the results obtained so far.

The predicted specific heat loss rates for the superinsulated and control houses, inclusive of a background air change rate of 0.2/hour, are 65W/K and 168W/K respectively. The results for the superinsulated houses are close to the predicted values, given the accuracy of the measurement, with the exception of house 1 which was incomplete when tested. The result for house 4 of 127W/K is some 25% less than the predicted value. This goes some way towards explaining the lower than expected heating costs for the control houses.

TABLE 4 Co-heating Results	Superinsulated Houses			Control
	House 0	House 1	House 3	House 4
Specific Heat Loss Rate W/K	68	81	73	127

251

Air Infiltration Experiments

In April 1986 a number of air infiltration experiments were performed in house 0 using an SF_6 tracer gas decay technique. The background air change rate was found to be ~0.1 ACH (air changes per hour), which is well within current performance requirements for Scandinavian houses. Experiments performed with the ventilation system operating in its various modes revealed the following air change rates; low - 0.6, normal - 0.9, boost - 1.1 ACH.

Some eighteen months after construction, house 0 was subjected to a blower door test to assess the level of background ventilation under depressurisation. The result was 1.43 ACH at 50 Pascals which is equivalent to less than 0.1 ACH under normal conditions. This result confirms that the choice of construction technique was wholly suitable for producing a very airtight house.

Air infiltration measurements carried out in the control houses revealed marginally higher background air change rates of the order of 0.3 ACH. This level was unacceptably low, and subsequent fitting of window frame ventilators has made provision for the air change rate to be increased to ~0.5 ACH.

RESULTS OF DATA ANALYSIS

t is only possible to report a limited set of results in the space available here. However, the final report on this project will be submitted to the Commission of the European Communities this year, and will eventually be published.

Energy Costs for 1986

Fig. 6 shows the total cost of all energy supplied to both groups of houses and the mean saving for the superinsulated houses is £102. These figures include the cost of electricity for running the ventilation systems in the superinsulated houses, at an average of £26 for the year. As reported by Cuyssevelt (1986) the ventilation systems achieved temperature efficiencies in the range 70-80% and CoP's ranging from 3 to 8.

The space heating requirements for the superinsulated houses illustrated in Fig. 2 show an average of £10 for the cost of gas, and a saving over the control houses of £56. £10 is very impressive. However, £56 would only allow an additional capital expenditure of some £650 if the savings were to balance the extra mortgage repayments by the fourth year.

Implications of Space and Water Heating System

The hot water production efficiencies are lower than expected, ranging as they do from 21% to 41% over a draw-off range of 93 to 442 litres/day. Over all the houses, an average of 38% of the total gas consumed is lost into the house uncontrollably. In the control houses this internal heat gain is useful for the majority of the winter, and serves to reduce the space heating requirement. However, in the superinsulated houses during the early and late winter this extra heat gain is surplus to requirements and leads to overheating, or additional venting of the houses to control internal

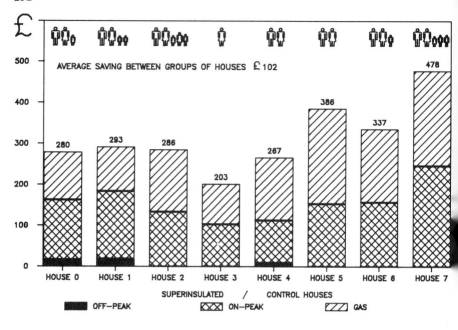

Fig. 6 Annual cost of all energy supplied to both groups of houses.

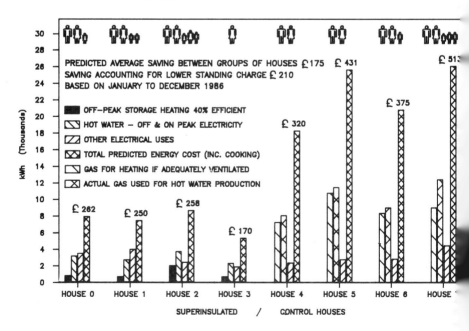

Fig. 7 Predicted costs for all electric superinsulated houses.

temperatures. Several steps have been taken during the course of the monitoring programme to improve the efficiency of the Cormorant systems, but only small increases of up to 5% have been noted.

THERMAL COMFORT

At the construction stage a number of temperature sensors were installed in, and underneath, the concrete floor slabs of one control house and one superinsulated house. Long term observations of these sensors show the superinsulated floor generally to be within 1-2°C of the room air temperature, whilst the control house shows differences of 4-8°C, producing less than desirable comfort conditions.

Since January 1987 globe temperature measurements have been taken in one superinsulated house and these show that the difference between the air temperature and wall surfaces temperature rarely exceeds 1°C. These results are reinforced by a number of thermographic images showing no difference in temperature between internal surfaces of external and internal walls. Whereas, with similar images of the control houses significant temperature variations were noted. The control houses also show a requirement for higher air temperatures, typically 1-1.5°C more than the superinsulated houses during heating periods. This tendency suggests the need to compensate for lower surface temperatures.

The major difference in comfort standards between the two groups of houses relates to the issue of condensation. There is a complete absence of condensation in the superinsulated houses. In contrast with this the control houses suffer severely with condensation on windows and some have small areas of mould growth on walls. In November 1986 attempts were made to alleviate this problem by installing the window frame ventilators that had been omitted from the control houses during construction. This measure has to a large extent failed to improve the situation and if the condensation continues at its present level the wooden frame windows will require substantial repair work.

ALTERNATIVE APPROACHES

Before drawing conclusions on the programme to date it is worthwhile considering the implications of alternative approaches to aspects of the design and construction of both the superinsulated and control houses.

Adequate Ventilation of the Control Houses

To examine the effect of adequately ventilating the control houses, the mean air change rate for each day of 1986 achieved by the ventilation systems in the superinsulated houses, was applied and the additional heating costs calculated. The mean saving on space heating in this scenario is increased to £85 for the year.

The cost of fitting a natural stack venting system such as that tested at TRADA (O'Connell & Pitts, 1983) would be approximately £200 in addition to the cooker extract and window frame vents. This modification would allow a more direct comparison to be drawn between the two groups of houses and would reduce the price differential to £1963.

An Alternative Space and Water Heating System

Given the significantly different requirements for space and water heating in the superinsulated houses, it could be argued that the systems should be completely separate. Further, an all electric house may have significant advantages in initial capital cost and any increases in running costs could be balanced by the absence of the gas standing charge.

Fig. 7 shows the calculated energy costs for the superinsulated houses using an all electric system, and for the control houses adequately ventilated and heated by the current gas system. The predicted average saving between the two groups of houses is £175 increasing to £210 if the reduced standing charges are taken into account. To arrive at these figures the superinsulated houses were assumed to be heated by a single off-peak storage radiator, taken to be 40% efficient. Hot water was obtained from a dual immersion heater, 210 litre cylinder, assumed capable of delivering up to 150 litres of water per day from an overnight charge. Demands above this level were taken at peak electricity rates. The extra cost of cooking by electricity was also taken into account in these calculations. The capital cost implications of an all electric house are significant, reducing the heating and hot water system costs to approximately £440. Taking into account builders profit the total extra cost of the superinsulated house would be reduced to £1625.

CONCLUSIONS

At this stage only preliminary conclusions can be drawn as changes to equipment will be made and further experiments will be performed to test some of the scenarios outlined in the previous section.

From the results available to date the superinsulated houses have exceeded expectations in terms of heating costs and standards of thermal comfort. There is a high level of occupant satisfaction with these houses. The same can not be said of the control group of houses where condensation is a major problem. Greater attention to the need for adequate ventilation at the design stage would have produced more realistic control houses. Retrospectively the superinsulated houses should have had an individually designed space and water heating system matched to the predicted loads.

REFERENCES

Dudek, S. and G. Valentine, (1986). Thermographic Field Survey Report, Milton Keynes Superinsulated Houses Project, School of Architecture, University of Newcastle upon Tyne.

Electricity Council, (1984). Energy Efficient Homes Seminar, Capenhurst.

O'Connell, P. J. and G. Pitts, (1983). Experiments with a Passive Ventilation System, Timber Research and Development Association, High Wycombe, Bucks.

Ruyssevelt, P. A., (1986). Ventilation and Heat Recovery in Superinsulated Houses, Proceedings of UK-ISES Conference C47, Superinsulation, London.

Heat Pipe Heat Exchangers for Boiler Flue Heat Recovery

N. H. Scurrah

Scurrah Hytech Products Limited
6 Market Street, Soham, Nr. Ely, Cambs CB7 5JG

HEAT PIPE HEAT EXCHANGERS FOR BOILER FLUE HEAT RECOVERY

Heat Recovery from the flues of Industrial Boilers has for many years been effected by placing Economisers in the flue stack.

Economisers first made an appearance onto boilers in 1845 to improve heat transfer for the products of combustion of large steam plant. They were used extensively on Lancashire boilers during the industrial heyday of the cotton and woollen industries.

The mid 1950's saw the gradual replacement of coal fired Lancashire boilers with multi-tubular shell boilers initially fired by oil, but more recently by natural gas. These boilers were considerably more efficient, 82%, than the old Lancashire 65% or 75% with economisers and so for a time there was little interest in economisers. The oil crisis of the 1970's and subsequent enormous increase in fuel costs coupled with the change in many cases to natural gas rekindled interest into use of economisers even on steam plant operating at 80 plus % efficiency.

Regarding a boiler as a cross flow heat exchanger with well mixed heating fluid is at uniform temperature, it follows that however efficient the heat exchange surfaces may be the products of combustion can never be cooled below the maximum temperature of the fluid and in practice the exit gases are well above the fluid temperature, 50°C at full load being typical.

It follows, therefore, that the practice of installing economisers between the pressure pump and the boiler limits the available additional heat transfer that an economiser can produce. Usually the feed water to the pressure pump has been heated to between 60°C and 80°C to reduce the dissolved oxygen within the feed water. Thus the temperature differential between the heat source, the flue products and the heat sink, the boiler feed water is relatively small, requiring large extended surfaces to effect adequate heat transfer.

The modern economisers consist of an array of finned tubes connected by headers at each end to effect flow and return of the water extracting heat from the flue gases. They can be very effective with clean fuels but do not lend themselves well to cleaning, being easily fouled up between fins. Corrosion of the surfaces of the heat exchanger is usually attributed to the temperature on these surfaces. If the flue gases are at all corrosive, as in the case with heavy oil, the cooling water being in direct contact with the heat exchanger can reduce the flue products to condense out acids. Equally well, the edges of the fins can be oxidised over a period of time, due to its temperature being much higher than the main wall of the tube. Finally because there are two headers top and bottom thermal stresses can build up if there is a lack of uniformity in the flow of the water through the tube. To overcome the problems, the design of modern economisers use expensive corrosive resistant materials having inferior thermal character- istics and are normally built into a bypass system such that on firing with fuel other than gas, the flue products are bypassed to waste in order that the heat exchanger surfaces do not become fouled.

Many boiler plants in Britain are converting to a dual-fuelling capability i.e. Natural Gas and Oil or Heavy Oil. British Gas Corporation offers a significant price reduction for consumers willing to have interruptible supplies i.e. at times of high gas demand the Gas Board will advise the withdrawal of supplies for a specified time period. In the past for some of their clients this withdrawal of supplies was for as long as eleven weeks during which time the clients boilers had to fire on fuel oil. During the oil price slide of 1986, the Gas Corporation were forced to reduce the cost of their supplies to dual fuelled customers in order to stay competitive, whereas those companies with sole supplies of gas saw a modest increase in cost.

The use of a standard economiser, which has to be bypassed during the periods of time the boiler is firing on oil, reduces the annual amount of heat which could be recovered.

There is a need, therefore, for an economiser that can -

a) Be used with all fuels, eliminates the need for bypass.

b) Can be easily cleaned.

c) By keeping the surface temperature high, eliminates cold surface, reducing or eliminating corrosion.

d) With less corrosion, copper with its thermally superior characteristics can be used resulting in a smaller more compact heat exchanger much reduced in weight and cost.

A heat recovery unit using heat pipes to extract the heat from the boiler flue products can be designed to have all the advantages itemized above.

Heat pipes are evacuated hermetically sealed tubes containing a volatile working fluid contained within a wick. When heat is applied to one end of the heat pipe the working fluid is vaporized taking in latent heat of vaporization and travels to the other colder end of the pipes where it condenses back into the wick giving out the latent heat of vaporization. The condensed liquid then returns to the source of the heat by capillary action through the wick. The heat pipe is in effect a passive heat transfer device which has no mechanical moving parts and can operate continuously with no degradation so long as the materials of the heat pipe are compatible.

When applied to the flues of boilers the heat pipes act as super heat conducting rods, part exposed to the flue products and part exposed to the cooling water, required to take away the recovered heat. But, whereas a normal economiser requires two manifolds to effect the flow and return of the cooling water, a heat pipe economiser or heat recovery unit requires a single manifold. This in effect means that the heat pipes are simply cantilevered out from the water manifold and it is a simple exercise to clean between the heat pipes in an array which make up an heat recovery unit.

The water in effect flows over the rods within a manifold and is separate from the flue products. Some heat exchange is lost, because heat has to be transferred into, along, and out of the heat pipe, so a heat pipe cannot compete with a cross flow tube within the boiler flue since the water in this case is heated directly by heat transfer through the tubular wall. But this can in fact be turned into an advantage.

The heat pipe by necessity is at a higher temperature than the cooling water, so much so that the occurrence of total condensing on the heat exchange surfaces is reduced.

Because the heat pipes in a heat recovery unit are not finned but plain it becomes even easier to effect cleaning, so much so, that there is no problem to fire on oil, residual fuel oil or even on coal.

HEAT PIPES v ECONOMISERS

In analysing the merits of use of an economiser the nature of steam generation and its uses should be inspected. Often in total condensate loss situations the feed water can be injected with process steam to eliminate dissolved oxygen in it prior to going to the feed pump and economiser (if fitted) to boiler. This is an application, which an economiser fitted to the boiler, but used in low pressure environment, can be used to eliminate (or at least reduce it considerably) the steam injection requirement.

The unit in this instance does not have to be manufactured to pressure vessel codes nor be subjected to regular insurance inspections. It also effectively provides more efficient heat transfer because the feed water is initially cooler and if heated in a counterflow manner can extract more heat from the flue gases for a particular size because the temperature differential between heat source and sink is significantly greater.

It is, therefore much cheaper to manufacture. These remarks hold equally well if a separate tank of process water is heated by a low pressure economiser. Only if there is full condensate return and no requirement for process hot water, should a high pressure economiser be considered.

Installation of a Morr Energy 'heat shuttle' in a pressurised situation as an economiser, recovers less heat than when installed in a low pressure heat recovery situation.

HEAT RECOVERED — KW

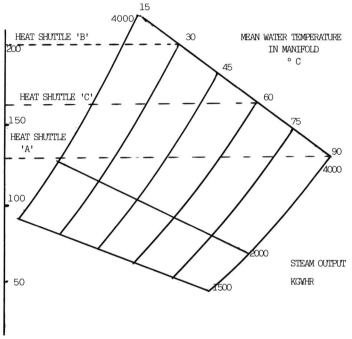

Fig. 1   shows the heat recovery using a heat shuttle at three
         firings at different manifold water temperatures

AT SHUTTLE 'A'

fers to a situation where there is 100% condensate return and in this case
e 'Heat Shuttle' is installed as an economiser i.e. on the pressure side
the feed pump thus requiring modulating water feed and fabricated to
essure vessel requirements involving insurance stamping etc. The
covery at a mean water temperature in the manifold of 40°C is 130KW.

AT SHUTTLE 'B'

fers to a situation where there is a total loss of steam and therefore
condensate return. In this case the 'Heat Shuttle' is used to preheat
e feed water tank. The unit can be fabricated to operate at low pressure
d does not need the full pressure vessel approval, nor the modulating
ed pump since a small circulator is adequate to maintain the flow of
ter through the 'Heat Shuttle' manifold to and from the feed tank.
cause the water temperature is much lower than 'Heat Shuttle 'A' the heat
ansferred is much higher due to the greater temperature differential
tween heat source and sink. The heat recovery at a mean water temperature
the manifold of 30°C is 195KW.

AT SHUTTLE 'C'

fers to a situation where there is partial condensate return and the heat
covered can be collected as in Case 'B' above except that in some cases
is might increase the feed water temperature too high. Where this
tuation can occur it is prudent to install a preheat tank which can be
ed to top up the feed water tank and any surplus water/heat used for
ocess requirements. If steam injection has been from 30 to 90°C then
KW have been injected costing a 88KW burn of fuel (assuming 75%)
ficiency of boiler. Thus, on a continuous basis application, in a total
ss situation, a heat recovery unit on low pressure into feed tank saves
,268 per annum more than a standard economiser applied to the high
essure side. This is not the end of the story, however, since the
plication of the economiser requires:-

(1) Full pressure vessel approval i.e. additional manufacturing
     cost.

(2) Modulating feed pump costing £2,500.

(3) Large size units cost more to install.

us application of economiser would cost £5,000 to £8,000 more and save
ss than a standard 'Heat Shuttle' of the same output.

ble 1 (overleaf) compares the costs and savings of the three types of
stallation.

TABLE 1

ECONOMISER OR HEAT SHUTTLE?

	(A) Economiser	(B) Low pressure into cold hotwell	(C) Low pressure into pre-heat tank to feed existing calorifier or for process
Cost of Unit (£)	8 000	8 000	8 000
High Pressure Manifold (£)	1 000	300	300
Cost of Modulating Feed (£)	2 500		
Cost of Pre-heat Tank (£)			1 000
Budget Installation Circuit (£)	3 500	2 000	2 500
By-pass for Low Fire and all Controls (£)	1 500	1 500	1 500
TOTAL COST	**£16 500**	**£11 800**	**£13 300**
Temperature of Water Entering Unit (°C)	90	30	60
Recovery KW	130	195	160
Savings per year (Continuous) (£) (Boiler efficiency 80%) Gas at 1.2p/KW	16 170	24 438	20 212
Therefore, Payback (Years)	1.02	0.48	0.65

Payback quicker (more savings achieved) if the heat recovered can be used for applications (B) and (C).

i.e. (B) £8 268 more savings than (A)

(C) £4 042 more savings than (A)

Where an installation is to be made on a boiler with dual fuelling, a standard economiser might be able to recover only £12,000 of heat, if the operator is required to fire on fuel other than gas for three months.

PROCESS WATER

The main application for the steam plant might be to provide heat to process water, for example, steam injection into laundry washing machines. The time taken to heat up the water dictates the process cycle time. If the water introduced to the washing machine is preheated by a low pressure economiser, not only is energy saved by preheating the water, but time is saved also, resulting in one case of the reduction in process cycle time of 10%. This meant the operation could effect one extra wash per shift. A high pressure economiser not only would have recovered less heat from the same unit, but would not have had the ability to improve productivity.

There are no hard and fast rules, it is essential that the process cycles are evaluated and the optimum implementation of heat recovery taken. In all cases the addition of an economiser will provide improved efficiency, but in most cases, a low pressure installation has superior results to a normally used standard high pressure economiser.

CASE HISTORIES

At the time of writing over 1500 'Heat Shuttles' heat recovery units have been installed, varying from small 100mm diameter domestic sized units to the largest so far manufactured, a unit for the 1m diameter flue of a 15000 Kg/hr steam boiler.

A programme to evaluate any boiler heat recovery has been developed which evaluates the heat recovered, the size and length of heat pipes and number of heat pipes per row and rows required give off infinite flexibility in design.

1) **EXHAUST FROM THERMAL FLUID HEATER**

Exhaust Temperature	515°C
Output	733 KW
Flue Diameter	381mm

15 row heat pipe heat recovery unit of 10 heat pipes per row each
of length 530mm recover 100 KW/hr on a 10 hour day.

Total installed cost of the equipment was £4,000 resulting in a
capital cost of £40/KW recovery.

HEAT RECOVERY DESIGN 86/03/33   DCA FROZEN FOODS

Exhaust Temperature Degree 'C'	515
Boiler size or output BTU/Hr	2500000
Diameter of Flue Inches	15
Carbon Dioxide Percentage	11
Fuel Used	Oil
Pipes per Row	10.0
Length of each pipe Inches	21.8
Number of Rows	15
Water Return Temperature Degree C	70
Efficiency	66.6%
Energy Wasted due to Latent Heat	225172.5 BTU/Hr
Energy Wasted (Sensible Heat)	1027703.1 BTU/hr

ACC.HEAT RECOVERED BTU/HR	TEMP. INPUT DEG. C	TEMP. OUT DEG. C	HEAT RECOVERED PER ROW BTU/HR
36894.3	515.0	496.5	36894.3
71848.0	496.5	479.4	34958.8
↓	↓	↓	↓
385023.3	350.6	342.5	20510.9
404780.1	342.5	334.7	19756.8

Reduction in Burner Firing   -   607635.8 BTU/Hr

Price of Unit =   £2,588.72

Photograph of the unit is shown in **Figure 2**

2) **EXHAUST FROM STEAM PLANT BOILER**

Exhaust Temperature	212°C
Boiler Output	13600 KG/hr
Flue Diameter	1250mm

15 rows of heat pipes, 25 per row, each of length 1220mm recover 350 KW/hr.

Total installed cost of the equipment was £15,000 resulting in a capital cost of £43/KW recovery.

HEAT RECOVERY DESIGN 86/04/44   STEPHENSON SOAP COMPANY

Exhaust Temperature Degree C	212
Boiler Size or output BTU/hr	30000000
Diameter of Flue Inches	39
Carbon Dioxide Percentage	9
Fuel Used	Gas
Pipes per Row	27.0
Length of each pipe Inches	51.9
Number of Rows	15
Water Manifold Temperature Degree C	50
Efficiency	81.6%
Energy Wasted due to Latent Heat	3.67m BTU/hr
Energy Wasted (Sensible Heat)	3.09m BTU/hr

ACC.HEAT RECOVERED BTU/HR	TEMP. INPUT DEG. C	TEMP. OUT DEG. C	HEAT RECOVERED PER ROW BTU/HR
116960	212.0 204	204.0	116960.7
↓	↓	↓	↓
1120305	143.3	140.0	54900.3
1162615	140.0	137.0	52310.1

Reduction in Burner Firing   -   1.4m BTU/hr

Estimated Annual Savings   =   4.8%

Price of Unit   =   £12,259.93

Photograph of the unit is shown in **Figure 3**

3) **EXHAUST FROM A BOILER IN A SWIMMING POOL**

Exhaust Temperature	325°C
Output	250 KW
Flue Diameter	300mm

20 row heat pipe heat recovery unit of 8 heat pipes per row each of length 500mm recover 40 KW/hr.

Total installed cost of the equipment was £2,500 resulting in a capital cost of £62.50 /KW recovery.

HEAT RECOVERY DESIGN 86/01/22   BILSTON SWIMMING POOL

Exhaust Temperature Degree 'C'	325
Boiler size or output BTU/Hr	850000
Diameter of Flue Inches	12
Carbon Dioxide Percentage	9
Fuel Used	Oil
Pipes per Row	8.0
Length of each pipe Inches	18.0
Number of Rows	20
Water Return Temperature Degree C	70
Efficiency	74.5%
Energy Wasted due to Latent Heat	68422.1 BTU/Hr
Energy Wasted (Sensible Heat)	221945.7 BTU/Hr

ACC.HEAT RECOVERED BTU/HR	TEMP. INPUT DEG. C	TEMP. OUT DEG. C	HEAT RECOVERED PER ROW BTU/HR
11621.0	325.0	808.0	11621.0
22263.8	308.0	292.8	10642.8
↓	↓	↓	↓
123903.6	171.2	167.3	3581.1
127310.0	167.3	163.6	3406.4

Reduction in Burner Firing	–	170800.3 BTU/Hr
Estimated Annual Savings	=	20.1%
Price of Unit	=	£2,285.04

Photograph of the unit is shown in **Figure 4**

FIGURE 2   —   HEAT RECOVERY UNIT WITH BUILT IN BY-PASS ON
THERMAL FLUID HEATER

FIGURE 3   —   HEAT RECOVERY UNIT ON 30,000 LB/HR
STEAM BOILER

FIGURE 4 — HEAT RECOVERY UNIT ON SWIMMING POOL BOILER

Energy Saving Measures in Pre-1919 Terrace Housing

A. J. A. Sluce* and D. Tong**

*BDP Energy and Environment, Sunlight House, Quay Street,
Manchester M60 3JA
** Building Use Studies Ltd., 13–16 Stephenson Way, London NW1 2HD

ABSTRACT

A sample of 44 households were monitored to establish their energy consumption
the temperatures they kept their houses at and various socio-economic factors.
Half of the sample were living in houses refurbished to the standards current
in 1985 and half had been refurbished to the same level but with the
following additional measures :

Internal insulation applied to external walls
Double glazing of downstairs rooms
Draught stripping of doors and windows
Extract fans controlled by humidistat.

The results showed that allowing for differences in use for cooking, hot
water and lights, the measures saved the occupants £45 a year or 10% of the
energy costs. No disbenefits were observed from the use of the measures and
they could all be installed satisfactorily in most situations. However,
supervision is needed to achieve success in all circumstances. The tenants
experienced problems in obtaining feedback on energy consumption in a short
enough time scale to be useful and in operating the controls provided by
manufacturers of thermostats, boilers and time clocks. The variability
between households' energy consumption was very wide and many examples of
wasteful usage were observed. In an attempt to reduce this variability an
energy awareness campaign has been initiated.

KEY WORDS

Energy saving; internal wall insulation; energy awareness; domestic energy
monitoring.

INTRODUCTION

The purpose of this Energy Efficiency Office, Energy Efficiency Demonstration
Scheme Project (supervised by BRECSU) was to demonstrate the benefit of a
package of energy saving measures applied to pre 1919, solid wall terrace
housing. The measures were as follows :

267

. Internal insulation applied to external walls
. Double glazing of downstairs rooms
. Draught stripping of doors and windows
. Extract fans controlled by a humidistat

They are appropriate to any solid-walled dwelling in both the public and private sector. The stock of pre 1919 dwellings is 4.75m of which 25% are in need of refurbishment. The replication potential for these measures is correspondingly high.

The actual benefit of the measures observed in this project will be affected by the behaviour of the households. To compensate for this source of variability the temperatures maintained in the houses and the socio economic circumstances of the occupants have been monitored. In view of the level of awareness of energy matters observed as a result of this monitoring an extension to the project was initiated to measure the effect of increasing the energy awareness of the occupants.

The host organisation was Merseyside Improved Houses (MIH), a housing association owning 11,000 houses in the Liverpool area. The samples examined were located in two separate areas in Kirkdale and Linacre supervised by the North and Sefton offices, of the association, respectively. The monitoring contractor was BDP Energy and Environment with Building Use Studies Ltd (BUS) sub-contracted to carry out the socio economic monitoring and the energy awareness campaign.

DESCRIPTION OF THE PROJECT

This demonstration was based on a comparison of the energy consumption of two groups of houses. The first, called the controls, consisted of houses refurbished to the current standards of the housing association in 1985 and the second, called the trials, consisted of houses refurbished to the curren standard but including additional energy saving features as described in the introduction. The energy consumption was established by recording the gas and electricity consumption in each house. This was carried out on a fortnightly basis so that fluctuations in the consumption could be observed and related to the activities in each house. No sub-metering was used so that the energy use for space heating had to be established by subtracting estimates of the energy used for cooking, water, heating and lights from the total. Because of the frequency of recording it was possible to make estimates of these loads.

The success of a demonstration of this type depends on obtaining two samples which are alike in all aspects except those aspects being tested. Given the highly variable nature of the building stock and its occupants this is difficult to achieve. To help to explain the undesirable variation between the samples, a number of other variables have been measured. These can be summarised as follows :

. Temperatures in the four principle rooms
. Outside air temperatures
. Principle dimensions of the houses
. A number of sociological factors

The measures adopted in the trial houses can also have an effect on the amount of maintenance required. This could not be ascertained during the span of the project. However, an important influence on the amount of maintenance required, both to decorations and to the fabric of the house, is the build up of condensation. In view of this fact attention was paid to the occurrence of condensation as a predictor of likely maintenance needs.

The measures being demonstrated should be capable of being installed under normal building site conditions. The installation of the measures, in particular the internal wall insulation, was observed by BRE personnel. The success achieved in reducing the amount of involuntary air infiltration was measured by pressure testing a sample of the houses.

The Physical Monitoring

The gas and electricity meters were read on a fortnightly basis. From the readings in July and August the consumption for water heating, cooking, lights and other electrical loads was estimated. The space heating load was calculated from the fortnightly readings minus the estimated base load.

Inside air temperature was recorded in four rooms in each house - two downstairs and two upstairs. The sensors were mounted in pattress boxes and wired back to Squirrel data loggers. In the control sample the wiring was surface mounted and the logger located under the stairs. In the trial sample the wiring was installed in conduits and the logger installed in a cupboard in the front room downstairs. The loggers were down loaded fortnightly when the meter readings were carried out. The result was a set of energy and temperature data which could be directly related.

At three locations the outside air temperature was measured with a single channel recorder in an enclosure on a garden wall. These were downloaded fortnightly when the house was visited.

The data from all the temperature monitoring was processed using the software provided by Grant and additional software written by BDP for creating average house temperatures, average difference between house temperature and outside air temperature and degree days occurring during the monitoring period.

In addition to the regular monitoring described above, a number of surveys were carried out. For example :

.   Observation of the installation of the measures
.   Pressure testing of sample of control and trial houses to establish benefit of draught stripping
.   Modelling of the benefit of the measures using the BREDEM model and the ESP program.

The Sociological Monitoring

It is now generally accepted that the behaviour of building occupants and their demographic characteristics are important influences on energy consumption. The purpose of the sociological survey was to monitor these influences and account for variations in consumption other than those resulting from the trial measures. It also assessed the benefits of the trial measures to householders in qualitative terms.

The survey comprised two interviews with tenants of the trial and control properties. The first interview was scheduled for the heating season of 1984/5, prior to completion of the trial modernisations, the second for the heating season, 1985/6. Additional data was collected fortnightly during the monitoring period in the form of an 'energy diary' for each household.

This was completed by the fieldworker when he visited each house to 'down load' the monitoring equipment.

The variables included in the survey covered the following :

. Household size e.g. number of adults, children and elderley
. Employment status of adults
. Self reported income level
. Previous method of space heating
. Satisfaction with space and water heating
. Use of central heating and methods of control, if used
. Demand for water heating (on scale 1-5)

For the above it is apparent that the ability of the field worker (Mr D. Bird was critical to the success of both the physical and the sociological monitoring. In the event he has achieved a very high standard; establishing good relations with the households and coping with the vagaries of the data loggers; which is a major contribution to the success of the project.

RESULTS

The installation of the measures.

With the exception of internal wall insulation the measures are suitable for installation without disrupting the household. The internal wall insulation is only appropriate to a major refurbishment as it requires stripping of existing plaster and replacing door and window linings, light switches and power sockets. Although acceptable finishes were achieved for the tenants the pressure testing indicating that there was considerable variation in the air tightness obtained from the measures. To reduce this variation would require considerable effort in studying each house to find the source of the leakage.

The average cost of the measures installed in the houses was as follows :

		£
Double glazing	:	170
Extract fans	:	90
Draught stripping	:	140
Wall insulation	:	712
Additional roof insulation	:	30
TOTAL :		1142

The total cost of the average refurbishment was £18,800.

Temperatures

The Squirrel loggers proved to be robust enough to get consistent sets of data from each house. There were occasional crashes and some sensors almost never recorded correctly. The effect of these lacunae was not significant as the houses exhibited consistent patterns of temperatures in the majority of cases. The temperatures were processed to create average house temperatures and average temperature difference between the inside and outsi temperatures per day (called the house degree day). These average temperatu have been plotted on graph 1 together with the annual energy consumption for space heating for each house.

The temperatures confirmed the pattern, for many of the households, of one room being heated and the remainder floating between outside air temperature and the temperature of the heated room. The average temperature in the control houses was less than that in the trial houses by 4°K. Within these averages there is considerable overlap but none of the control houses were able to get as warm as the warmest 30% of the trial houses (see graph 1). The energy saving measures allowed the tenants with the life styles of this sample to be considerably warmer. This was confirmed by the social surveys where the majority of the tenants said they felt that they were warm in the trial houses.

Energy Consumption

The energy consumption was based on fortnightly meter readings. Although some meters had to be replaced during the study little data was lost. The meter readings provided surprisingly regular patterns of energy use which could be interpreted into the loads for cooking, hot water and lighting.

For each household the values interpolated for non-heating purposes could be confirmed by the social surveys. There were obviously some atypical results but most families with children used more energy than the single, elderly households for example.

The average trial household used less energy than the average control household (4768 kWh, a £92 saving). The social survey had highlighted the fact that control households were probably going to use more energy for non heating purposes because of family composition. Allowing for the non heating purposes the space heating saving is then £42 on the average, taking the consumption in the controls of £160 down to £118 for the trials. This represents a saving of gas and electricity; the gas consumption reducing by 2417 kWh and the electricity by 286 kWh.

The measures have obviously made significant savings for the trial households but not of the level predicted before the project. The reason is the way the houses were occupied by the predominantly low income households. Very few were using the central heating for whole house heating. The preferred method was to heat one room. This alters the way the heat is lost from the house. The traditional approach has been to assume that heat is lost through the fabric and by ventilation. A saving in either of these losses is then taken as a benefit. In practise, with a single room being heated the heat is lost from that room to the outside and to the rest of the house. The heat going to the rest of the house can be the major loss from the occupied room but is a small quantity of heat which can easily be removed by the air changes in the rest of the house without any heat being lost through the fabric. In this case the insulation to a large part of the external envelope yields little benefit in saving heat loss.

During the study it was found that the trial houses enjoyed considerable benefit from the higher temperatures obtained in the unheated parts of the house due to solar gain. So the wall insulation does benefit the household, but does not appear as an energy saving in the comparison carried out here. A more accurate comparison would have been to heat the control houses to the same level as the trial houses and see how much extra energy was needed to achieve these levels. From graph 1 it is apparent that considerable additional energy would be required in some houses.

Sociological Factors

The surveys were conducted successfully and benefited from the rapport which had been built up by the field worker. It is important to select the right person for this job. David Bird was a social science graduate from Liverpool University and lived locally. Other surveys which have made use of people from other disciplines and not from the locality have run into difficulty due to getting negative responses from the households being surveyed. Table 1 lists a sample of the results from the surveys and gives a feel for the type of data available.

The trial group were coming, in large part, from unsatisfactory accommodation and had little experience of central heating. In the first winter they have made little use of the central heating on time clock control. The reason given for this, in many cases, was fear of high bills. The controls having been in their houses longer had more idea of what heating bills they would receive and had adopted methods for budgeting for this in many cases. Even so, only half the group used the central heating under time clock control. Two messages emerge from these facts. One was that the controls on the heating systems are not easily understood and people rely on using the on/off switch to control the system when they think they need it. The other was that many households would benefit from more direct feedback about their energy consumption so that they could purchase the heat they could afford. Quarterly billing allowed large bills to be run up which the tenants had difficulty paying, in some cases.

Despite the lack of use of the central heating the trial households reported that they had found their houses warm. From the temperature readings it is clear that they were warmer. 60% of the trials did not heat the bedrooms as against 40% of the controls, which reinforces the benefit the trials were getting from the measures.

From the survey it was clear that the heating systems should be matched to the needs of the tenants. Central heating was clearly not what they required. The preferred mode of operation seemed to be a gas fire in the occupied room which could be switched on and off as required. Many households used the immersion heater for heating water which is more expensive than using gas. The true cost depends on the timing and quantity of water used. The reason for the use of electricity was the fact that the control was easily understood. An instantaneous gas water heater would probably suit the needs of many of the households. The combination of gas fires and an instantaneous gas water heater would be an acceptable provision provided the capital and maintenance costs were less than the conventional boiler, radiators and water storage cylinder. To further test these ideas an energy awareness campaign has been initiated starting in January 1987.

CONCLUSIONS

The method adopted for monitoring using four channel squirrel recorders and fortnightly visit for downloading of recorders, meter reading and social survey worked well. It represents a cost effective approach, provided a suitable field worker can be found. The households found the level of intrusion acceptable in general. Cash grants were made to them to cover the inconvenience and redecoration costs.

The pressure testing highlighted the leakiness of the houses. Further work is needed to interpret the figures. The absence of condensation in the houses can be attributed to the number of air changes. It is interesting that in a sample of similar houses draught stripping has led to condensation in a number of cases.

The measures increased the comfort of the occupants; saved them money; reduced the risk of condensation and gave occupants the ability to choose what heating level they wanted to pay for. In particular, the internal wall insulation fitted the houses for changes in expectations about thermal comfort which are bound to occur in the next thirty years.

Households of the type studied here have difficulty in using the controls on equipment. Examples were found of boilers which could not be relighted because of the contortions required and controls which could not be set because of the reduction in manual dexterity caused by age. There is a need for manufacturers to produce controls which are as easily understood as an on/off switch. Many of the heating systems were used in this way despite the controls which were available. The difficulty of using the controls led to electricity being used for water heating when gas would have been cheaper. In view of these problems and the wide range between apparently similar households an energy awareness campaign has been initiated.

The project is due to finish in July 1987 and a fuller discussion of the results will be possible then.

ACKNOWLEDGEMENTS

Our thanks are due to the following :

The representatives of the host organisation who have been most involved with the project
Mr S Porter   Director, North Area of MIH
Mr J Lennon   Development Manager, North Area of MIH
Mr A d'Henin   Development Manager, Sefton Area of MIH

The project officer at BRESCU (Building Research Energy Conservation Support Unit)
Mr A D Russell

The representative of Pilkingtons plc the supplier of the internal wall insulation
Mr J Alderson

REFERENCES

Porter, S (1987) MIH/BRESCU Energy Demonstration Project, Voluntary
          Housing, March 1987
BRESCU   (1986) Project Profile No 209, BRESCU, Watford
Ward, D   (1987) Building Research Establishment private communication

274

GRAPH 1 - Average Temperatures and
Annual Energy - July 1985 - July 1986

Space Heating
Annual Energy Consumption
in kWh (x1000)

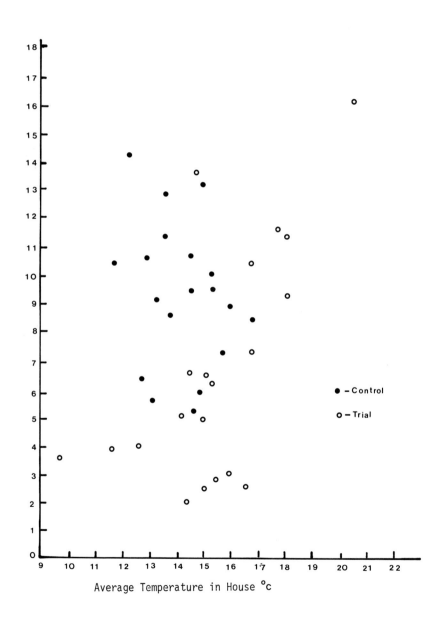

Average Temperature in House °c

TABLE 1 - Sample of Results from the Social Surveys

House Number	Annual Energy kWh	Numbers in Household			Number of Radiators	Employment Status
		Total	Children	Elderly		
CONTROLS						
1	42348	3	2	0	3	PT
2	30537	4	2	0	4	-
3	24939	4	0	2	3	FT
4	23473	3	0	1	7	FT
5	22340	2	0	1	3	-
6	21831	1	0	1	6	-
7	20448	1	0	1	4	-
8	20371	2	0	1	8	PT
9	19334	2	1	0	3	-
10	19280	2	1	0	3	PT
11	18189	2	0	2	4	-
12	17569	5	3	0	4	-
13	17359	4	2	0	7	-
14	16984	2	1	0	3	-
15	16239	2	0	2	7	-
16	15429	1	0	1	3	-
17	14959	4	2	0	4	FT
18	13762	2	0	1	3	FT
19	13116	1	0	1	3	-
20	12881	4	2	0	3	-
TRIALS						
1	39411	4	2	0	3	FT
2	24031	3	1	0	7	FT
3	23869	4	2	0	3	FT
4	23485	4	1	2	7	-
5	19906	5	3	0	8	-
6	17658	2	1	0	0	-
7	17586	5	0	0	3	-
8	15394	2	0	2	3	FT
9	13405	3	0	2	3	-
10	13322	2	1	0	3	-
11	13310	3	0	0	3	-
12	12625	2	1	0	3	-
13	12543	1	0	0	7	PT
14	12473	2	1	0	7	FT
15	12168	2	0	2	7	-
16	11957	1	0	0	3	-
17	11776	1	0	1	3	-
18	10905	1	0	0	3	-
19	10790	1	0	1	3	-
20	10043	2	1	0	7	-
21	9032	3	0	0	6	-
22	7316	1	0	1	0	-

This is a technology package some one could use. Several industrial examples

Combining Thermal Storage and Heat Pumps in Industrial Cooling Systems

M. F. Söderström

*Division of Energy Systems, Department of Mechanical Engineering
Linköping Institute of Technology, Sweden*

ABSTRACT

Industrial cooling and heating loads may vary substantially with time. This may lead to systems having complex control equipment, high investment cost and low COP for heat pumps in situations with partial load. Including thermal storage in industrial cooling systems to decrease the power variations results in improved system performance and reduced cooling machine/heat pump investment.

A computerized analysis method is used to find the best combination of thermal storage and heat pump to reach a good profitability. The method is applied to on site collected data from some industries. It is shown that profitable applications of thermal storage are mostly found on the cold side of the heat pump.

KEYWORDS

Thermal energy storage; industry; heat pump; sizing; net present value; analysis method; cooling system.

INTRODUCTION

Applications for thermal energy storage (TES) may be found throughout industry (see e.g. Söderström, 1986 and Lambrecht and Söderström, 1987). Here the interest is focussed on the combination of thermal storage and cooling machines/heat pumps in industry.

Using heat pumps in industrial cooling systems results in less variations due to the climate and decreased water costs compared to ordinary systems. Heat output may be used for tap water heating and for the heating of premises thus saving primary energy for these needs.

277

Varying loads create problems on the heating side as well as on the cooling side of the heat pump. The partial load problem on the cooling side can be solved by splitting the maximum cooling power or two or more heat pumps. On the heating side an extra boiler may be used to cover a deficit, in a surplus situation the energy may be cooled off to atmosphere or an external water flow. This way of solving the problems leads to complex control equipment, high investment cost and low COP in situations with partial load.

Using TES as a buffer before the evaporator of the heat pump decreases the cooling power needed in the heat pump and thus the investment. There might be a need for a flow rate buffer as well if the flow rate variation is large. On the heating side the TES plays the role of moving excess heat produced by the heat pump from production hours to non-production hours.

PERFORMANCE OF A THERMAL STORAGE UNIT

When a thermal storage is used as a buffer the goal is to achieve a non-varying output from the storage (Fig. 1a). To reach this a certain storage capacity and power transfer is needed. Decreasing the storage capacity has the effect shown in Fig. 1b. When decreasing storage capacity, the "saw tooth" is moved to the left. The influence on output power from decreased charging and discharging power is shown in Fig. 1c. The lower the power transfer the more will the power level out of the storage vary.

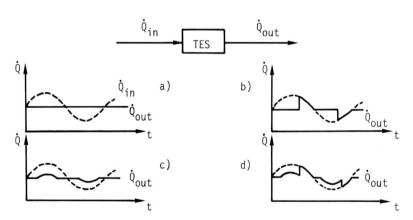

Fig. 1.   The influence of different buffer capacities and power transfer on the power level from storage, $\dot{Q}_{out}$. a) Ideal situation, b) Decreased capacity, c) Decreased power transfer (charging and discharging) and d) Combined effects.

The combined effect of decreased storage capacity and charging /discharging power is shown in Fig. 1d. Power levels and temperature differences determine the heat exchange area of the

TES. For a given heat exchange (charging/discharging fluid to storage material) geometry this in turn determines the capacity and thus the investment.

ANALYSIS METHOD

The thermal storage must be designed to have the right storage capacity, charging and discharging power levels with respect to the heat source power variations and the cold sink power variations.

Power levels from the source and the sink are used to evaluate the need for storage capacity and storage power. The input (heat source power) is to be changed to an output (cold sink power demand) by the use of an intermediate storage unit.

In a first run the capacity and power requirements to fulfil these needs are calculated. Succeedingly limitations in capacity and power levels for the storage are included to simulate the influence on system performance.

In Fig. 2 the method is schematically shown. A thorough description is given by Söderström (1986).

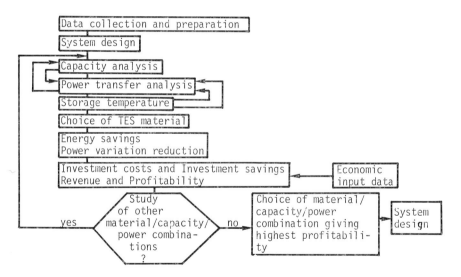

Fig 2.   Analysis method. Schematic structure.

To analyze the profitability the net present value method is used. Annual revenue, R, of a certain installed storage capacity, C, could be written as:

$$R(C) = N \cdot p \sum_{Q=0}^{C} Q \cdot Pr(Q) + N \cdot p \sum_{Q=C}^{I_{max}} C \cdot Pr(Q) \qquad (1)$$

IEE-S

where N is the number of cycles annually, p the energy price, Q the capacity needed for every cycle and Pr(Q) the probability that this capacity is needed, I_{max} is the largest capacity needed in the actual application.

The investment cost as a function of storage capacity and/or power transfer is compared to the net present value of all costs and revenues during the life time of the project as a function of the storage capacity and power levels. The profitability, P, for a certain capacity (C) - power transfer (Ċ) combination is:

$$P_{C,\dot{C}} = NPV(R_{C,\dot{C}} - K_{C,\dot{C}}) - INV_{C,\dot{C}} \qquad (2)$$

where K is the annual operation and maintenance cost and INV the investment.

In the buffer storage case (cooling side of the heat pump/ /cooling machine) the revenue is calculated as the decreased heat pump investment due to reduced cooling power variation.

CASE STUDIES

Three industrial cooling systems have been studied; One in a plastics industry, one in a foundry and one in a heat treatment shop.

Plastics Industry

Cooling water is used to cool the moulds and hydraulic pumps of a set of injection moulding machines. The existing system includes a cooling tower to cool the water. This leads to large variations in cooling water temperature during the year, which affects the quality of the products.

Today the cooling water temperature to the machines varies around 18 ^{o}C. With a new system this temperature will be 8 ^{o}C and the temperature after the machines will vary around 14 ^{o}C.

The maximum cooling power needed is almost 200 kW and the average is 120 kW. Using a buffer storage to effect this reduction would need a capacity of 730 MJ and a charging and discharging power of 75 kW.

The storage material will be a latent one having a phase change temperature of about 14 ^{o}C. Investment cost for the thermal storage is 50 kSEK at 70 SEK/MJ. The reduced cooling power requirement will lead to a reduction in investment of 160 kSEK.

Presently the heating of premises and tap hot water requires 2.1 TJ annually, while 2.7 TJ is lost via the cooling tower.

The heating power from the heat pump at 60 ^{o}C will be 160 kW (COP = 4). The heat pump will cover the heat demand of the premises (transmission and ventilation) down to -4 ^{o}C during

daytime (6 am to 10 pm) and down to -11 °C during nighttime
(10 pm to 6 am). The annual savings is 0.5 TJ corresponding to
42 kSEK at an energy price of 0.3 SEK/kWh.

An energy storage could cover the energy demand of the premises down to -4 °C during weekends. This would require a
thermal storage capacity of 16.2 GJ saving about 310 GJ annually. Using surplus energy from nighttime for use during daytime the annual savings is 16 GJ requiring a storage capacity
of 220 MJ. Thus, the durations are far too short to make a
thermal storage investment for the heating of premisis economically possible.

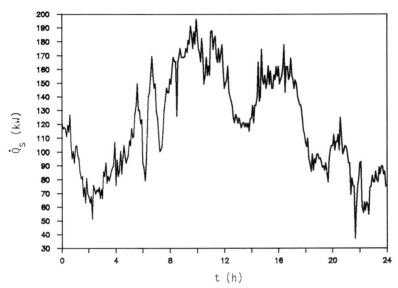

t (h)

Fig. 3. Cooling power variation in a plastics industry.

Aluminium foundry

In an aluminium foundry employing die casting the cooling
water flow rate is about 60 m³/h. The cooling power varies
during a week according to Fig. 4. The present system uses a
water-to-air heat exchanger (for the heating of premises) and
an outdoor pond for the cooling of the water. This leads to
unsufficient cooling and sometimes contamination of the water.

In an improved situation a closed circuit system with a heat
pump/cooling machine is used. Fig. 4 shows that cooling power
varies between 0 and 800 kW during a week. The average cooling
power is 385 kW. The temperature varies ± 8°C around 22 °C.

The required storage capacity, C, to reach at a stable average
cooling power is 12.5 GJ. The utilization of such a storage is
shown in Fig. 5 together with the utilization curves for 11 and
10 GJ respectively. The application seems to be a good one as

the storage capacity will be quite evenly utilized over all the inventory (I) range. The inventory is the accumulated energy stored in the TES.

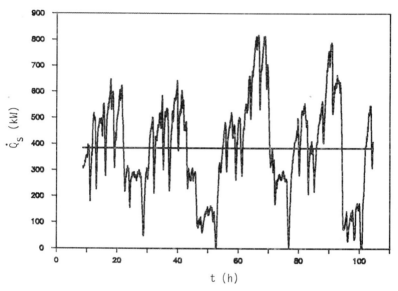

Fig.4. Cooling power needed in the cooling water system.

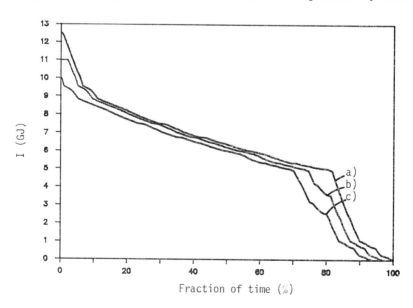

Fig. 5. Capacity utilization for C = 12.5 GJ (a), 11 GJ (b), 10 GJ (c).

The maximum charging power used is 434 kW and the maximum
discharging power is 415 kW. Using a storage capacity of
12.5 GJ. The power duration curves for charging and discharging are shown in Fig. 6 and 7 respectively. The charging
power duration as well as discharging power duration are far
from the ideal as regards duration time at high power level.
The length of the studied period (in this case a week) is used
as time base for these curves as well as for the capacity utilization in Fig. 5. The added charging and discharging power
duration is close to 100 %, i e the storage is working almost
all the time.

Reducing power transfer levels has a severe impact on the
power out of storage. A charging and discharging power of
350 kW will result in a power out of storage varying from 320
to 475 kW at a storage capacity of 12.5 GJ. At a capacity of
12 GJ and a power transfer of 434 kW the power out of storage
will vary from 385 to 640 kW.

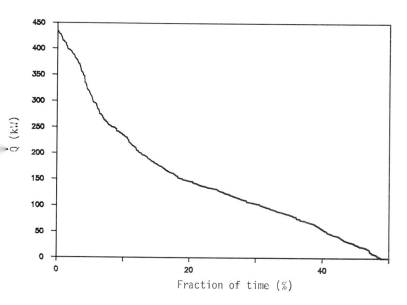

Fig. 6.   Charging power duration for C = 12.5 GJ.

s storage material a phase change material is chosen with a
hase change temperature of 22 °C.

he profitability, P, of the investment is shown in Fig. 8 as
 function of storage capacity, C. The storage cost is
50 SEK/kWh and the cooling machine cost is assumed to be
EK 2 000 per kW of cooling power. The maximum profitability
s 235 kSEK at C = 12.5 GJ.

he cost of storage materials and the investment cost of the
ooling machine influence the profitability. Decreasing the

284

heat pump investment to 1.4 kSEK/kW the profitability becomes
zero. Increasing the storage cost to 280 SEK/kWh gives the
same effect. Using the heat from the cooling machine to heat
premises and tap hot water would save about 150 m^3 of oil at a
value of 0.3 MSEK, increasing the profitability substantially.

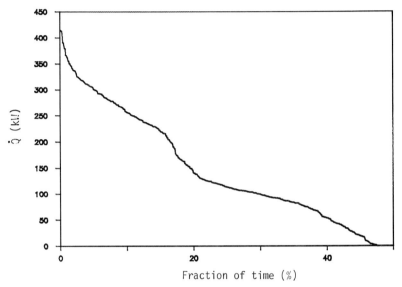

Fig. 7. Discharging power durations for C = 12.5 GJ.

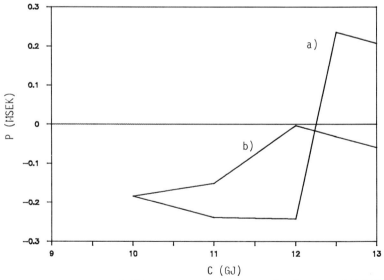

Fig. 8. Profitability as a function of storage capacity for
two power transfer levels a) 434 kW and b) 300 kW.

The temperature duration for low temperatures during non-production hours is too short to make a TES on the heating side
of the heat pump economically possible.

Heat Treatment Shop

The quenching fluids are cooled using circulating water.
Cooling power varies between 125 and 275 kW. The cooling power
variation is built up of flow rate and temperature variations.
To reach the average cooling power level, 174 kW, a flow rate
buffer as well as a temperature buffer will be needed.

The flow rate buffer will consist of a water tank having a
volume of 10 m^3. Incorporating this tank will have the effect
on cooling power variation shown in Fig. 9. The power varies
between 150 and 220 kW.

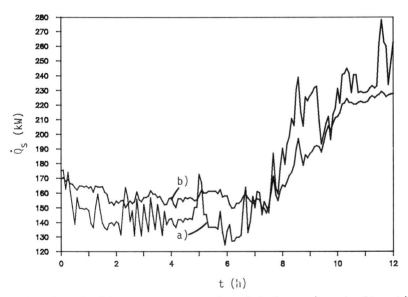

Fig. 9.   Cooling water power level before a) and after b)
          installation of the flow rate buffer.

The investment in the water tank is 30 kSEK and the profitability about 70 kSEK (reduced heat pump investment).

A temperature buffer to levelize the temperature to 20 °C is
not profitable. Such a buffer would be working with very small
temperature differences at both charging and discharging.

In this case the heat from the cooling machine could not be
used within the plant as the needs for heating of premises is
very small (only some holidays during winter). The heat could,
however, be used by a neighbour plant, which could take care

of all the heat the year around (for the heating of processes).
This would save 6 TJ annually corresponding to 400 kSEK.

CONCLUSIONS

The profitable use of thermal storage in combinations with
heat pumps seems to be on the cold side of the apparatus. Here
the pay back period is very short as the investment in the heat
pump is directly reduced by the thermal storage investment.

On the heating side the annual number of cycles are too few,
at least if the heat sink is the heating of premises. This is
of course due to the heat pump alone being able to satisfy a
large part of the heating demand from the premises.

The author wishes to thank the Swedish National Board for
Technical Development for financial support of this project.

REFERENCES

Söderström, M. (1986). Evaluation Method of Industrial Thermal
   Storges. Linköping Studies in Science and Technology.
   Dissertations. No. 146.
Lambrecht, J. and Söderström, M. Einsatzmöglichkeiten von
   Wärmespeichern in der Industrie. (Heat Storage Application
   Possibilities in Industry.) To appear in the July issue
   (1987) of BWK (Brennstoff Wärme Kraft).

Saving Megabucks by Saving Megawatts

Andrew Warren

*Association for the Conservation of Energy, 9 Sherlock Mews,
London W1M 3RH*

The use of differing criteria between investment in the efficient use, and investment in the supply of energy, has led to a gross mis-allocation of resources in Britain. It is vital to establish a satisfactory mechanism for ensuring preference for the energy resource which involves its provider, and therefore its ultimate consumer, in the least cost.

The first report of the newly established House of Commons Select Committee on Energy, published back in 1981, stated:

"We were dismayed to find that seven years after the first major oil price increases, the Department of Energy has no clear idea of whether investing around £1,300 million in a single nuclear plant (or a smaller but still important amount in a fossil fuel station) is as cost-effective as spending a similar sum to promote energy conservation."

The objective of this lecture is to answer the question whether investment in energy conservation can be comparable with, and be a viable alternative to investment in new supply capacity; and if so, how this can be achieved in Britain.

Why is this issue important? The question of whether conservation and new supply can be compared is not esoteric debate over recent years: the equivalent of between 40-45% of the UK's total industrial investment has been committed to major energy supply projects, and one energy industry alone - electricity - has announced that to serve England and Wales, 9 new power stations will be required within the next dozen years at a cost of some £17 billion. I shall therefore be addressing today primarily the role of the utility in this regard.

In order to ensure that investment monies are committed to the most cost effective resource, it is vitally important both in the national interest and in that of energy consumers to know whether the deliberate promotion of energy conservation programmes offers a lower cost alternative to investment in new supply.

This is not a theoretical concept. There is a growing body of expertise and experience, both in North America and now in Europe, which demonstrates clearly that such conservation programmes can be introduced in a practical way, which can offer a reliable and cost-effective return on capital invested by the sponsoring utility. As George Manteatis, executive vice-president of the world's largest integrated private sector utility, Pacific Gas and Electric has concluded:
"Conservation programmes are considerably less expensive than the cost of adding new capacity, and are clearly less risky from an investment perspective".

This paper sets out to establish first, whether there is any misallocation of resources which would justify making such comparisons in Britain; second, if there is a prima facie case to suggest such misallocations, how such comparative programmes can work best; and third, what are the requirements for introducing such activities in this country?

First, is there a misallocation of resources? Surely market-forces should succeed in ensuring that energy consumers receive the correct price signals so that they purchase only the fuel they require, investing for themselves in any equipment necessary to ensure the greatest efficiency in the energy they use? In this scenario, in competitive markets the rational choices of consumers would lead to an efficient allocation of society's resources. Under such conditions, investment plans would automatically be least cost plans.

Do we live in such a world? It has to be admitted that there have in the past been Secretaries of State for Energy who have believed that rational efficiency decisions would flow from rational price signals. Unfortunately, the marketplace for energy goods and services remains an imperfect one - incomplete information, institutional limits on competition, contrasting industry structures - thus ensuring some significant impediments to the free interplay of market forces in the energy sector.

It has become manifestly clear over recent years that such a simplistic approach to energy supply and demand equations does not produce optimum results. Considerable over-capacity and over investment can ensue: witness the position in Scotland, where the South of Scotland Electricity Board currently possesses nearly twice as much generating capacity as its consumers require. I cannot but believe that had the criteria I am advocating been adopted in Scotland, there would not now be quite the same embarrasse de richesse currently enjoyed.

Witness also the battle going on within British Gas, between the technicians - determined to raise sales within four years by 16% to 20 billion therms per year by 1990, and the accountants who know that in many of its markets gas is having to compete with falling oil prices, and should therefore having either be pulling out of these markets (hopefully temporarily) or risk selling at a loss. I should like to believe too that these accountants are also looking further ahead into the early 1990s, should the 20 billion therms target be reached, and, appreciating that finding further supplies to replace the present short-term abundance will be an expensive operation, are looking to encourage demand management specifically to address that problem. I should like to believe that because it would provide a positive example of how we too in Britain could adopt the 'least cost resource acquisition' policy with benefit. I should like to believe it, but I still reckon that unless the expansionary Zeitgeist of the industry is eliminated by either the genuine pressure of the shareholders or, more likely, by a more inventionist regulatory authority than OFGAS, then such economic logic will not occur. I shall return to this later.

Most utilities have traditionally seen themselves as suppliers of a commodity, and like many other enterprises, strive towards increasing profits by increasing sales of their commodity. This has historically been accomplished by constructing new power plants or investing in new gas fields. Now of course none of the big energy suppliers would ever be so unpublic-spirited (or for that matter, so crude) as to suggest that they were not all of them absolutely supportive of the British government's 20% savings campaign. When energy conservation has been considered, it has essentially been perceived as a passive concept - something that may well increase in the fullness of time, but not a matter concerning which a utility should initiate independent policies to stimulate. And as has been shown to be a sad trait in human nature, the individuals at the top might have wondered if empire-diminishing, not building, was really in their interests. They could appreciate that, whatever the economic sense cutting back on new supplies might make both to their customers and themselves, it wasn't automatically the best recipe for carving their names in the history books.

It is now accepted by our Government that the UK is performing relatively badly in exploiting cost-effective energy efficiency opportunities. Much play has been made of our intention to rise from down the bottom of the European energy efficiency league to the top by the mid

1990s, by saving at least 20% of our current £36 billion annual consumption. It is incidentally also the target of the entire European Community to improve energy efficiency by 20% between 1983 and 1995, which would mean in effect that even if we were successful in terms of efficiency league placings we should be running very hard only to stand still.

But grasping that success will be difficult. Much has been documented about the stringent investment criteria which private industry adopts regarding discretionary expenditures like energy efficiency, as opposed to those in the mainstream 'business development' category. There is still a tendency to use short payback measures, requiring not just ultra-swift returns on capital but disregarding the continuing benefits over future years.

It is perhaps a greater indictment of government that such simplistic investment criteria also continue to be used regarding energy efficiency in the public sector, which still occupies around half the building stock.

Decisions relating to energy efficiency investment in public buildings are almost exclusively related to departmental 'cash limits', and has been well documented elsewhere - obsessive concern with controlling the Public Sector Borrowing Requirement has led to a bias against capital investment programmes in favour of revenue expenditure.

One example: the Comptroller and Auditor General found that health authorities, faced with limited resources and many calls upon them, favoured energy saving measures with a payback period of 5 years or less (equivalent to a minimum real rate of return of roughly 15%). Indeed he reported on a range of energy efficiency projects with an average payback period of 1.9 years which had still not been undertaken.

That minimum real rate of return of 15% - achieved on a payback period of 5 years - should be contrasted with that required for energy supply side investments since the 1978 White Paper, specifically that the Test Discount Rate chosen for project appraisal should be consistent with the objective of achieving the 5% real rate of return after risk on the investment programme as a whole. There would appear to be prima facie grounds for suspecting a misallocation of investment resources within different parts of the public sector.

More generally, there is evidence of a wider political consideration here and institutional bias as well. Answering Commons Select Committee's criticism on this point, the Government agreed that:-

"investments in energy supply and energy use ought to be determined according to similar criteria",

but continued,

"This does not imply that all investments with a realistic prospect of earning a 5% return will go ahead. In allocating resources to different activities, Government makes judgements on priorities between them and on priorities within the programme. Particular investments must be justifiable in all the circumstances and the balance between different objectives must be held"

This was not a distinction which Sir Frank Layfield felt able to recognise in his report on the proposed Sizewell B power station:

"The construction of Sizewell B should not preclude cost-effective public sector investment in conservation if the Government's stated project selection criteria are consistently applied. In principle, I should assume that there will not be capital rationing. If there is no capital rationing, it should be assumed that public capital will be available for those public projects which pass the Government's primary test for project selection, at present the 5% internal rate of return. It follows that public capital would be available for investment in public conservation measures which pass the 5% test."

Sir Frank did however have the grace to add:

"In practice, however, I am less certain that public conservation measures would obtain the same priority as investment in new generating capacity."

The Select Committee concurs, by proposing that the Department of Energy produce an annual list ranking the anticipated rates of return on all projects, supply and demand, with the Government's purview. The Government has not yet obliged.

I now move to answer my second question: how can this misallocation be reduced? The Department of Energy's present view, as provided in evidence to the Sizewell B Inquiry, that

"The question of whether conservation investment saves capacity or primary energy, and in what amounts, cannot be answered quantitatively in ways useful to planning."

It is the contention of this section of this lecture that the precise opposite is true. Conservation can be meaningfully forecast, and its cost can be compared to the cost of new supply capacity. Considerable expertise has been developed concerning the ways in which this fifth fuel resource can be assessed. At the Bonneville Power Administration covering the Pacific North-West, computer programmes have been developed to assess the effect of certain conservation measures upon different aspects of the known building stock. The next step is to assess the penetration rate, the rate at which savings can be achieved. The data is used in comparison with fuel price forecasts and economic indicators to drive the price response components of the BPA's load forecasting models. They also estimate the amount of conservation which would have occurred without the various incentive and promotional programmes on offer, the 'freeloader' effect.

"Conservation supply curves" are used as the basis for comparing conservation with other supply resource supply curves. An important factor in developing conservation and other supply curves is the question of when the resource will be available: new electric generating plants have a discreet construction period - some longer than others - and will only produce power when completed. Conservation, likewise, takes time to 'build', but unlike a generating plant from the outset it produces resources up to completion i.e., saturation of the conservation measures. With conservation, the 'rate of construction' penetration rate is as important as the total project length.

BPA have developed a linear programming analytical model called a least cost resource mix model. This considers annual and peak requirements of the system, and then estimates the mix of resources - whether conservation or conventional supply - which can meet these needs at the lowest cost to the region. The inputs include the latest regional load forecast, including the level of conservation assumed to occur without any extra incentive, and price induced conservation. Conservation which BPA can choose to acquire - a discretionary resource - is included in the conservation supply curve rather than the load forecast, since this is conservation which is one of the options the least cost resource mix model can choose in order to meet demand at the least cost.

Obviously such an embryonic science is still inexact - but no more so than conventional supply/demand forecasting - and it is interesting to note the growing pressure upon the US Department of Energy from state utility commissions for support for information regarding improved methods for estimating the impact of new end-use technologies.

In Europe there is only one other country which like Britain has a wide variety of indigenous energy resources, and that is Norway. It is therefore particularly instructive to look at this process as undertaken by Oslo Lysverker (literally, the Oslo City Light Company).

Since 1982 they have been running a conservation programme with a year 2000 target of reducing their gross 1980 consumption by 15%. This savings target was reached by comparing the estimated cost of saving one kilowatt hour compared to the long run marginal cost of new supplies. On a conservative basis they estimated that 15% of 1980 production would be an economic target which would be acheivable, with investments which would always be below the long run marginal cost of certain new supply options. Other supply options - like for instance two new gas fired power stations currently under consideration - were at this stage still regarded as economically feasible.

However in 1986 Oslo Lysverker's proprietors, the City Council, asked the utility to examine again the comparison of the costs of conservation and the long run marginal costs of new supply, in order to see whether the savings target could be raised above 15% - which they were achieving with some ease - and still remain cost-effective.

Conservation is now explicitly included in Oslo Lysverker's annual plan as a 'supply' option, rather than as purely an implicit element in the demand forecast as had previously been the case - and is still the case in the UK. The concept of regarding energy conservation as the 'fifth fuel', rather than just a good cause, has been established.

However a very relevant question to ask is how Oslo Lysverker actually achieve their energy conservation targets. How do they, and their North American counterparts actually acquire the fifth fuel?

Oslo Lysverker promote conservation investments through a system of grants and loans implemented through trained and qualified energy auditors. These grants and loans apply to all classes of customers - industrial, commercial or residential. The important point is that through the use of various types of incentives and various levels of incentive, Oslo Lysverker is succeeding in cost effectively acquiring the energy conservation resource that they have planned.

In the words of the Tennessee Valley Authority, the largest publicly owned energy utility:

"Conservation programmes are treated as a power supply option, since the impact of conservation programmes can be controlled by TVA in the process of planning the power supply system."

The evidence accumulated from 10 years of gas and electric utility experience with energy conservation programmes in the USA overwhelmingly supports this statement. North American utilities promote energy conservation investments through grants and loans, which have been demonstrated to be successful in stimulating predicted levels of conservation investments. The Pacific Gas and Electric Company of Northern California, for example, has offered residential customers free home energy surveys, no interest loans or cash rebates, and incentives to encourage the purchase of energy efficient domestic appliances. These schemes have been highly successful both in encouraging applications, but also in encouraging action from the utilities' customers. Some 220,000 consumers undertook comprehensive energy saving work on their homes through the cash rebate scheme in 1985 alone, while the zero interest loan scheme has attracted more than 440,000 participants.

Commercial and Industrial customers have been similarly encouraged, with similar success. The example I have quoted here is the Pacific Gas and Electric Company, but this utility is by no means unique in the United States. I could have mentioned Brooklyn Union Gas, or Florida Power and Light or Northern States Power or Potomac Electric Power Company. All these utilities use energy conservation as a 'fifth fuel', and all have experience that demonstrates that the effectiveness of conservation programmes can most certainly be predicted with sufficient accuracy for the purposes of utility planners.

Demand-side policies reduce utility planning uncertainty and risk. These investment options are small, modular and incremental in nature. Compared with conventional coal or nuclear fired power plants or new gas fields, they have shorter lead times, low capital requirements, and offer the utility a quick return on its investment. In an era of highly uncertain demand, utilities are finding that conservation and load management investments offer a unique opportunity to improve load factors, increase velocity of cashflows, reduce capital expenditure, and the financial risks associated with excess generating capacity.

We come now to the third and final of the questions I raised at the start. If a misallocation of resources is occurring, and if it can be shown that energy conservation can be delivered at a cheaper cost than alternative conventional supply sources, then why does not either the utility, or those who oversee the utility, ensure that this happens? How do we make this happen in Britain?

In answering my first question regarding potential misallocation of investment resources, I had reason to cast doubt upon the wisdom of some of the investment policies of our utilities. This is not to suggest that the captains of our fuel supply industries are uniquely short sighted. To be frank, I do not believe that the whole concept of promoting energy saving in preference to new supply sources would ever have happened in the United States had it been left to the suppliers alone. However, in America there is another critically important player in this game: each state has a public utility commission, the role of which is to act 'pro bono publico', to ensure that the private company, or even the public Corporation, which has been given the monopoly energy franchise for an area does act in the long-term interests not only of its shareholders, but also of its consumers and in the overall public interest.

Their role, and their relative effectiveness, differs from State to State. But suffice it to say that their attitude is personified by the resolution unanimously adopted at the 1984 Annual Convention of the National Association of Regulatory Utility Commissioners. This stated that:

"It is necessary that demand and supply planning be fully integrated to ensure that resources are used prudently, and that our energy needs are met at the least possible cost.'

These regulatory commissioners can have the power to forbid proposed tariff increases, to disallow injudicious purchases being charged to their customers and, to require positive demonstrations that any utility is indeed pursuing the 'least cost option' of the US regulatory system. Suffice it to say it is by no means as bureaucratic nor as long-winded as some of its partisan critics allege. But one conclusion we can draw from studying these organisations is that the more committed the regulatory authority is to ensuring efficient investment policies are adopted, the more likely it is that the utility it is overseeing will also act efficiently in the service it provides both for its consumers and ultimately its shareholders. The more toothless the watchdog, the more likely it is that the monopoly franchise will take injudicious decisions. Britain's one energy equivalent, OFGAS, was greeted with derision by the media when inaugurated - the Economist magazine dubbed it precisely as a "regulatory dwarf": a view obviously held by Sheffield Foregemasters who have chosen to bypass OFGAS entirely. And certainly OFGAS' initial remit appears to prevent it from considering the wisdom of British Gas' purchasing policies. As these account for well over half the ultimate costs to the user they are of considerable importance, and one that I have argued earlier is likely to grow if their 16% market growth target by 1990 is achieved.

Historically it has been the sponsoring government department which has overseen the wisdom of proposed purchases or investment projects. I have already quoted the negative attitude of the Department of Energy regarding the concept of comparing £1.5 billion on Sizewell with demand management programme. I am informed that, despite the earlier strictures of the Commons Select Committee, this was not an issue considered when British Gas sought to spend £20 billion on Sleipner gas. All too often it has been a case of the civil servants in the sponsoring division empathising all too closely with the aspirations of the utility they ostensibly oversee - capture of the 'regulators' by the 'regulated.

However, I am encouraged by research undertaken contemporaneously with the Department published work within the UK Department of Energy's Policy Unit, and involving senior executives from all the major supply interests. This was made publicly available only via leak to the Commons Select Committee on Energy, and its research was promptly terminated.

Nonetheless it did make the following recommendations:

"There is an imbalance in the national investment resources being devoted to energy supply and use. If the nation is to receive the maximum benefit from the investment resources it expends, then a greater emphasis needs to be placed on conservation."

"There are significant and continuing national benefits to be gained from increased conservation investment .... in the longer term, increased conservation investment should reduce the call on public funds for energy supply."

"In the absence of additional funds, it would be worthwhile examining whether some of the Department's current expenditure on R and D on longer term sources of supply could, in the short term, more advantageously be used to promote conservation."

That was in 1983. If we had acted upon these recommendations then, we would have been in the vanguard of least cost resource acquisition. As it is, subsequently, nothing has happened.

How do we solve this problem of institutional inertia? I think the answer lies substantially in the proposal made by the Select Committee as part of its 1985 gas depletion report, which stated:

Improved energy efficiency will be critical to prolong the life of United Kingdom continental shelf gas resources. The Government must direct its policies towards ensuring that our energy needs are met at the lowest resource cost to the nation.

This requires a thorough assessment of the relative economic advantages of investment in supply and demand. The energy utilities may have a statutory requirement placed upon them to invest in energy efficiency."

This statutory requirement to assess alternative supply and demand investments was specifically featured in several manifestos in this summers election.

What is needed is for the Government to extend the powers of OFGAS, so as to require the gas industry to undertake such comparisons, with a failure to do so leading ultimately to a refusal to approve rate increases. Similarly the Government should so instruct the electricity industry in its turn to do likewise, with a failure on their part leading ultimately to the Treasury withholding approval for supply expenditure. Above all, Government must move to a situation whereby the evaluation of energy efficiency investments in the public domain are judged by precisely the same criteria as those applied to supply investments. controlled by the Government.

We need to ensure not only that the relevant sums are done, but that they are seen to be done, so that they are open to public inspection and challenge - if necessary in the kind of public interest forum provided by American utility regulators.

I anticipate that the United Kingdom will increasingly come to appreciate that not only can the supply versus demand calculations be undertaken with a considerable measure of accuracy, but that it is in all our economic, social and environmental interests for those sums to be done, and the conclusions reached and acted upon.

Strategies for Improving Energy Efficiency and Energy Management in Smaller Businesses

D. J. White, C. Eng.

Sears plc

ABSTRACT

The paper relates to the establishment of an energy management programme to cover a large number of retail shops with special reference to small units.

Reference is made to the initial survey and the work done with evaluation of the results, and the subsquent development of a programme to cover both people and equipment.

A number of simple points which can have an important effect on energy use are considered with a review of the present and future opportunities.

Some ideas on technical developments and innovation are included with consideration of the increased importance of the role of the Energy Manager.

The results against original targets with future forecasts are included.

KEYWORDS

Targets; Energy Managers; Lighting levels; Environment; Control Equipment; Future developments.

BACKGROUND

The work encompassed in this paper results from a programme that has been undertaken during the last 2 years in the Sears Group by an energy management team. It arises from the case presented by the Secretary of State for Energy, Peter Walker, in which he claims savings of the order of 10% are feasible from nearly all energy consumed, providing the user takes practical and realistic steps to prevent wastage and eliminate or replace inefficient methods of energy consumption.

Within the Sears Group we have over 4,500 shops located generally in the

"High Street". The major departmental stores unit is Selfridges in Oxford Street London, with a further 10 large stores operated throughout the U.K., by Lewis's. We also incorporate the manufacture and retail of shoes through some 2,500 outlets, operated by the British Shoe Coporation, with ladies fashion clothes sold through Wallis and Miss Selfridge. Mens outfitting includes Foster Brothers Companies incorporating Adams, Dormie, Esquire and Millets, whilst sports goods are dealt with through Olympus.

In the jewellery sector we have Mappin and Webb and the Crown Jewellers, Garrards.   There is also a large motor distribution division, and within the William Hill organisation there are nearly 900 licensed betting shops in the UK. When the office warehouse and manufacturing facilities are also included we have a total Group energy bill currently in excess of £16M. per annum.

If departmental stores are excluded from our reported energy figures the remaining smaller units have an average energy bill of some £3,000 per unit with a distribution spread from the smallest at about £750 per annum, whereas by comparison, our largest departmental outlet has an energy bill of some £1.7M. Our Group operates with the autonomy of each division being considered as a prime motivator in achieving performance targets.

From this background information the engineering division of Sears was requested by the Group Chairman to establish a programme of energy management and conservation that would reflect the Group policy also recognising the existing varied trading methods inherent in a diverse organisation. At the same time the programme had to be capable of achieving the 10% target saving which had been set by the Sears Group.

PLANNING

The initial energy consumption data that was collected from our Companies covered a three year period, and after it had been analysed in detail it provided a sound basis from which to plan the work. In particular it was discovered that over 80% of our total energy was provided by electricity and our initial analysis was therefore concentrated into this area. It also became apparent with our medium and smaller sized shop units that Sears were moving into a sector where little energy management work had been undertaken by U.K., organisations including the Department of Energy and there were no relevant "case histories" that could be consulted.

The co-operation and support of the divisional managing directors was seen to be the key to the success of an energy conservation programme and a visit was made to each to discuss the ideas that had been formed. From these first meetings emerged a clear pattern of support and it was agreed with all divisions using in excess of £1M per annum that they would appoint a fulltime Energy Manager, whilst those using less should include energy management in the job description of a senior employee who was closely involved in the mechanical, electrical and servicing side of the business. In all cases it was agreed that one of the senior divisional directors would also be made specifically responsible on a normal day to day basis, with the Energy Manager reporting directly to him.

Because of the increase in workload in the Engineering Division it was decided to appoint a highly qualified engineer with experience in energy management at the centre to control and co-ordinate the work being done. The initial establishment of the programme was in place by December 1985 and provided a platform from which it was possible to move ahead.

IMPLEMENTATION

From the initial stage the conflict between the requirements of the retailer and the energy manager became increasingly apparent and the reason for our large increase in the energy consumed over recent years was identified. In particular 3 areas were of importance.

1. Increased lighting levels were seen by the retail and particularly fashion sales personnel as necessary for the effective display of goods, with the firm belief that by increasing the amount of light that sales would increase. As a result they were introducing extra lights for both general display and also spotlighting purposes.

2. The ever increasing need to improve the shop environment to attract customers was becoming more important. In particular the desire to include some form of environmental control including air conditioning was seen as a priority. However, the increased lighting was a major cause of problems due to the heat gain being experienced from increasing spot and general lighting systems.

3. The conviction in the mind of many shop managers that trading could only be fully effective if all doors were left open in normal trading hours in order to remove any barrier that might be seen by customers from free entry into the shop.

To enable these specific points to be dealt with effectively required a detailed technical assessment which caused us to divide our plan for energy management into 2 separate and clearly defined areas.

A. Energy management, related to good housekeeping where we seek to involve everyone in the organisation to play their individual parts to minimise wastage or unnecessary use of any of the energy supply sources. In general we now see this being of particular importance in our older premises where there is little existing control equipment available, or in those cases where equipment has been fitted and it may have fallen into disrepair.

B. Improved and more effective use of the energy we purchase to ensure that it is adequately controlled and only used as necessary for our business. This approach is particularly related to our new or re-furbished shops and enables us to seek improvement well in excess of the planned 10% saving resulting from housekeeping. We now have a target saving of 25% over past practices firmly set and we have been able to achieve this level in our more recently built units.

ENERGY MANAGEMENT THROUGH GOOD HOUSEKEEPING

From the beginning of our programme we recognised the need to include the small operations in order that we were not seen to ignore those cases with an annual energy bill of less than £1,000 per annum. In such outlets it was always accepted that low capital cost measures and good housekeeping through education and training would be necessary. Underlying the work we must face a number of basic facts relating to the persistent apathy amongst staff that has to be overcome.

There is a general tendency common amongst many people to do nothing if it is not possible to do everything; similarly there remains a tendency to measure and record nothing because it is not possible to accurately or precisely measure everything. For the same reason small investment in controls and equipment is not made because they are not considered to be the complete and ultimate answer.

During the past 12 months the reduction in the price of oil and the consequent effect on general energy prices has led to further complacency associated with an awful naivety regarding future trends. People readily forget the need for future competitiveness and the problem that will arise when escalation of energy prices recommences as will surely happen if world demand moves ahead.

To tackle these problems is not helped by the common view that energy management is seen as an unglamorous and undynamic activity which appears to be very remote from either satisfying the customer, or assisting to compete with other high street outlets. As a result if it is made the responsibility of any one person it is all too often someone whose own personality reflects the apathetic attitudes commonly shown to the subject.

Recognising the problems to be encountered in the shop units must be a key to the success of any attempts by management to succeed in an exercise which consists almost entirely in the motivation of people, and ensuring that their own awareness and co-operation in their own actions is crucial to the success of a programme. Inevitably this means we have to record, monitor and report to the senior management and employees the progress that is being made.

In seeking a catalyst to ensure that "energy management" becomes a topic considered each and every day by every employee we should also recognise that:-

1. Within the working environment each of our employees on a broad average uses some £250 of energy per year.

2. Every employee will be aware of the substantial costs that they have to meet outside of the work environment in paying for the energy used at home for heating, lighting and cooking. Beyond this area most people also have to recognise the substantial energy costs required for petrol to be able to travel in their car. There is no doubt that relating energy savings at work to the opportunities at home makes the subject more interesting and of greater relevance to most people.

3. There is need to preserve as far as possible the discrete supplies of fossil fuel that are available. Currently the energy stored over many millions of years is being used at an alarming rate with no readily acceptable substitute so far discovered.

OPERATION

Within the area of good housekeeping there are a number of specific areas which can be readily identified as having a substantial effect on the energy consumed. A few examples can readily be made to indicate the potential to save in excess of 10% of energy usage with minimum costs.

The need for a much wider recognition of the Law which is laid down in Statutory Instrument No. 1013 dated 1980, this when paraphrased with an earlier document No.2160 dated 1974 states, subject to some specified provisions:

"No person shall use, cause or permit the use of electricity or fuel for the purpose of heating premises so as to cause the temperature of those premises to exceed 19°C which is equated to 66.2°F."

The failure to recognise the law with it's legal limit is widespread and can have a significant effect on energy consumption, as the increase in temperature by 1° centigrade will require approximately an additional 8% of heating energy at normal winter levels. In addition to the possibility of the equipment operating at incorrect settings there is the added possibility that thermostats may be located in areas where the temperatures to which the instrument responds, does not relate to the environment of the people working in the area. It is not unusual for thermostats to be located near to doors giving rise to peculiar air streams, with the fresh air entering the shop effectively controlling the thermostat.

Nearly all heating systems are controlled by some automatic device such as a time clock which can frequently be found with both the settings and the actual time or day set incorrectly. Examples of older time clocks are often seen where the mechanical adjustment to set trip levers is extremely inaccurate and as a result even with best endeavour an accuracy of half an hour per day is very difficult to achieve, and this can have the effect of increasing consumption by 6% based on the normal opening period of eight and a half hours.

Those shops that are equipped with air conditioning installations need to have thermostats reset from the 19°C winter level to 21°C or perhaps 22°C during the summer a potential saving of 10% which is frequently forgotten especially in managed units where the staff is particularly retail orientated and is non-technical.

The significance of two simple control systems of the type frequently fitted in small shop units with the thermostat and time switch are shown above and demonstrate how easily some 20% of excess energy can be saved by simple management control. Important as the setting of controls may be it must also be remembered that security or locking of equipment to prevent any tampering by staff or customers is equally essential to prevent both accidental and deliberate changes being made.

Considerable energy can also be wasted by faulty or inadequate maintenance of plant and equipment, particularly where heat exchangers, filters and fans are involved. It is reported in the "Businessman's Energy Saver" manual that failure to clean filter can have the effect of doubling the warm up period necessary to reach normal room temperature and at the same time increase the load on plant and equipment, and this can represent a further 10% excess energy use.

In addition to the above opportunities which are totally controlled by people, further savings can be made by taking advantage of new technical innovations and developments where even for retro-fitment into old premises many pieces of equipment can show a pay-back in the order of 12-18 months.

NEW BUILDINGS

The effectiveness of new equipment is now sufficient for us to expect a 25% energy saving in a re-furbished facility, based on the original operating expenses, without accepting any deterioration in the environmental conditions that existed, in fact we expect that in addition we will improve the ambiance or general acceptability in the premises. We also are finding an increasing opportunity to take advantage of new developments to improve the technical performance and effectiveness of capital equipment with the benefit of improved reliability. It is our view that with the proper selection of equipment to suit the particular application there may well be no necessity to incur additional capital expense in order to provide a considerable reduction in energy consumption.

The relationship of an energy efficiency programme to the specification of electrical and mechanical services cannot be under-estimated and its relevance to energy management personnel is essential. It is only through the involvement in the decision making process regarding the specification of equipment, and the decisions with regard to contracts - that it is possible to ensure that adequate attention is being given to energy management.

New premises are expected by our customers to exhibit higher standards than in the past, and in particular they need to have an environment which is equally acceptable in both the summer and winter. The demand for increased lighting levels in recent years has been met in many high street shops through the increased use of conventional spotlights for colour display purposes with the associated problems of a high level of heat generation.

A detail study will reveal in many cases that the total heat load of a shop can be met by the lights for all reasonable winter conditions without any additional heating. In the past the air conditioning plant may have been required to operate throughout the year, and few examples have been seen where this heat source was taken into account at the decision stage. As a result normal systems that have been fitted may be over specified for heating but under capacity for cooling in those cases when the chilling facility has been included.

Recent developments in the field of display lighting that can yield very much higher levels of light output for a given consumption of energy (with consequently less heat output) can be made to look very attractive. However, each would appear to have serious drawbacks at this time. Our experience has been of numerous cases of lamp life substantially below that specified by the manufacturers and although we all understand the reason for the failures no acceptable solution has been forthcoming from the lamp makers. In another example of compact fluorescent systems we have an acceptable light source but a power factor which is so low as to make energy savings insignificant, although the product has the associated benefit of long lamp life.

In the absence of a solution which is relevant to our requirements, and almost certainly that of the majority of high street retail shops, we have found it necessary to seek our own solution to the problems. One unit now being introduced represents an example of a product developed and manufactured to our requirements and specification, this comprises a small electronic device that is able to reduce the electrical voltage to all incandescent lamps by a pre-set amount which we currently vary between 5 and 7% according to the surrounding environment. When used in conjunction with a soft start up, through means of a reduced supply voltage for approximately two seconds we should double the life of all lamps by reducing the thermal shock. This will save both the cost of new lamps and also the labour involved in replacing failed lamps whilst saving energy. On a theoretical basis the design lamp life of 2,000 hours or eight months of average operation will be extended beyond one year.

Sample shops where we have undertaken tests have not revealed any adverse comments from either the shop management or staff about reduced lighting levels. In an attempt to establish reaction we have deliberately retrofitted a section of an existing store with the unit to everyone's complete satisfaction.

Much work still has to be undertaken on the broader economies of energy saving measures, especially in the lighting field with such a major part of our cost being provided by what is really a consumable item. New light sources are providing an acceptable colour rendering and overcome some of the above difficulties, but each seem to have their own shortcomings. One of which is the large light output from a small source which causes problems in establishing even light distribution especially for display purposes.

Other electrical control equipment is becoming available to perform a variety of tasks relating to the control of equipment, and it extends from the time clock which may be used for operating the lighting and heating systems to the elaborate and sophisticated energy management systems capable of monitoring and controlling all energy related services in any size of operation. Increasingly we are automatically controlling all lighting and the period when maximum lighting levels are permitted is limited to the period of shop opening hours. Outside this period lights that are sufficient for the appropriate safety of staff and the work being done, such as cleaning, is clearly defined and controlled. In a similar way plant and equipment related to the heating and environment is controlled with an optimising facility, which by simple micro-processor control will vary the start up time of equipment to take into account both internal and external temperatures in order to calculate the time the building requires to respond and so ensure the pre-set temperature is achieved at a time related to the store opening.

Such devices can be most effective in operation and are well known to energy managers because of the large variety that are available. By suitable selection for each application we believe that a payback in the order of one year can be achieved.

Other more specialised equipment is becoming available from some of the smaller manufacturers which by careful selection may be found to be very effective. Typical of such devices is a small unit which has undergone extensive testing by us and shows considerable economy when fitted to individual small electric motors operating from a single phase supply such as refrigeraters and deep freeze units where 24 hour operation is necessary and frequently the motor operates below the design Horse-power capability. The device simply reduces the power sent to the motor by sensing the change in certain characteristics of these motors when operating under part load and it is able to ensure that sufficient power is made available to ensure the motor continues at a constant speed. Other equipment of a similar concept recently evaluated for escalators shows a good payback due to the large percentage of time such equipment has to be running at a low utilisation rate and is well below the design capability and hence the capacity of the motor.

DEVELOPMENT

Future work which we are now considering is based on an even more basic review of our objectives in shop design, in order to establish exactly what criteria need to be satisfied so that we can ensure that both our customers and employees find the general surroundings comfortable and attractive. To this end we have to study the shop ambiance which may well reflect more of the customers' home atmosphere and to enable them to see potential purchases in conditions that they desire to have in daily life. Probably this will reflect something of the theatre and it will almost certainly be spectacular for reasons of careful planning rather than the stark brilliance which has been the trend in recent years.

Environmentally we have to face the difficult problems that exist between the needs of our staff, who even in winter, may be dressed in light clothes and that of the customers who may be wearing heavy outdoor clothing and who enter into the shop environment which may easily represent a temperature difference of 20° or 30°C when taking into account the wind chill factor outside. This is a particular technical problem and we are undertaking a test programme to establish how the environmental temperatures in the shop may be influencing the occupants. Apart from providing a more acceptable set of conditions we are convinced that we may also be able to save energy. We are continually reviewing the latest technical papers and undertaking development work ourselves. Test work is being undertaken to enable us to understand more about shop environment and in particular the effect and influence of the various sources of heat that exist.

It is necessary to understand that comfort is not solely dependant on ambient air temperature because radiant heat is an important factor on the comfort conditions of any working or selling area. We know that the atmosphere within the shop unit is a subtle combination of the air space temperature and the radiant temperature and we know how each is generated. What is not understood is the significance of the air and radiant temperatures in an environment where the staff substantially remains stationary and the customers move about. Air temperature is measured by the familiar dry bulb thermometer but it can be modified to take into account the impingement of radiant energy and the convective currents that are liberated from the conductive surfaces and any inter action could profoundly affect the energy consumption within the space. Reference has been made earlier in the paper to the effect of small adjustments to the statutory 19°C. When radiant energy is also taken into consideration this adjustment may become even more significant.

Light energy input shows itself primarily as heating and secondly as light. Most of the heat liberated transmits itself primarily as radiation and secondly as convective currents, when the radiant energy strikes objects in its path these absorb the energy which then retransmit again as radiant and convective heat components whilst some of it is conducted away. Normal design practice does not take into account fully the radiant effect of light sources and the influence on comfort of the occupants of the shop, nor is it recognised sufficiently as a mode of heating in the energy balance.

Decorations, wall, floor and ceiling colour, the placement of light sources, the air distribution and diffusion air all play their part in setting the comfort conditions. It is our expectation that test work will reveal considerable scope for energy reduction and at the same time make a more satisfactory trading area environment, all in addition to reducing costs to the benefit of the Group profits.

When the heating requirement of a typical shop unit is examined with insulation values for the building that are specified by the current building regulations it will frequently be found that the heat loss approximately is balanced by the heat input from the lighting system on the specified design day. The use of electricity for lighting still remains our major source of energy and this will continue for the foreseeable future.

RESULTS

Sears clearly benefits from its size which has enabled us to appoint six full time energy managers each of whom relates directly to the central energy team at the Engineering Division. Regular meetings ensure that the experience and results of tests are readily available to all companies and we are able to prevent duplication of effort.

Clearly the observations made in this paper refer specifically to the experience gained from the energy programme which has effectively been in operation for just over 12 months. However, because of the variety of outlets we have in the High Street and because of the similarity of the results that have been achieved there must be every reason to believe that similar opportunities exist in nearly all retail areas.

Work done earlier by other companies with large departmental stores have been reported through Department of Energy demonstration schemes and already show the effectiveness of a specific energy programme. Even though it must be anticipated that such outlets would already have been more efficient in the choice of equipment due to the design requirements for the effective plant facility appraisal initially necessary to support the request for large capital sanctions. If this view is correct it confirms the tremendous opportunities that remain in a substantial number of smaller retail outlets.

It is especially disappointing that so little work has been done by the Department of Energy to generate enthusiasm by a direct approach to the retail sector, where the Secretary of State has already claimed some £140M. a year can be saved. Current official information and case studies are particularly related to the large retail stores, and of course this only acts as a further factor to persuade the smaller user that he is unlikely to achieve any realistic returns.

The complacency of the consumer has often been matched by the lack of interest shown by the equipment manufacturers whose product range is frequently inappropriate to our requirements. We require commercial equipment, with capacity and capability, between the normal domestic units and the large consumer areas. The area where energy usage is between

£1,000 and £5,000 per annum is typical of the majority of shop units and requires controls that will achieve a direct 10% saving in use, whilst having a maximum of 18 months payback. Some of the equipment that is becoming available which is necessary to upgrade shop environment frequently originates only from overseas sources and even many domestic products are dependant on imported parts. Unless the UK manufacturers respond to our requirements we will almost certainly need to use further imported items of equipment.

Whilst the work will continue on the original premises within the Sears Group of Companies we expect that the major returns will come from any facilities that are subject to building design work where the demands of the merchandise display and sales departments have influenced architects to respond with an increased use of the various services but with little recognition of the significance that such changes will have from the viewpoint of energy management or consumption.

From the beginning Sears recognised the importance of a technical approach to energy management and care was taken to ensure that each of our energy managers was technically qualified and experienced. We are now beginning to see the benefits of energy management being further integrated into our Companies as part of the normal development or evolution of the business. We expect this to continue with an increased accountability being given to them, to handle all of the analysis work necessary regarding heating and lighting prior to specifying the service equipment that is to be chosen.

To ensure the effectiveness of this work is followed through they are also becoming involved in the commissioning and acceptance testing of new facilities to ensure they will achieve the planned performance. In turn this is likely to lead to further investigation into the maintenance programmes that have been established for the mechanical equipment and especially to ensure that fans and filter have been checked and maintainted in accordance with manufacturers requirements.

Innovation for energy efficiency on our existing units meant for us a return to the basic principles. They must be the foundation for any programme and in our case are summarised below:

Commitment
Programme - Planning - Implementation
Measurement - recording
Control
Review - monitoring - recording

In addition to the saving of costs through reduced use of energy we have the additional benefit coming through of increased lamp life and improved appraisal of service equipment which with the benefits of scale add up to a further bonus from the substantial benefits that we are well on the way to achieving

From our original target of a 10% saving in energy we have developed a programme which in the third year will save some £2M. It is the intention to continue the early initiative, and the efforts of the energy management team enables us to forecast that the programme will now contribute some £10M. to group profits over the first 5 years of operation. At that time we shall be saving some 17% of our original energy bill as a result of the progressive introduction of new equipment into those units being built or refurbished.

Need education & implementation
Buy in for

Energy-Efficient Participation by Building Occupants in Environmental Control: Research and Design Considerations

H. B. Willey

University of Auckland, School of Architecture, Private Bag, Auckland, New Zealand

ABSTRACT

The compatibility is discussed of some naturalness in the internal environment of buildings, and of occupant access to the control of that environment, with energy efficiency. Cybernetic theory is used to explore a range of alternative environmental control strategies, including various forms of hierarchical control. The importance of providing the occupants with a clear understanding of how their actions will affect the environmental performance of the building, and the importance of rapid response and feedback when they take such action, are both emphasized. An approach to computer modelling of occupant control actions is outlined, with both research and design applications.

KEYWORDS

Energy efficiency; building occupants; environmental control; cybernetics; hierarchical control; computer modelling; fuzzy algorithms.

INTRODUCTION

The desirability of reducing the energy consumption of environmental services in commercial and institutional buildings has led to two main responses. On the one hand, deep-plan buildings with tightly-controlled environments have been designed, and the apparent dissatisfaction of some occupants has been noted. On the other hand, new building forms, such as those embodying an atrium, and new forms of control, such as those now available in lighting, have been proposed and the ideas are being worked out both in theoretical studies and in design practice. These developments would proceed with greater depth of insight and greater clarity of direction if the trends in environmental control were analysed in such a way that some of the issues at the heart of the changes being made were brought into better focus.

THE PROVISION OF OCCUPANT CONTROL OF THE ENVIRONMENT

One particular issue that has yet to be given more than token recognition by
the majority of researchers and practitioners is the role of a building's
occupants in the environmental control system. Consideration of occupants in
research is usually restricted to their role as receivers of the environment
and research papers discuss the fine tuning of the comfort conditions to be
established by a building's plant. This pattern of thinking dominated the
energy conferences of the 1970s where there was a concentration on improving
the plant. Despite the growing interest at that time in systems theory, the
occupants were ignored and building fabric characteristics were seen only as
parameters to plant operation. Little change of pattern is evident in even
the more recent conferences, where discussion has centred on making plant more
efficient, rather than the whole system of building plus occupants. The
regrettable tendency for most reported research or design practice to
concentrate on efficiency of plant design and operation may well reflect that
research funds are more readily available for determinate, engineering-
orientated projects and that buildings designed in this manner lead to more
authoritative and objective description.

Automatic control of the environment for energy conservation purposes has been
implemented with an increasing tightness of that control, accompanied by a
minimizing of air change rates in completely sealed buildings. The facade is
often designed for thermal opacity, this having the advantage of decreasing
the perimeter zone for air conditioning and increasing the floor area able to
be served from central plant (Baum, 1984). The problems that have been
encountered with tightly-controlled environments include: concern that
laboratory-derived control set-points do not take account of many factors that
influence environmental comfort (McIntyre, 1982); significant occupant
dissatisfaction with the tightly-controlled environments and with the lack of
remedial access to system controls; the production of a bland, soporific
environment from the uniformity of temperatures and lighting and the low air
change rate; and the emergence of a 'sick building syndrome' (Ashley, 1986a)
in which some occupants become chronic suffers of headaches, sore throats,
lethargy and other ailments.

A commonly recurring complaint amongst building occupants is the lack of
involvement in environmental control, either through individual control or an
effective complaints procedure.   Lack of access to normal system controls may
be expressed sometimes in behaviour to beat the system (wedging doors open,
making illicit keys to open windows intended for cleaning access only,
tampering with sealed thermostats, etc), reinforcing the view that, one way or
another, energy users play a key role in effective energy conservation.

The uniform internal environments produced by the designed unresponsiveness
of building envelopes to the external environment plus the normal uniformity
of the plant-generated environment have been counteracted in a handful of
cases by an artificial simulation of a 'natural' environment. Experiments
with a controlled, time-varying thermal environment have shown that slow small
changes in temperature can provide stimulation and increase satisfaction
(Gerlach, 1974). Other techniques have been tried (such as introducing eddies
into air movement (Dowson, 1977)) but some factors cannot be provided. For
example, people prefer to sit by a window not only for the perceived qualities
of the daylight but also because of the view and the status associated with
such a position.

Two issues are highlighted by the preceding: a perceived preference for some
naturalness in the internal environment and a desire by occupants for some

access to control of the environment. Ideally the naturalness should be achieved by designing buildings to be suitably responsive to and admissive of the external environment, and by less tight control of the internal environment by the building services system. The usual argument against this is that such a strategy would lead to increased energy consumption. A similar argument opposes surrendering some measure of environmental control to the occupants. It is also argued that occupants would be less likely to achieve 'optimum' comfort conditions, although general studies by environmental psychologists (Ittelson and others, 1974) suggest that occupant access to controls is more important than the actual physical state of the environment in ensuring that the occupants are satisfied with their comfort state.

If the strategies of naturalness and occupant control are followed, two major approaches to building design can be identified:

a) Shaping the building to respond to and make efficient use of ambient energy. This can lead to a largely 'free-running' (as distinct from plant-determined) internal environment, with some degree of occupant control of both adjustable building fabric (blinds, sun control devices, openable windows, etc) and the limited plant required.

b) A largely plant-determined environment without tight control and with occupants having either (or both) of some individual controls (task lighting, individual air supply, etc) or some access to general controls (ambient lighting, thermostats, building management, etc).

The key to the success of either approach is ensuring the effectiveness of the occupant control actions and their compatibility with the overall energy efficiency of the building. Effective occupant action requires their understanding how to use the controls, which implies an understanding of the performance of the whole environmental control system. It also requires that the occupants achieve satisfaction with the effects of their actions, which has implications for the provision of tangible feedback and for the response time of the system to each potential control action. If adequate understanding of system performance or adequate feedback cannot be practically provided in the building design, occupant control actions may have to be provided through a hierarchy of controls, with control tasks shared between the occupants and automatic sensors/effectors (and possibly some provision for either to override the other). These considerations will be discussed shortly.

A secondary factor in the successful implementation of these approaches to building design is an adequate prediction of the performance of the system at the design stage. The environmental control system, embodying both occupant and automatic control over a 'natural' internal environment, is much more complex than the tightly-controlled, plant-determined environments for which good simulation models are available. An approach to modelling the more complex systems will also be discussed.

VARIETY IN THE ENVIRONMENTAL CONTROL SYSTEM

Variety and its Reduction

A significant attribute which occupants bring to the control of the environment is an ability to deal with systems, such as the natural environment, which display high variety. The cybernetic concept of *variety* is a measure of the complexity and variability of a system (Ashby, 1956).

Each state of the environmental system can be represented by the set of value of all the system variables for that state. The variety of the system will b the number of states in which it may exist. This number will depend upon the range of values of each of the variables and on the number of possible combinations of these values which in turn depends upon the amount of interaction between the variables.

In the case of a system, such as that of the physical environment, where the variables are continuous within their ranges rather than adopting discrete values, a measure of variety might be obtained by imposing a stepped scale on the range of each variable, analogous to the rating scales used in psychometric studies. If the number of steps in each scale corresponds to the number of separate intervals of values of the variable able to be discriminated by the occupants of a building, this measure of variety would become particularly meaningful.

When the variety of a system is less than the maximum possible number of combinations of values, the system is exhibiting constraint. The *variety reduction* which this constraint brings about may be through a decrease in th ranges of values which the variables take or through a decrease in the interrelatedness of the variables. In general, environmental comfort requir only variety reduction in the form of a restriction of the ranges of variabl so that they lie within the 'comfort zone'. Comfort *per se* does not depend upon a reduction in the interrelationships between environmental variables. Nor does it depend upon such a restriction of the ranges of environmental variables that a single mode of system behaviour is sufficient; that is, that no adaptive control action is necessary (such as switching on a light or opening a window). Where variety reduction involves a reduction of interrelationships between variables or the elimination of adaptive behaviou it will be the result of a requirement or side-effect of the control measure employed.

Subsystem Structure of the Environment

It can be demonstrated (Ashby, 1960) that a system which displays adaptive behaviour must consist of weakly joined subsystems in order to reach equilibrium in a reasonable adaptation time. The terrestrial environment i observed to consist of sparsely joined subsystems but much weaker joins are found in an internal environment which is controlled by plant. This latter is assisted by the relative responsiveness of the building enclosure to eacl variable including the blocking of some inputs and the provision of an artificial substitute. (Some relationships are more difficult to sever tha others: even efficient modern light sources still produce some heat.)

Complementing this, both the building occupants' physiological sensory mechanisms and constructed monitoring devices display dispersion but much m so in the latter case, where the dispersion is also much clearer. (This is not only where the stimuli are physically different but also where a stimul is being monitored for different purposes, as in a temperature control thermostat and a fusible link in a fire door stay.) The connections that exist in people's sensory mechanisms produce a much richer repertoire of behaviour.

It may be possible to approach this richness of response with computerized electromechanical controls but the independence of monitoring subsystems ensures a specialization which should result in greater efficiency for each monitoring device. The appropriateness of this specialization is, however dependent upon the maintenance of constraints on the environment. When this latter condition ceases to be met, the potential efficiency of highly specialized electromechanical control is unlikely to be realized; in fact, such control may cease to be effective at all. A classic example is the thermostatically controlled heater which continues to pour heat into a room whose windows or doors have been left wide open to provide ventilation, with a resultant high rate of ventilation heat loss. Paradoxically, the appropriate control action, which should be apparent to an occupant of the room, is often never taken by that person, for reasons which may become clear shortly.

A formal way of stating the above is that, in general, electro-mechanical controls possess responses of very low variety. Since a system displaying a particular amount of variety can only be effectively controlled by a system displaying an equal or greater amount of variety, it is clear that low variety controls can only effectively handle a low variety environment. Conversely, the high variety 'natural' environment preferred by a building's occupants can be appropriately controlled by the high variety responses of those occupants, or by a combination of occupant and automatic control which retains the high variety contributed by the occupants.

Nevertheless, building occupants are often denied any role in environmental control. This decision is usually on the grounds that the occupants will either cause the building and its plant to perform less efficiently (especially in energy terms) or make their own task performance less efficient by this distraction. However, this takes a very narrow view of the way this role might manifest itself. In principle, the control model derived from cybernetics indicates no negative effects from occupant interaction, only the positive advantage of increasing the variety of responses available. The application of cybernetics concepts will need to be extended to see how such occupant interaction might appropriately be achieved.

CONDITIONS FOR OCCUPANT INVOLVEMENT IN ENVIRONMENTAL CONTROL

Control as a Non-distracting Secondary Task

Environmental control will become a conscious action only when triggered by an environmental stimulus reaching a threshold level. It has been observed in field studies (Haigh, 1981) that preoccupation with a primary task makes this threshold level variable but usually higher than suggested from laboratory studies. It is a matter of good building design to ensure that action is required infrequently and that the actions available are not time-consuming. (For example, the action might involve a switch or dial accessible at the work station; to make direct or indirect adjustments to building fabric or environmental services, or even to contribute feedback to a computer controlling fabric and services for analysis and potential action.) The time taken for such actions should be compared with the effect of increased satisfaction when analysing the occupants' task performance.

An alternative approach to this considers the hierarchy of basic needs described by Maslow (1954) in which lower level needs such as survival, safety and physical comfort must be satisfied before higher level needs can become regular sources of behaviour motivation. This can be taken as an argument to free people from the need to exercise any personal control over their environment but it is not at all certain that an occasional need for people to take steps to modify an already near-satisfactory environment is a significant inhibiter of higher level activity: indeed, to withdraw altogether the opportunity for people to perform some form of fine tuning may produce dissatisfaction. A perhaps more important implication of the hierarchy of needs is that an occupant's environmental control action may take into account its effect at higher levels in the needs hierarchy. In that sense the occupant's control action could be said to have a higher level goal than that of standard electromechanical controls. This implies a requirement for a matching level of information. In other words, the information must display a complementary hierarchical structure and be sufficiently comprehensive and clear to indicate the consequences of any control action based upon it.

Some insight into action thresholds can be gained from the now classic studies of human operators in industrial process control (Edwards and Lees, 1974). However, environmental control is not a primary task and these studies must be used with some care. One situation which has been studied in which environmental control is an important secondary task is teaching (Haigh, 1981). Teachers develop a strategy for control of classroom environments so that it is appropriate for teaching and learning without becoming a distraction for either teacher or pupils. However, the strategies are not necessarily the same for all teachers. Studies have shown that similarly-designed and orientated classrooms with similar class programmes display widely different energy performances (Perkins and Powell, 1979). There is a clear need for more research to develop an environmental system which will lead to a control strategy that is similar for all occupants and also energy efficient. It would be fundamental to this work to ensure that the environmental system was understood fully by the occupants.

Energy Efficient Control

The fact that we are not physiologically equipped to sense energy consumption together with a failure to comprehend the environment as a system, means we remain largely ignorant of the consequences in energy terms of the control actions we take. In addition, it has been demonstrated that the set of possible states for environmental comfort will rarely be a subset of the states for satisfactory performance of a primary task (Wyon, 1973). In other words, comfort does not imply satisfactory performance. Both of these factor suggest that for energy efficiency and satisfactory task performance to be achieved simultaneously with comfort in a building where occupants have some control over their environment, the environmental control system will have to be designed either to ensure that occupants have the necessary information or to have a restriction on the achievable states. The latter may be implemente through a hierarchial control system, as will be discussed shortly.

The ability of occupants to understand how their environmental system works has an interesting history. The changes in the subsystem structure of environmental control, from traditional forms to those employing modern plant in which there is a central source and a ducted or piped distribution network have created problems in the design of buildings which were not faced by earlier generations of architects. These architects had only to handle the articulation of fabric and the integration into that fabric of individual heating or lighting elements such as coal or gas fires and gas or electric lighting. There was a resultant clarity of expression in the environmental control and a physical immediacy in the adjustments made by the occupants to the system. More recently, architects and engineers have handled the probler of modern plant with two polarized solutions; namely, fully exposed environmental servicing and the concealment of servicing above a suspended ceiling. Neither of these solutions answers two basic questions: how can th occupants of a building be shown how the environmental control of their building works as a whole system, and how can the relationship between the adjustments they may make and the system as a whole be best expressed in the design.

Although to some extent the recent buildings which have incorporated extensive passive control can be seen as a reversion towards traditional modes of control, it does not follow that a complementary clarity of system expression will be regained automatically. It has been pointed out that the physical forms can no longer be the same and the demands being made of the controls will be more sophisticated (Dowson, 1977). At the same time the opportunity exists here for using advanced technical design to provide built solutions that are simpler and more easily responded to by people.

Information Flow in the Environmental System

In cybernetic terms, the key problem with modern forms of the environmental control system is that the flow of information within the system is poorly managed compared to the flow of energy. This statement recognizes that the occupants are part of that system. In most present systems the flow of information is rudimentary, given the potential complexity of the system. In turn, the energy flows are forced to be simple, imposing a subsystem structure which relegates occupants to being recipients of the environment rather than involved with its control.

The provision of information to the occupants, and the complementary access to appropriate controls, has implications for both the form of the building and the form of the controls on plant and fabric. Allowing a degree of occupant control requires that the occupants be provided with information not only of the disturbances that need to be responded to but also of the appropriate response to make under the circumstances. The clarity with which the occupants perceive the control system is all the more important because of the higher level at which occupant control may operate, as noted earlier.

Control action decisions taken by the occupants will take account of two important criteria: *anticipated response time* and *anticipated satisfaction* to be achieved. Some systems provide a rapid response (switching on lights; increasing fan speed) while others can be very slow (some heating systems). However, what is important is the time taken for information about the response to be received. An adapting system must obtain information on the effectiveness of its behaviour. An occupant will not be satisfied until that information indicates a successful control action. The information need not be about the variable being controlled. The information from a slow response heating system could be visual (a red glow) or aural (a humming sound), inducing satisfaction from the *anticipated* effect.

Implicit in the anticipated satisfaction resulting from any control action (with fast or slow response) is a recognition of secondary effects of the action taken. Opening a window in winter may produce air that is fresher as well as cooler. Anticipated satisfaction may be increased by a more elaborate control action which may take longer to implement. The two criteria may therefore need to be counter-balanced in any control action decision by an occupant. Together they form an *effectiveness* criterion which is complemented by a third basic criterion, that of the *efficiency* of the human input to the system (the human energy input compared to the size of change produced in an environmental variable or to the length of time that variable is maintained at a preferred value). Efficiency may be enhanced by a form of hierarchical control.

One of the particular characteristics of occupant control is that the complexity of the system and its information inputs does not allow it to be treated in the determinate manner of electromechanical controls. Rather, it requires a probabilistic treatment where different occupants with different past experiences may respond in somewhat different ways to given environmental conditions. Individual occupants must also be assumed to arrive at a particular control action in a non-determinate manner in response to a particular combination of information inputs.

A further important characteristic is that occupant control is adaptive and uses information which cannot be 'built into' the system by a designer. As has been pointed out (Beer, 1966; Lerner, 1972), a cybernetic approach enables systems to be devised which are capable, in principle, of going beyond the range of actions foreseen by the designer. A similar effect is achieved in what are otherwise intended to be deterministic control systems by the informal intervention of affected people. Just as the low variety of the decision-making process at the formal company board meeting is augmented by the high variety interaction at committee meetings, at the Friday evening 'happy hour' and through informal contacts (Beer, 1966), so too can low variety environmental controls be augmented by the highly redundant responses of the building occupants to match the level of variety in the environmental disturbances.

HIERARCHICAL FORMS OF ENVIRONMENTAL CONTROL

The informal intervention of occupants in the control system may ensure that environmental performance of buildings satisfies their criteria but does not necessarily ensure the satisfaction of ancillary criteria such as minimizing energy consumption. It would be possible to achieve the latter by the imposition of controls on energy consumption but this may then conflict with the qualitative criteria. Hierarchical controls could ensure the simultaneous satisfaction of both sets of criteria, and four forms of hierarchical control can be identified for this purpose.

Superordinate Restriction of Control States

Both the situation state (the set of values of the environmental variables) and the control state (the settings of the control mechanism) can be evaluated as satisfactory or unsatisfactory. An unsatisfactory control state could be one that involved a high energy consumption, even if it was able to achieve a satisfactory situation state. Occupants are unlikely to seek such a satisfactory control state unless they are paying for the energy in some way. The control of energy consumption may, as an alternative, be sought superordinately by constraints placed on the available control states through the design of the building fabric and the plant, or through an override place on the controls (as in some present forms of light switching). This hierarchical control system recognises the difference between a person's motivational orientation and value orientation (Parsons and Shils, 1951) as bases for control action decisions. The former can be relied upon to achieve satisfactory states of the environment; the latter is superseded by a superordinate restriction on the control states which can be achieved in the process.

Amplifying Regulation

An alternative hierarchical control system seeks primarily to optimize the
efficiency of the human input to control (defined above) by replacing active
adaptation with a process of selection; that is, using a small amount of
regulation to achieve a large amount of regulatory system behaviour.
Cybernetically, this "amplifying regulation" (Ashby, 1956) takes the form of
a master controller performing a small amount of selection and leaving one
or more subcontrollers to carry out the detailed interactive control (Pask,
1968). This places the occupant of a building in a superordinate role with
respect to automatic control. Energy conservation would then be achieved by
the automatic control (the subcontrollers) in their programmed responses to
the commands of the occupants (singly or collectively the master controller).

Computer-based Control Interface

A combination of the above forms of hierarchical control would see the direct
link from occupants to subcontrollers replaced by an indirect link via a
computer. The computer would respond both to the occupants and to a
programmed set of criteria (including energy-use goals), and would take
account of both before instructing the subcontrollers. (By comparison with
the others, this could be termed 'coordinate' control.) One attraction of
this approach is that the occupants do not need a detailed understanding of
the functioning of the subcontrollers. Their input to the computer could be
information about their comfort or desired changes to environmental variables.

A request to a computer can be seen as a sophisticated form of switch control.
A trend in environmental control from adjustable fabric to switch-operated
plant has been noted elsewhere, with a simultaneous trend to denying building
occupants access to that control which is paradoxical because a switch is
potentially a more accessible and efficient form of control (Willey, 1985).
The computer interface restores the control access without the perceived
disadvantages of occupants using plant inefficiently in energy terms.

Rudimentary desk-top devices for transmitting 'comfort votes' have been used
in field studies of thermal comfort. More sophisticated equipment could be
developed, perhaps geared to the increasing use of networked desk-top computer
terminals which could be linked to a building management system in an overall
"intelligent building" approach. Verbal statements such as "I am a little too
warm" can be interpreted by an application of fuzzy set theory (MacVicar-
Welan, 1976; Mamdani, 1977) and algorithms developed for a fuzzy logic
controller to provide appropriate responses. The system can be designed to
be self-organizing with adaptive fuzzy control developing new or better
algorithms on the basis of the successfulness of initial algorithms (Procyk,
1979). Such self-adaptive control will be able to take account of any higher
level motivation for the feedback from occupants. It should be able to
acquire the variety of responses inherently available in occupant control but
without direct occupant intervention (although this latter may be a source of
dissatisfaction with this approach).

Local and Background Control

To provide occupants with some hands-on fine-tuning of their immediate
environment, control can be separated into two components: occupant control
of local systems and automatic control of a background system. This is
already well-established in task/ambient lighting installations (Alexander and

others, 1982) and offers exciting prospects as a form of low level air supply
providing both personal supply and displacement room ventilation (Appleby,
1986). The local/background concept avoids using energy to maintain comfort
conditions through the whole volume of a space. In addition, it should be
feasible to design the background system to sense the use of the local systems
and to adapt the more explicitly energy conscious background system in such a
way as to reduce energy-inefficient use of the local systems.

The satisfaction derived from direct contact with controls and from the
immediacy of the effect produced will be greater if the local system is
floor-sourced, employing a raised floor, rather than ceiling-sourced,
necessitating remote controls and sometimes less immediate effects.

MODELLING ENVIRONMENTAL PERFORMANCE

It has been suggested that the present computer evaluations of energy
consumption in buildings may, in the near future, be required to be applied to
simulations of the environmental performance of proposed buildings to ensure
compliance with energy standards (Ashley, 1986b). If so, there is a very real
danger that the acceptableness of designs will be determined by what can be
adequately modelled. Occupant participation in environmental control may be
precluded by the limitations of the testing procedure.

Modelling occupant control actions is difficult within the conventional
theoretical framework. As is typified by the common resort to either a fully
exposed or a concealed environmental services system, it appears that in order
to solve a problem which is not well understood the solution space is usually
reduced substantially by a constraint imposed from outside the system, from
considerations of energy consumption, from aesthetics, or even perhaps
arbitrarily. To avoid this, a much sounder theoretical framework is required
which considers the dynamics of the system comprising not only the building
and the physical environment but also the actions of the building's occupants.
One approach to this more general theoretical base has been outlined above.
A more general but more detailed discussion is provided elsewhere (Willey,
1978).

Without a simulation model which recognizes that the occupants are an integral
part of the control system, the search for new forms of environmental control
is restricted to the working out of ideas in architectural and engineering
practice. Research is restricted to fine-tuning non-occupant forms of
control. The key to simulating occupant control actions lies in developing
adequate algorithms in a subroutine of the computer model. As with the
occupant-computer interface described earlier, this is best done with fuzzy
set theory. The use of fuzzy algorithms and the testing of fuzzy models has
been well documented (Maiers and Sherif, 1985) and the application to
environmental models has been discussed elsewhere (Willey, 1982).

PROSPECTS

The immediate goal is to develop design strategies for producing 'natural'
environments in buildings with a degree of occupant participation while still
ensuring energy efficiency. Basic cybernetic concepts in communication and
control and a number of hierarchical control systems have been outlined as
one response to this. Such systems have the advantage of requiring less
understanding by the occupants of how the whole environmental system works
with the potential for dissatisfaction at the lack of a fuller involvement.

he simulation model necessary to evaluate the alternative systems could be
sed to explore a wider range of control systems, including less
lant-intensive forms.

s well as simulation studies, the insights gained from setting up the model
ould be used to examine building regulations with a view to their
estructuring to recognize the interrelations in a general environmental
ontrol model. This would ensure that the regulations did not impose a
tructure on a designed control system which was in conflict with the
natural' control system. If a comprehensive environmental model can be
eveloped to guide both design and research, only its application to research
ould be unfettered so long as the design was constrained by regulations and
rofessional lore (including the present demarcation of professional tasks)
hich were based on concepts no longer relevant to the control situation. The
onstruction of the model must therefore be accompanied by the investigation
f ways of ensuring that the results of the research may be readily
ssimilated into design practice.

REFERENCES

lexander, D.K., V.H.C. Crisp, G.T. McKennan, and C.M. Parry (1982). Localised
lighting - a low energy alternative to uniform lighting in offices. *CIBS
National Lighting Conference Proceedings*, University of Warwick, pp.112-125.
ppleby, P. (1986). Low level supply: application and design. *Building
Services, 8(11)*, 59-62.
shby, W.R. (1956). *An Introduction to Cybernetics*. Chapman and Hall, London.
shby, W.R. (1960). *Design for a Brain* (second edition). Chapman and Hall,
London.
shley, S. (1986a). Sick buildings. *Building Services, 8(2)*, 25-30.
shley, S. (1986b). Computer modelling and building services. *Building
Services, 8(11)*, 40-41.
aum, R.T. (1984). Contemporary energy conservation design considerations and
methods. In H.J. Cowan (Ed.), *Energy Conservation in the Design of Multi-
storey Buildings*. Pergamon Press, Sydney. pp.29-40.
eer, S. (1966). *Decision and Control*. Wiley, London.
owson, P. (1977). Offices. *The Arup Journal, 12(4)*, 2-23.
dwards, E., and F.P. Lees (Eds.) (1974). *The Human Operator in Process
Control*. Taylor and Francis, London.
erlach, K.A. (1974). Environmental design to counter occupational boredom. *J.
Arch. Res., 3(3)*, 15-19.
igh, D. (1981). User response in environmental control. In D. Hawkes and J.
Owers (Eds.), *The Architecture of Energy*. Construction Press, Harlow.
pp.45-63.
ttelson, W.H., H.M. Proshansky, L.G. Rivlin, and G.H. Winkel (1974). *An
Introduction to Environmental Psychology*. Holt, Rinehart and Winston, New
York.
rner, A.Ya. (1972). *Fundamentals of Cybernetics*. Chapman and Hall, London.
cVicar-Welan, P.J. (1976). Fuzzy sets for man-machine interaction. *Int. J.
Man-Mach. Stud., 8*, 687-697.
Intyre, D.A. (1982). Chamber Studies - reductio ad absurdum? *Energy and
Buildings, 5*, 89-96.
iers, J., and Y.S. Sherif (1985). Applications of fuzzy set theory. *IEEE
Trans. Syst., Man & Cybern., SMC-15(1)*, 175-189.
mdani, E.H. (1977). Applications of fuzzy set theory to control systems - a
survey. In M.M. Gupta, G.N. Saridis and B.R. Gaines (Eds.), *Fuzzy Automata
and Decision Processes*. North-Holland, Amsterdam. pp.77-88.
slow, A.H. (1954). *Motivation and Personality*. Harper Brothers, New York.

316

Parsons, T., and E.A. Shils (1951). *Towards a General Theory of Action.* Harvard University Press, Cambridge, Mass.

Pask, G. (1968). *An Approach to Cybernetics.* Hutchinson, London.

Perkins, M., and J. Powell (1979). The assessment of energy utilization in primary school buildings. Hampshire County Council internal paper.

Procyk, T., and E.H. Mamdani, (1979). A linguistic self organizing process controller. *Automatica, 15,* 15-30.

Willey, H.B. (1978). *A Theoretical Framework of Environmental Control in Buildings.* PhD thesis, University of Cambridge.

Willey, H.B. (1982). Fuzzy perception, fuzzy modelling and fuzzy control of the environment in buildings. *Arch. Sci. Rev., 25(3),* 75-80.

Willey, H.B. (1985). An occupant orientation for architectural science. In J.D. Kendrick (Ed.), *Design and Science,* Proceedings of the 18th ANZAScA Conference. University of Adelaide. pp.11.1-11.10.

Wyon, D.P. (1973). The role of the environment in buildings today: thermal aspects. *Build Int., 6(1),* 39-54.

Addendum

Energy Efficiency—the Role of Government

B. D. Emmett

Director-General of the UK Energy Efficiency Office, Thames House South, Millbank, London SW1P 4QJ

INTRODUCTORY REMARKS

Chairman, Ladies and Gentlemen, this is my first public engagement after eleven working days as Director-General of the Energy Efficiency Office and I count it as a matter of good fortune that it should be in Newcastle and at this Conference. The Conference must assess my following remarks accordingly. The general good fortune of being here today is reinforced by the leading role of Newcastle in the field of energy efficiency. In this particular sector of affairs - you Mr. Chairman, will be more familiar with examples in other spheres - Newcastle and the North-East have given a lead not only within the UK but also within Europe. Moreover, I think the Conference falls at an interesting stage in the UK's development and implementation of energy conservation and energy efficiency policies.

SOME GOVERNMENT STRATEGIES

Later today I shall be visiting the Newcastle Neighbourhood Energy Action offices to hear about the progress that has been made there in reducing heating bills whilst improving living conditions for the underprivileged and older citizens of Newcastle. Industry and commerce have also taken up the energy efficiency message vigorously here. Shortly, I will be presenting the regional NIFES/Powematic Energy Manager of the Year Award to just one of the many companies in this region which are reaping the benefits of improved energy management practices. The energy management groups in the North East are also very active in promoting energy efficiency - last year the Tyneside group was the British Gas Energy Management Group of the Year. But perhaps most important for the future of energy efficiency in the region, is the study which is currently underway, under the auspices of the Northern Region Councils Association with assistance from a wide range of companies and organisations, to analyse energy patterns and the potential for savings across all sectors. The study is funded, like this Conference, by the CEC and I am sure that Dr. Fee will be able to tell us more about this.

All this activity and more adds up to an impressive record in energy efficiency for the region. It has taken place against the background of continuing efforts by the Energy Efficiency Office. Our overall approach has been clear, to tackle the barriers to the free operation of market forces which have prevented the UK from realising its full potential energy efficiency. For example, to tackle the problems of lack of management attention, the EEO held Breakfast Meetings between Ministers and senior executives all over the country, with the result that many companies have, for the first time, appointed a manager to be directly responsible for their energy costs. Many of you will remember the very successful event here in the Civic Centre last September, which features local companies like Lonrho, which have implemented a comprehensive programme of energy management measures, and like Newcastle City Council, whose excellent work I have already mentioned. The EEO has backed up high profile activities like these with a programme of advertising, promotional events and literature, and advice booklets to ensure that awareness of energy efficiency is maintained through all sectors of the community.

To help tackle the lack of relevant skills, the EEO has developed Monitoring and Targetting programes in over 15 sectors to date (including industries like metals and chemicals which are very important for this region) to introduce techniques for the accurate and effective measurement of energy and setting of targets - the key to good energy management. Through the programme, a wide range of firms in industries as diverse as textiles and water have been able to make savings of up to 18%, often with very little initial investment. We also encourage the Energy Management Group movement, to help energy managers develop the skills they need from the groups' seminars and meetings. There are over 75 groups throughout the country actively promoting energy efficiency - including the four excellent groups here in the North East. And the Demonstration Scheme and our R&D programme has encouraged the development and replication of new energy efficiency technology and techniques, throughout industry, commerce and the public sector. We also work to overcome institutional and financial barriers to energy efficiency, and I hope that there are innovative approaches in this field also.

Contract energy management offers one route for organisations in the public and private sectors to draw on outside resources of finance and expertise to improve their energy efficiency at no risk to themselves. You will be hearing more about this subject later. One of the recent tasks of the EEO is to produce guidance, along with the Treasury, to clarify the public expenditure treatment of such schemes and remove any obstacle to their development in the public sector.

So our work takes many forms - last but not least, our Regional Energy Efficiency Officers take the work of the EEO right into the heart of the community. Arthur Hoare, our North East Regional Energy Efficiency Officer, has played a considerable role in promoting energy efficiency in the region, and indeed has helped to organise this Conference. These are just some of the methods by which the EEO works with others in the community to encourage greater energy efficiency.

The EEO has had considerable success over the last few years in promoting energy efficiency through cost effective collaborative activities. But I believe that we must now question this, and other strategies, and whether new approaches are needed to take energy efficiency forward into the future.

CONCLUDING REMARKS

If the nation's work on enhancing energy efficency is to progress and prosper, I think those of us involved have to look hard at what we do, the environment in which we are operating and the relevance and attractiveness of our programmes. There have been changes in recent years in the wider energy, social and economic environment. When I am acting professionally as DG of the EEO and assessing the VFM of the EEO's expenditure, defending its programmes, fighting for its budgets, etc. how do I deal with changes in world energy prices and demand, changes in energy availability, the consequences of the switch from energy conservation to energy efficiency, the real economic benefits of energy savings attributable to Government expenditure, etc. Some of the schemes I am now responsible for managing are defended on the basis that £1 of public expenditure leads to £30 per year savings in energy costs - on that sort of return I personally will not be satisfied until for each £1 of public expenditure there is a £1 recovery for the public purse from the beneficiaries. Energy efficiency may be linked with apple pie and motherhood; energy efficiency may also be caring, conservationist, constructive commonsense, even communataire; but is it the automatic flavour of the month that it used to be? I think we need to think hard about all these things - there are some stock answers but some of those brought down from the THS shelves look decidedly dusty to me. We need to be sure that the EEO's work is fully adjusted to the situation, possibilities and personalities of 1987 and beyond.

Fortunately, this Conference and the performance of Newcastle and the North-East in energy efficiency, provide many of the questions, if not all the answers, and I am sure that all of us concerned with this Conference will be better brief as a result of our participation. (At this stage in the proceedings Mr. Emmett presented the regional NIFES/Powrmatic Energy Mangement Award for 1987 to Mr. Colin Baker of Lonrho Textiles).

C. E. C. Initiatives in the Development of Energy Efficient Plant and Processes

Dr. D. Fee

Energy Directorate, C.E.C. Brussels, Belgium

INTRODUCTION

Since the first energy saving Resolution was passed by the Council of Minister in December 1974, the Commission has implemented a series of initiatives which have been aimed at assisting the realisation of the Council's energy saving objectives. The Energy Council of December 1974 also set energy objectives to be attained by 1985. The energy saving objectives set by the Council called for a 15% reduction in energy consumption in the Community by 1985 using the 1973 consumption level as a base. This target was surpassed, and figures for 1985 show that overall energy consumption dropped by 20% during the period 1973-1985.

The industrial sector of the EUR-12 accounts for 30% of the final energy consumption and therefore ranks second to the residential and tertiary sector in energy use. The industrial sector also accounts for approximately 30% of the Community's Gross domestic Product. However, a large share of G.D.P. in other sectors of the economy (tertiary, transport, services, etc.), stems directly from industrial activity. The industrial sector is therefore a very fertile field for energy efficiency initiatives since savings there will have impacts at several levels of the national economy.

The initiatives undertaken by the Community are not concentrated in any one area, and try to effect savings from improvements in technology, infomation or in the provision of finance.

The Council of Ministers set energy objectives for 1995 at their meeting in September 1986. These objectives call for a further improvement of 20% in the rational use of energy by 1995. In the current energy market situation of relatively low fossil fuel prices this objective will be diffilut to obtain. However, the maintenance and strengthening of the measure already taken by the Member States and the Community should go a long way to the achievement of the Council's objectives.

Actions at Community Level

Introduction

In view of the importance of making efficient use of energy in undertakings, the Council, acting on a Commission proposal, took several steps in the 1970s to promote this aim, particularly in industry. (See, for instance, the Counci resolution of 17 December 1974 and the recommendation of 25 October 1977).

In its resolution of 9 June 1980 the Council sought to step up efforts in the Community to save energy and recommended that the Member States adopt certain guidelines on a basic energy savings programme. It issued further guidelines in its resolution of 15 January 1985.

Alongside this coordination, the commission initiated a number of Community actions described below.

Sectorial audits

With the cooperation of industrial circles, the Commission set out to make and publish energy audits for different industries with a view to making a major contribution towards energy efficiency in industry.

Each report covers a specific industrial sector and comprises the following sections : a brief description of the industry and its energy profile, the technologies already available or still in the course of development and those being demonstrated as capable of promoting a more rational use of energy.

Various consultancies in Member States have carried out the studies and these have been published with the agreement of the European professional associations concerned.

Technically speaking, the audit reports are very useful,particularly in sectors characterized by numerous small and medium-sized undertakings. The reports enable small businesses to check their own consumption against a reference level and benefit from the description of the available technology in order to improve their own performance. For management, the reports are a guide to be used in making decisions on RUE investments.

The information is also directed at banks which will be called upon to finance such investments. The audits are published in the "Energy" series by the Directorate-General for the information Market and Innovation (DGX111).

The Commission has made provision for regular updates.

The following audits have so far been completed :

(a) Published audits

No 1 :   STEEL   (Nuova Italsider)

No 2 :   ALUMINIUM   (Aluminium Federation Ltd.)

No 3 :   PAPER   (P.A. Management Consultants Ltd.)

No 4 :   GLASS   (Comite Permanent des Industries du
                                        Verre de la CEE)

No 5 :   BRICKS AND CLAYS   (British Ceramics Research
                                        Association Ltd.)

No 6 :   CHEMICALS   (Serete)

(b) Audits completed and in course of publication

No 7 :   GRAINS   (SEMA)

No 8 :   MILK   (Birch & Krogboe)

(c) Audits planned or in preparation

No 9 :   CERAMICS   (MacHale Associates, Dublin;
                                        European Studies, Athens)

         SUGAR

         NON-FERROUS METALS

         ABATTOIRS

         BREWERIES

         CANNERIES

         STEEL   (Update of the first audit)

         CONSTRUCTION MATERIALS

         METAL GOODS

         TEXTILES, LEATHER, CLOTHING

         RUBBER

Non-nuclear research and development programme

Aware of the contribution of new technologies towards the achievement of
the energy policy objectives, the Council in 1975 approved a first
Community energy R&D programme managed by the Directorate-General for
Science, Research and Development (DG XII).   The programme included a
large section on energy saving which in fact covered the following aspects:
improving insulation in buildings, the use of heat pumps, urban transport,
the recovery of residual heat, materials recycling , energy from waste,
industrial processes and energy storage.

More than 100 research contracts were negotiated with Community institutions and undertakings and the budget for the programme exceeded 22 MECU, half of it paid by the Commission.

The programme identified a huge range of techniques of use to industry. The results were announced at the conference on "New Ways to Save Energy" held in Brussels in October 1979.

Given the success of the first programme, the council decided in 1979 TO ADOPT A FURTHER PROGRAMME. The call to submit projects attracted some 600 proposals relating to energy saving, 160 of which were selected. The total cost of the second energy saving programme is some 50 MECU, 25 million of which is being paid by the EEC. The results of the second programme were presented at the conference on "Energy Conservation in Buildings" held in November 1984 at the Hague and at the conference on "Energy Conservation in Industry" held in February 1984 in Dusseldorf.

In the first two programmes research connected with industry covered the following topics :

- general research and development on industrial heat pumps (particularly in the second programme);
- controlling and improving combustion;
- advanced heat exchangers;
-heat recovery in industrial processes-
- organic Ranking cycles;
- the optimization of metallurgical processes;

- recycling of raw materials, plastics, fibreglass, etc.;
- developing catalysts for the chemical industry;
- devulcanization of rubber (used tyres);
- process integration in the textile industry (impregnating damp fabric, etc.);
- improving various processes in the cement and paper industries and in preparing grain for human consumption, etc.;
- the use of microwaves in industry.

The invitation to submit projects for the third R&D programme was closed in July 1985.

The Commission received 376 proposals on energy saving but the budget available for contracts is only some 23.9 MECU. This has meant that the Commission and the advisory project selection committee, made up of government officials, have had to drop many valuable prosposals.

The Commission has given priority to projects in which several research centres or industries are involved, and in particular to topics on which work was done at earlier stages (e.g. basic research on combustion). This has resulted in the launching of a global harmonized research project on the subject.

Since the general call for tender of 1985 the Commission has concentrated the research effort on perfecting new industrial processes making major energy savings. Several specific calls for tender were made where the 1985 call for tender and the smaller specific calls for tender have led to a total of 116 projects being accepted in the energy saving sector. Research tends towards multidisciplinary activities and integrating advanced technologies developed in other contexts. Examples of fields worth considering are ; applied superconductivity, the applied development of new diagnostic methods, applied MHD (Magneto Hydro Dynamics).

Energy Demonstration programme

Everyone recongises the importance of real-life demonstration of technologies resulting in the more efficient use of energy if the penetration of these technologies into the market is to be speeded up.   :

In 1977 the Commission therefore proposed that the Council should set up a programme covering various aspects of energy that were of importance to the community as a whole. In June 1978 the Council adopted Regulation No 1303/78 on the granting of financial support for Community demonstration projects resulting in a major improvement in energy efficiency. Financial support for such projects has been renewed successively every year since then. Nevertheless there has been a change in the type of project accepted and in the clauses concerning repayment of the subsidy in the event of the project proving successful.

In the section on energy saving in industry, 224 projects were adopted between 1979 and 1987. To this should be added 57 projects connected with the energy industry and 26 projects on heat and power. By the end of April 1987, 309 contracts providing Community support for demonstration projects in energy saving have been concluded. Of these 84 have now been completed. A list of current projects broken down by subject and the technology involved is given in Annex I.

The Commission had an analysis of the programme results made by independent experts (See COM (85)29/2). Their verdict is that the section on energy saving in industry can be regarded as the section of the demonstration programme which has the best prospects of success.

The number and quality of the projects, their profitability, the large amount of energy saved by each project and the potential for reproducing these savings on a larger scale are the main reasons for the overall success.

The Energy Bus programme has been extended from October 1985 to September 1987 but its aims have been reframed. It will concentrate on a few sectors where there is a high energy saving potential (ceramics, abattoirs and the preparation of cold cuts, dairies, breweries and malt-houses, textile finishing, the leather industry, industrial washeries).

In each case the methodology used to analyse a sector is worked out by someone familiar with that sector.

The Commission's departments plan to use a sample of the completed audits to assess how far undertakings have actually carried out the investments recommended as a result of the audits.

Loan systems

The European Investment Bank grants loans for certain types of investment
in energy saving, including in industry.

Between 1973 and 1986 the EIS lent nearly 15,500 MECU to energy sector
and 3,800 million of this went on the rational use of energy. Some of
the projects (involving over 1 000 MECU) received support from the New
Community Instrument (NCI).

The Bank's decisions on lending, in accordance with principles enunciated
in the Treaty of Rome and laid down by the Committee of Governors of the
Central Banks, are based on economic and financial criteria including, as
an essential factor, the solvency of the applicant and the prospects of
the project being profitable. Major projects are examined by the EIB
itself. Smaller projects are assessed by the financial intermediaries in
the Community countries which have Links with the Bank (global Loans)-
subloans ranging from 20 000 ECU to 7,5 MECU are granted under EIB rules.

Over 400 projects have been financed in this way in Italy, France, the
United Kingdom and Denmark and have resulted in savings of 2.5m toe/year.

The major projects mentioned in an EIB Report relate to the development
of geothermal resources, the modernization of oil refineries, global
loans for investments aimed at achieving the Community's energy goals,
combined heat and power generation projects, etc.

The financial terms for obtaining EIB loans are very attractive in
certain countries which have had high rates of interest, such as
Italy, Greece, Ireland, the United Kingdom and France.

New methods of financing

Firms wishing to make RUE investments, whether they are users or
producers of equipment, have various financing methods at their
disposal.

Firstly self-financing may be mentioned; this requires little comment
other than to say that few firms can do this in practice. Even where
they could, they may prefer to devote their funds to development
projects with a higher risk, but entailing greater prospects for
eventual expansions or redeployment.

Several Member States provide partial subsidies or tax relief for energy saving
investments. However, firms will have to turn to the money market for the
major part of their investment capital.

Banks will sometimes be reluctant to grant loans on favourable terms
(Long-term, Low interest rate) where the overall financial position of the
firms is not strong, even if the investments proposed present little risk
and are very profitable. Such firms will not obtain an EIB loan, as they
cannot offer sufficient guarantees.

A credit insurance mechanism, the conditions for which remain to be worked
out, could provide a valid solution to this very common financing problem.

One of the effects for firms, apart from stimulating investments, would be lower interest rates and access to a type of preferential loan on the Financial market.

Present systems of subsidies and tax reliefs could usefully be supplemented or replaced by a credit insurance system. The Commission plans to study such a system for equipment manufacturers in a post-demonstration stage. The Member States could also consider a system of this kind for users of RUE equipment.

Another way of financing RUE investments is for users to turn to an outside body which would study, implement and finance the investments on behalf of the firm.

This system, known as Third Party Financing (or "performance contract" or financing through savings") began in the United States and Canada around 1980 and has developed rapidly, especially in commercial, services (hospitals, schools, etc.) and office sectors.

The financial intermediary is repaid from the energy savings made over a period which is agreed on, and is responsible for the management of the equipment. The user may receive a refund on his energy costs.

After the agreed period, the equipment normally becomes the property of the user, who has not had to commit any funds at the outset.

The penetration of Third Party Financing (TPF) into industry has been more recent. So far it has been restricted to non-processing sectors such as boiler rooms, air conditioning of premises, remote control, combined heat and power generation, etc.

The main reasons are the complexity of industrial processes, which do not give much scope for standardization; the reluctance of industries to let outsiders know about their manufacturing methods; the lack of experience of TPF bodies and uncertainty concerning the repayment of the sums advanced (working life of machines, etc.).

In Europe, companies offering this type of service have been slower to emerge. They are linked with large groups in the energy sector which have sought to diversify or with design and engineering consultants. A considerable inital capital stake is required for the first five to six years of operation, after which repayments tend to catch up with the rate at which new investments are made. However, TPF bodies have easier access to the financial market han RUE equipment users, as the risk is distributed over a portfolio of projects whose quality of design is above the average normally ound i industry. It is certainly a great advantage to belong to a group which has a reputation for financial soundness (such as an oil company), because the profitability of TPF companies depends on the proportion of the burrowings which they can mobilize.

The Commission is studying ways of assisting industry to use TPF in collaboration with the public and private bodies involved. In 1986 a study on the possibilities for TPF in the Community was published.

The Commission has also prepared model Third Party Financing contracts for use in a European context. These model contracts are accompanied by guide books which explain their operation. A seminar on Third Party Financing which will be held in Luxembourg in October 1987 will examine the wider application of this technique in Europe and will present the work already completed by the Commission.

Where equipment manufacturers are concerned, the Commission is at present considering whether to present a proposal for a Council Regulation providing for credit insurance for products for the efficient use of energy. The aim of this proposal would be to facilitate access to the financial market for RUE equipment manufacturers by providing partial coverage of the risk for banks in case of bankruptcies, losses, etc.

One of the various financing possibilities would be for roughly half the sum to be insured by a surcharge on the interest rate, with a subsidy from the Community after the projects had been assessed by the Commission.

This type of system could have an important leverage effect and lead to considerable industrial activity and energy savings (several m toe/year by the year 2000) using limited resources, which careful selection of projects should keep within strict bounds.

Access to this Community mechanism would be reserved for projects of sufficient scale, but Member States could provide for a similar mechanism on a national or regional scale for smaller development projects.

Publications

The Commission has set up an extensive system of publications which deal with the activities mentioned above.

The demonstration projects are the subject of regular publications on the contracts in progress and of leaflets describing each successfully terminated project.

Industry audit reports are published in the form of brochures intended primarily for firms and federations involved, government departments,etc.

The project carried out under the Research and Development Programmes are presented in the form of brochures.

The final reports of projects are available from the Publications Office in Luxembourg. Certain confidential information is sometimes recorded in unpublished annexes. In addition, video cassettes on some projects are available for loan.

Conferences and seminars

Several very well attended conferences have been held to make known the results of the Research and Development Programmes. On-site information workshops are more usual for demonstration projects and this practice should expand considerably over the next few years.

Energy saving associations

Energy saving associations are very important collaborators in
disseminating knowledge of RUE technology.

These bodies also contribute to a better understanding of energy problems
and can draw the attention of public authorities and the Commission to
trends in industry.

National federations of consultants and energy managers, under which most
private or semi-public energy consultancy bodies are grouped, recently
formed themselves into a European Federation at EEC level, called the EFEM
(European Federation of Energy Management Associations).

The Commission welcomes this initiative and will endeavour to involve
this federation as much as possible in any RUE work which it may initiate.

The energy saving associations can play a dynamizing role in large sectors
of the economy, especially in industry. Some of these associations are in
fact mainly or exclusively made up of energy managers in business, and engage
in wide-ranging exchanges of information on actual and possible developments
in their firms, regions or sectors. In a number of countries they have
organized themselves into regional clubs.

This activity appears to confirm that energy problems are a permanent concern
for many sectors of industry and should be given practical encouragement
by the Member States.

Data Banks

The commission has set up two data banks on energy saving.

The EDSES system at the joint Research Centre in Ispra, Italy, provides
support to teams carrying out energy audits under the Energy Bus
Programme.

Data are transmitted anonymously by participants and then processed. The
system provides statistics on consumption and potential energy savings.

The SESAME data base contains information on research and development and
demonstration projects and on hydrocarbon technology projects in the
Community. It has been open to all the Commission departments since the
beginning of 1984. For a trial period, the section containing data on
demonstration projects and hydrocarbon technology projects has now been made
available, by means of international teleprocessing systems, to authorities
responsible for energy questions in all the Community Member states. At
the end of the trial period , information on Community technological projects
(energy demonstration projects and hydrocarbon technology projects) will be
made available to the general public through the major data bank centres
which supply the information market in Europe. It is planned to open them to public
access by the end of 1987. From this date onwards, anyone wanting this
information will only need to interrogate one of these centres to obtain
data on projects managed by DG XVII. It is invisaged that users of the Prestel,
Bildschirmtext and Minitel systems will also have access to SESAME at a later
date.

SESAME now contains all data on the (roughly 1,000 projects managed by DG

XV11. There are new projects every year and the range of information on each project is to be extended. SESAME should progressively become the core of a Community centre for information on RUE technology and, in collaboration with European manufacturers, on equipment and techniques available on the market.

Energy Efficiency—Promoting Action in the North East

David Green

Energy Adviser to Association of Metropolitan Authorities
Newcastle upon Tyne, UK

KEYNOTE ADDRESS

Many commentators have recognised the benefits that energy efficiency can bring - yet to many others it is still essentially a marginal area of activity. What can be done to tackle this situation, and, significantly, what future role can energy efficiency play in areas such as the North East?

In answering this question one needs to first examine, in broad terms, the nature of energy demand in the North East and the factors which influence its use and that is why the Northern Regional Council's Association have been pleased to support the current regional energy study in association with the European Commission.

Change is clearly taking place. It is a truism to say heavy engineering no longer has the role it once had in the region. This is not the place to set out the pros and cons of this trend, but rather to point out the effect of this has upon the nature and level of energy use in the North East.

Newcastle the City Council has found, for example, that the companies and activities it supports, when established, tend to be in product areas that are less energy intensive, whilst, in Newcastle there is a clear shift to the service sector which will have a futher effect upon the level and nature of energy demand. We would now expect more energy consumed in the commercial sector and less in the manufacturing sector, more for shops and offices, less for factor premises and so on. The intensity of energy used per employee will probably have declined significantly.

Alongside this is the nature of demand in the domestic sector. The North's largely aging population will produce a particular requirement for space heating, and the extent of the public housing sector compared to other regions will also mean certain solutions will be more suited to action here than elsewhere.

All this means we need to build throughout the region an energy efficiency strategy which:

*   encourages new and existing industry to monitor and plan their energy use in the most appropriate manner.

*   enables public owners of housing to seek effective investment options designed to boost the energy efficiency of the properties they own.

*   mobilises the interests of institutions in achieving good value for money and targets energy management efforts accordingly .

What does this mean in practical terms?

Here I turn to the example of Newcastle upon Tyne, an authority which has acted to achieve the efficient use of energy since its first initiatives in 1968.

In outlining what Newcastle is doing, it is important to recognise that any local authority does not naturally have a focus in the energy field. However it does have a concern with:

*   ensuring its tenants live in a reasonable environment.

*   stimulation of the local economy in an effective manner.

*   investment in measures designed to improve the operation and performance of the authority.

Such matters in turn affect an authorities response to the opportunities for the efficient use of energy. Hence in Newcastle our approach to energy demand management reflects such concerns and has evolved to cover a number of policy areas.

Firstly; in the housing sector; the Council's Housing Committee has maintained a substantial insulation and heating programme. This work has drawn heavily upon "third party finance", and used the Council's resources to improve insulating and heating in most of the Council's 43 multi-storey blocks, together with a significant number of the Council's 20,000 or so worst heated houses - out of a total stock of around 50,000 (about 40% of the city total housing stock).

This work has in turn, at the height of the programme, supported some 300 jobs, largely in the private sector.

Secondly; members of the Council have recognised the need to ensure local companies do become efficient users of energy. Future jobs may depend upon it. Hence our efforts have ranged from the constuction of low energy use factory units for new businesses, to the pioneering work of our Energy Information Centre and its support for the private sector lead Energy Task Force for industry.

Thirdly; as a Council Newcastle's own energy bill runs to millions of pounds. A clear incentive therefore exists to promote efficiency. Our City Architects Team has lead the way in this task, using dedicated funding from our central capital budget to achieve savings running at well over a million pounds a year on a recurring basis. Awards have been won from British Gas and the government's Audit Commission amongst others.

Yet a key part of all this work is to ensure that the techniques of energy efficiency are used by and are relevant to the Community. The Energy Information Centre has a crucial role to play here, and is a central access point to many local energy efficiency services, having handled well over 12,000 direct inquiries in the two years of its existence. Allied with the work of our local home insulation project for lower income households, Keeping Newcastle Warm, the practical benefits of better use have been (literally) brought home to many householders.

Such work has clearly put the region on the map. In the domestic sector alone, well over 7,000 people are now employed by the network of 400 community insulation projects initiated from Newcastle under the aegis of Neighbourhood Energy Action.

Taken alongside the expertise of the academic institutions and industry in the field of energy use, it is the Council's view that the region is well on the way to become a clear centre of excellence in the field of energy efficiency.

So what of the future?

There are a number of opportunities we must now be prepared to take forward:

- the scope for partnership.

  with the private sector being given a leading role in redevelopment, it will be vital to ensure that the scope this may offer for investment in energy efficiency is adequately promoted and developed.

- the inner city.

  enterprise in the inner city is not just a catchphrase; we need to build up practical approaches and targetted energy efficiency improvements could bring many practical benefits to industry and households alike.

- neighbourhood initiatives.

  practical participation of community agencies is now happening. We need to make sure this is adequately supported, and recognised throughtout major institutions.

Finally, we must also look for a new relationship with the privately (or soon to be privately) owned fuel utilities. Corporate advantage can be in energy efficiency, and much more could be done by some utilities to recognise this and respond in a creative manner.

So, in conclusion, whilst oil prices may go up and down, what is clear is that energy efficiency is here to stay and many opportunities are now open to make sure that the northern region effectively leads the way to ensure that the UK does demonstrate that energy efficiency is not just fine words, but essentially about practical and popular action.

A Computer Based Fault Diagnosis System for Refrigeration Plant

N. P. Daniels

*WS Atkins Management Consultants, Woodcote Grove, Ashley Road,
Epsom, Surrey KT18 5BW*

ABSTRACT

To assist the operators of refrigeration plant a microcomputer based fault diagnosis system is being developed that can be operated by plant operators, inputing information manually, or be fully automated gathering data from the plant with permanent instrumentation and data links.

The importance of running refrigeration plant is ever increasing along with the demands of management to reduce the labour related maintenance budgets. This results in plant that runs and still apparently produces its product being neglected. This leads to increased running costs and can lead to major plant failure, with the inevitable costs to rectify the faults exceeding the costs of maintaining the plant properly.

This short paper discusses the development of a computer based fault diagnosis system for refrigeration plant.

A Survey on the Energy Performance of the Turin Building Stock

G. V. Fracastoro*, V. Giaretto**, M. Masoero** and G. Pavoni**

*Istituto di Fisica, Università della Basilicata, Potenza
**Dipartimento di Energetica, Politecnico di Torino

ABSTRACT

The paper describes the results of a survey performed on 300 buildings of Turin (Italy) in order to improve the understanding of the energy performance of the city building stock and evaluate the feasibility of district heating or of an energy retrofit campaign. Data on the dimensions, construction and maintenance condition of buildings and heating systems were collected on-site according to a well tested survey methodology. A simulation model, developed by the authors, was then employed to calculate the main thermophysical and geometrical features and the annual energy consumption of each building. For each feature a correlation has been established with the data available from the 1981 National Census (volume, age, number of floors, and whether the building is detached or not). Based on these correlations, the annual energy consumption and installed power was estimated for each building, obtaining a good agreement with the data derived from the detailed investigation.

KEYWORDS

District heating, energy conservation, building survey, energy estimating methods.

INTRODUCTION

This paper presents the results of a field inquiry on space heating demand at the urban level, which has been completed within the framework of a research program on energy policy issues in the area of Turin, a city of about one million inhabitants situated in the cold Northwestern part of Italy (latitude 45°N). Activity of the work group, which includes the major local gas and electric utilities as well as the regional and town administration, is coordinated by the Department of Energy Engineering of the Polytechnic University of Turin; funding is provided by ENEA (Comitato Nazionale per la Ricerca e lo Sviluppo dell'Energia Nucleare e delle Energie Alternative), under research grant no. 206/84 (research coordinated by Prof. E. Lavagno).

The work group was formed in 1981 with the goal of investigating the technical feasibility and the economics of a set of different strategies aimed at optimizing the use of energy in a large, industrialised urban area. Options include the transformation of an existing power station into a cogeneration plant, to be connected to a district heating network that would serve roughly one fifth of the Turin population (200,000 people). Altogether the system would have a thermal capacity of 576 MW, and an electric capacity of 136 MW. Alternative or complementary plans call for large-scale retrofitting of buildings, in order to reduce heating demand, and an extension of the natural gas distribution network. A plan of the south part of Turin, indicating the district heating grid layout, is shown in Fig. 1.

340

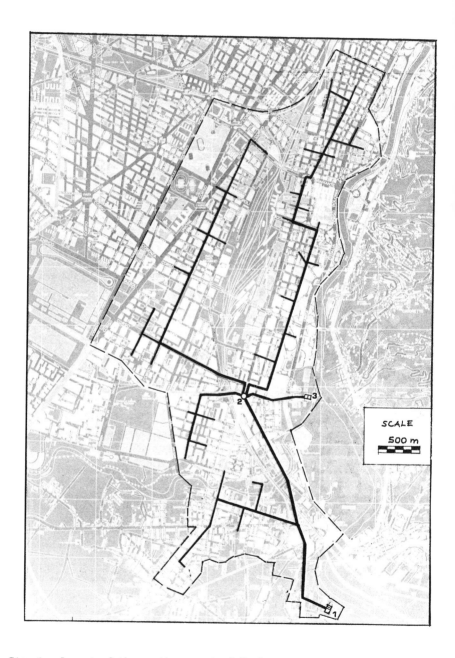

Fig. 1. Layout of the southern part of Turin
1. Cogeneration power station (Electric 136 MW, Thermal 226 MW)
2. Distribution station
3. Auxiliary thermal plant (350 MW)

The research activity within the Politecnico has been focused on three areas: the preliminary design of the distribution grid, the analysis of the energy end use, and the environmental impact issues associated with the various energy production, distribution and utilisation technologies.

The aim of the investigation on the end use was to build up an accurate and detailed map of the spatial distribution of heating demand in Turin. Knowledge of such distribution was required on one hand in order to carry out the feasibility study of district heating, on the other to evaluate the technical and economical feasibility of retrofitting of the building stock.

This study has been performed within a larger framework, consisting of investigations at different scale levels:

1.   Information on the entire building stock (e.g., size and year of construction of buildings) was initially obtained from the 1981 Census data base (Pavoni, Calì and Ruscica, 1984).

2.   A questionnaire was distributed to a sample of roughly one tenth of the population (10% of the questionnaires were returned compiled), which provided information about size of apartments, occupants' habits, heating costs, etc. (Carpignano, Filippi and Giaretto, 1986).

3.   A field survey was performed on a sample of 300 buildings (1/60th of the entire building stock), concerning the characteristics of the building structure and services (results are presented in this paper).

4.   Finally, 30 buildings of the sample of 300 will be instrumented and monitored to provide indoor air temperatures, electric energy use profiles and heating consumption data.

This set of investigations should allow a step-by-step extrapolation of the information collected, with a high level of accuracy, from the smaller scale samples all the way up to the entire building stock.

A number of similar investigations carried out in the past in Italy and abroad (Socolow, 1981; Fels, 1986; Pagani et al., 1984; Boffa, 1984) have provided the background for our study. The distinctive feature of the present study is the integrated methodology which makes use of field investigations, energy simulation models, and extrapolations based on rigorous statistical criteria.

METHODOLOGY

The first priority which was set forth at the beginning of the inquiry was to obtain results which could be readily compared with those provided by similar surveys presently being carried out in Italy. For this reason, a well tested survey methodology, developed by the National Research Council (CNR) of Italy, was adopted (Boffa, 1984).

External contractors were asked to complete a surveying form with data on the dimensions, construction characteristics, and maintenance condition of the building structure and heating system.

The investigation aimed at deriving the spatial and time distribution of energy demand for building heating (E) and design heating power demand (Q), i.e.:

$$Q = f(x,y)$$

$$E = f(x,y,t)$$

(1)

The reference time interval (t) adopted was the whole heating season (15 October - 15 April) and the corresponding climatic data were those of the typical reference year (CNR, 1982). Data on the entire building stock were obtained from the 1981 Census Data Base (CDB). Among all the data contained in the CDB, only four of them, which will be referred to as "Reference Parameters" (RP), have been retained as relevant pieces of information about the building energy behaviour. The RP are:

> $P1$ = building gross volume (V)
> $P2$ = building year of construction or reconstruction (An)
> $P3$ = number of floors above ground (Np)
> $P4$ = an index specifying whether the building is
>   detached or not (Ic)

Therefore, the set of discrete spatial functions

> $P1 = f(x,y)$
> ....
> $P4 = f(x,y)$

(2)

is known from the CDB.

However, the RP alone do not provide sufficient information to calculate Q and E. To do so, one needs to know the relationship linking the RP to the thermal and geometrical parameters which influence the building thermal behaviour. Such quantities (wall surface areas, U-value and orientation, envelope air permeability, solar gains, etc.) are called "Thermophysical Parameters" (TP).

A proportional sample of 300 buildings was therefore selected in order to determine the link between RP and TP; among all possible TP, the following turned out to be the most important:

> $p1$ = surface to volume ratio (S/V)
> $p2$ = average U-value of the building envelope (Um)
> $p3$ = effective free heat gains per unit volume (Ag)

As the TP were known from the field investigations (directly or after pre-elaboration), the following "Interpolating Functions" (IF) were determined, together with the asociated errors:

> $pi = f(P1 ... P4, pj \neq i)$

(3)

Finally a Simplified Model (SM), based on the degree-day approach, was used to determine Q and E using the TP and some of the RP:

> $Q = f(P1 ... P4, p1 ... p3)$
>
> $E = f(P1 ... P4, p1 ... p3)$

(4)

To check the accuracy of the whole procedure, the results of the SM were campared to those provided by a Detailed Model (DM), which is basically a simulation tool making use of all the data collected from the field survey.

The flow chart of the whole methodology is shown in Fig. 2. The solid lines indicate the steps of the simplified procedure.

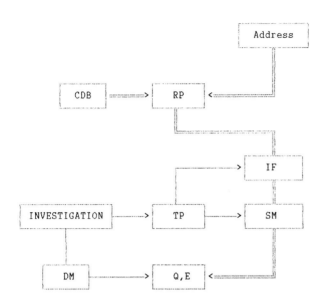

Fig. 2. Flow chart of the methodology.

BUILDING SURVEY

Surveying of the buildings was carried out with the aid of a set of data collection sheets, including the following sections:

1. **General information:** contains information on building location, year of construction, use, size, construction type and heating/ventilating system, energy consumption.

2. **Building component description:** includes a description of construction type, surface areas, orientation and shading of each wall constituting the envelope of the building.

3. **Building component description:** is a collection of forms in which the thermal properties of each component in the building envelope are described (U-value, thermal capacity, solar factor, etc.).

4. **Maintenance condition:** this section includes a checklist of possible problems in the building envelope (e.g.: moisture problems, thermal bridges, leaky windows, etc.) and heating system (poor boiler performance, leaky pipework components, unbalancing in the ambient temperatures, etc.).

The data contained in the collection sheets were transferred on magnetic tape for all successive handling by computer.

BUILDING SAMPLE SELECTION

The sample size was based on a tradeoff between the need to achieve a good statistical representation of the universe and the financial cost of the investigation.

According to the CDB classification of the RP, the building stock is divided into 96 classes, resulting from four volume categories, four age classes, three number of floor categories and two values for the detached/non detached index (4 x 4 x 3 x 2 = 96). Each of these 96 classes should in principle contain as many buildings as it is required to provide a reliable estimate of the energy consumption of a building belonging to that class.

The minimum number of elements required to restrain, with a certain level of probability, the real average within a certain confidence interval around the sample average may be estimated with the t-Student theory (valid for normally distributed populations). According to this theory, the minimum number of elements (n) is given by:

$$n = (2 \ t \ \sigma \ / \ p)^2 \tag{5}$$

where

t = a function of n and of the required probability level (Student's function)
σ = standard deviation of the universe with respect to the selected variable (i.e., energy consumption per unit volume)
p = confidence interval

Adopting a 95% probability level, knowing from previous researches (Fracastoro and Lyberg, 1983) that the standard deviation of energy consumption for residential buildings is typically σ = 20-25%, and assuming a 15-20% confidence interval, one obtains:

$$n = 20 \div 25$$

An inspection of the composition of the 96 classes in which the building stock was subdivided, showed that a large number of classes was almost void. Consequently, it was decided to restrict the investigation to a number of buildings such that each of the 16 "aggregated classes", formed by combination of age and volume categories (presumably the most important ones), contained on the average 20 elements. Therefore, the required sample size became:

$$N = 16 \times 20 = 320$$

This figure was subsequently reduced to 300 buildings.

Once the size of the sample was decided, its composition was still to be determined. The adopted criterium was that of proportionality, that is the frequency distribution of volume, age, number of floors and "detachement" in the sample should be as close as possible to that of the universe (i.e., the Southern quarters of Turin which would be served by the district heating network). The frequency distribution of sample and universe (the entire city of Turin) is given in Tables 1A÷1D. Deviations in class distribution is justified by the fact that the composition of the building stock of Southern Turin does not exactly coincide with that of the whole city.

TABLE 1A   Volume Distribution

Volume (m³)	< 1750		1750÷2800		2800÷6000		> 6000	
Sample	68	22.7%	32	10.7%	6	22.0%	134	44.7%
Universe	7100	38.6%	3810	20.7%	5546	30.1%	1921	10.5%

TABLE 1B   Age Distribution

Age (year)	> 1971		1946÷1971		1919÷1945		< 1919	
Sample	18	6.0%	188	62.7%	61	20.3%	33	11.0%
Universe	471	2.6%	9875	53.7%	4340	23.6%	3691	20.1%

TABLE 1C   Number of floors

Floors	0÷2		3÷6		> 6	
Sample	50	16.7%	190	63.3%	60	20.0%
Universe	4165	22.7%	9945	54.1%	4267	23.2%

TABLE 1D   Detached/Non detached

Type	Detached		Non-detached	
Sample	49	16.3%	251	83.7%
Universe	2779	15.1%	15598	84.9%

ENERGY ANALYSIS

The Simplified Model (SM) consists of the following equations:

$$Q = V (Um\ S/V + 0.35\ n) (Ti - To) \tag{6}$$

$$E = V \{24\ DD\ [Um\ S/V + 0.35\ n] - Ag\} \tag{7}$$

The Interpolating Functions for Um, S/V (see also Fig. 3), and Ag are the following:

$$Um = \begin{cases} 1.67 & \text{if } An < 1950 \\ 0.20 + 0.009 \ (An-1800) & \text{if } An \geq 1950 \end{cases} \qquad (8)$$

$$S/V = A + B/V + C/Np \qquad (9)$$

where

A = 0.3261; B = 182.33; C = 0.3267   (detached buildings)
A = 0.3498; B = 76.48; C = 0.3057   (non detached buildings)

$$Ag = 0.9168 + 0.4277 \ S/V \qquad (10)$$

The meaning of the symbols in Equations 6÷10 is the following:

Q = design heating power (W)
E = annual heating energy demand (Wh)
V = gross heated volume (m3)
Um = average U-value of the envelope (W/m²°C)
S/V= surface to volume ratio (1/m)
n = number of air changes per hour = 0.5 Vol./h
Ti = indoor design temperature = 20°C
To = outdoor design temperature = -8°C
DD = heating degree-days = 2570 °C·d
Ag = effective free heat gains per unit volume (W/m3)
Np = number of floors

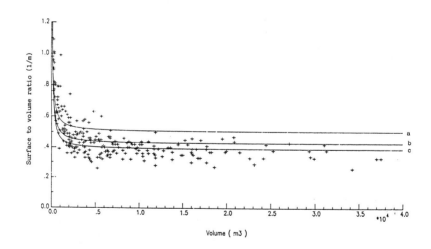

Fig. 3.   Interpolating Function for the Surface to Volume ratio
          Number of floors: a = 2;   b = 4;   c = 8.

The Detailed Model (DM) is of the steady-state type, but keeps track of the
most relevant dynamic phenomena by means of the so-called "coefficient of
utilization of heat gains." It determines the heating demand of the building
on a monthly basis according to the equation:

$$E = [Ec + Ev - f \ (Es + Ew + Ei)] \qquad (11)$$

where

Ec = conduction heat losses through the building envelope
Ev = ventilation heat losses
Es = solar heat gains (opaque walls)
Ew = solar heat gains (windows)
Ei = internal heat gains
f = coefficient of utilization of the heat gains

The meteorological input data used by the model (outdoor temperature and solar radiation) are those of the typical refernce year (CNR, 1982). The coefficient of utilization f is calculated according to the procedure illustrated by Agnoletto, Brunello and Zecchin (1979) as a function of the dimensions and thermal properties of the building.

The results of the energy analysis are summarized in Fig. 4 and in Table 2. The first graph of Fig. 4 shows the ratio of the heating power estimated with the SM to the power calculated with the DM. The three horizontal lines indicate the amplitude of the ± σ interval, σ being the standard deviation of the population. Similarly, the second graph shows the ratio of estimated (SM) to calculated (DM) annual heating energy demand.

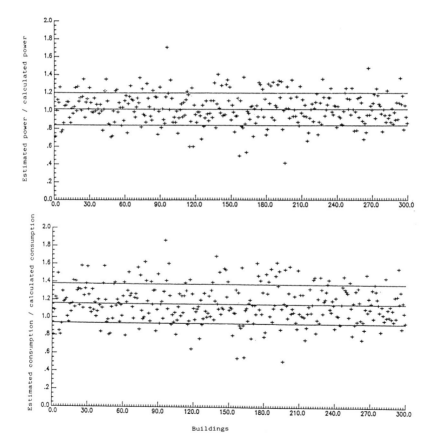

Fig. 4. Ratio of estimated to calculated power (top) and estimated to calculated heating energy (bottom) for the sample of 300 buildings.

TABLE 2 FREQUENCY DISTRIBUTION OF PERCENT ERROR

HEATING POWER	FREQ.	CUM. FREQ.	PERC. FREC.	CUM. PERC.
5|*******************************	70	70	23.3	23.3
10|*******************************	70	140	23.3	46.7
15|**************************	58	198	19.3	66.0
20|*************	30	228	10.0	76.0
25|*************	30	258	10.0	86.0
30|*********	19	277	6.3	92.3
35|***	7	284	2.3	94.7
40|**	5	289	1.7	96.3
45|*	3	292	1.0	97.3
50|*	2	294	.7	98.0

```
   +---------+---------+---------+---
   0        20        40        60
           Frequency
```

Arithmetic mean ....:	14.05
Weighted mean (extremes excluded) ....:	11.55
Modal class ....:	1
Modal value ....:	5
Median class ....:	3
Median value ....:	10.86
Standard deviation ....:	14.51
Coefficient of curtosys ....:	23.62
Coefficient of asimmetry ....:	3.77

HEATING ENERGY DEMAND	FREQ.	CUM. FREQ.	PERC. FREC.	CUM. PERC.
5|*****************************	48	48	16.0	16.0
10|*******************************	52	100	17.3	33.3
15|******************************	50	150	16.7	50.0
20|**********************	38	188	12.7	62.7
25|***************************	47	235	15.7	78.3
30|*************	23	258	7.7	86.0
35|************	20	278	6.7	92.7
40|*********	16	294	5.3	98.0
45|*	1	295	.3	98.3
50|*	1	296	.3	98.7

```
   +-----+-----+-----+-----+------
   0    10    20    30    40
          Frequency
```

Arithmetic mean ....:	17.15
Weighted mean (extremes excluded) ....:	16.03
Modal class ....:	2
Modal value ....:	8.33
Median class ....:	3
Median value ....:	15.00
Standard deviation ....:	12.69
Coefficient of curtosys ....:	6.33
Coefficient of asimmetry ....:	1.66

Table 2 shows the main statistical features of the frequency distribution of the percent error (in absolute value), which is defined as:

|Estimated Power (SM) - Calculated Power (DM)| / Calculated Power (DM)

or

|Estimated Energy (SM) - Calculated Energy (DM)| / Calculated Energy (DM)

Inspection of Table 2 shows that, in general, a better agreement is found in the heating power values than in heating energy demand. For heating power, the average value of the error is equal to 14%; for two thirds of the buidings the error is less than 15%, and in 90% of the cases it does not exceed 28%. For heating energy demand, the average value of the error is equal to 17%; for two thirds of the buidings the error is less than 21%, and in 90% of the cases less than 33%.

CONCLUSIONS

The statistical analysis of the results provided by the survey on the energy demand for space heating in Turin has confirmed the validity of the approach which has been adopted. By combining field surveys, simulation models and simplified prediction methods, a fairly accurate map of the energy behaviour of a large-scale urban system was obtained. Obviously, it must be stressed that the correlations presented in this paper are valid for the building stock of the city of Turin only. Nevertheless, the methodology seems flexible enough in its structure to be applicable to residential communities in general.

REFERENCES

Agnoletto, L., Brunello, P., and R. Zecchin (1979). Simple methods for predicting thermal behaviour and energy consumption of buildings. Proceedings ASHRAE-DOE Conference on Thermal Performance of Exterior Envelopes of Buildings. ASHRAE, New York.
Boffa C. (1984). Il riscaldamento di oltre mille edifici italiani. HTE - Habitat, Tecnologia, Energia, Anno VI, no. 29.
Carpignano P., Filippi M., and V. Giaretto (1986) Indagine sul riscaldamento ambientale nell'area metropolitana torinese: Risultati e commenti. Proceedings of 30th Congresso Nazionale ATI.
CNR (1982 Dati climatici per la progettazione edile e impiantistica. PEG, Milano.
Fels M. (1986). Monographic issue of the Journal Energy on "PRISM" (Princeton Scorekeeping Method).
Fracastoro G., and Lyberg M.D. (1983). Guiding Principles Concerning Design of Experiments, Instrumentations, and Measuring Techniques. Swedish Council for Building Research, Stockholm.
Pagani R. et al (1984). Energy conservation on existing buildings in Milan: Retrofit/maintenance associated program on 2000 dwellings. Proceedings of ACEE 1984 Summer Study on Energy Efficiency in Buldings, Santa Cruz (California), American Council for an Energy-Efficient Economy.
Pavoni G., Cali' M., and G. Ruscica (1984). Un modello per le strategie di conservazione energetica a Torino. Recuperare - Edilizia, Design, Impianti, n. 13, September-October 1984.
Socolow R., editor (1981) Saving energy in the home. Princeton experiment at Twin Rivers, Balliger.

Lots of Simple ideas work IF you want them to!

District Heating and Combined Heat and Power—a Major Contributor to Energy Conservation in Denmark

Fleming Hammer

Bruun & Sørensen Energiteknik AS, Aarhus, Denmark

INTRODUCTION

5 million Danes live in a fairly rough climate: Cold, wet and always windy.

More than 40% of the total heat demand is covered by district heating, which is to the benefit of Denmark's balance of payment and to the benefit of the environment as we do not import/use too much primary energy.

In fact, within a few years more than 50% of our heating demands will be covered by DH, and the major heat supplier will be a number of existing and future CHP-stations.

Today I would like to talk about various aspects of DH and CHP in Denmark. I will try to give you an impression of what we have achieved through pragmatic cooperation between producers and consumers of heat and mutually between consumers.

DH & CHP is certainly not a new invention, but it has developed differently in Denmark than elsewhere. Historically we have always been short of indigneous fuels, so efficient use of imported fuels has always been important in Denmark.

When the oil embargo appeared in 1973, the DH coverage was quite large, but most of it was based on fuel oil, and only a minor part on CHP. Due to DH's flexibility it was possible to reduce the dependency on oil very fast, and today less than 25% of the DH-production is based on oil; CHP based on coal being the major contributor with 50%. (Fig 1)

Looking at Denmark's total demand for heating, you will see how the oil contribution, which was more than 90% in 1973, will be reduced through enhanced use of mainly CHP at the end of this century. (Fig 2)

First generation CHP-plants consisted of a large number of diesel engines in local power stations around the country. When electricity generation was concentrated on few larger plants during the 50s and 60s, the DH-plants were kept on line by boilers only. Of the 1st generation schemes only a few in the largest cities, now based on steam turbines, were left in 1973. (Fig 3)

Second generation-schemes were started in the mid 70s. The power companies realized that co-generation was also a viable option, and a number of new schemes were established. To name some of them:

In Aabenraa a 3.3 km transmission line from Enstedværket to the city's existing DH-scheme was established, and a bleed was installed in the turbine. The same thing happened in Vordingborg, although 360 m's of the pipeline had to be underwater. Varde was connected to the Esbjerg-scheme through a 15 km transmission pipeline.

In Kalundborg the situation was different since there was no DH-scheme available. However, Denmark's largest power plant was converted, and a new DH-scheme was installed. I shall turn back to this scheme and to the Herning scheme where a new power plant was built and connected to the existing DH-networks. These plants were initiated shortly after 1973 and commissioned before 1982.

New schemes to complement the existing ones in and around our large cities like Copenhagen, Odense and Aarhus, and a large regional system connecting 4 cities around the Littlebelt were approved a few years later. These schemes will be operational and fully extended respectively within a few years and contribute heavily to the increase of the CHP-share.

An important factor in the development of the large CHP schemes was the existence of a great number of local district heating schemes which could be integrated in the large systems, whereby the desirable shift from oil to coal based CHP has been possible.

Many of the small local systems were initiated in the 1950s by a group of individuals, who formed a co-operative to take care of the planning, building and operation of the common district heating plant to the benefit of the members connected to the system. In the larger cities, the schemes were usually built by the municipal authorities.

During the 1960s many new schemes appeared. Today we have 350 DH-utilities of which 60 are municipal while the rest are co-operatives. A good deal of these, whether private or municipal, are today integrated in the large schemes buying heat from the power companies. (Fig 4)

Third generation. Very recently it has been agreed between the electrical utilities and the government that a large number of socalled "decentralized CHP-stations" shall be installed throughout the country and be based on domestic fuels.

In total 450 MW_e shall be installed before 1995, and some 30 minor and medium-sized towns with a minimum of 1000 homes and an existing DH-system are now being considered for these plants.

Natural gas, straw, wood chips, waste and biogas are the fuels to be taken into account.

CASE STORIES

Taking the development in three selected cities as models, I will now talk about some technical, operational, financial, administrative and environmental aspects of the CHP-concept.

Herning

Denmark's first dedicated district heating station based on lignite was established in 1950 in the town of Herning, and took over the heat supply of 140 houses through a 5 km network. Thereby, the houseowners could give up their own stoves or central heating boilers. The system was expanded during the 1950s, and by 1963 it was so big that the waste heat from a new incinerator burning household waste could be fed into the system as an all year base load. In the neighbouring town of Ikast and in the suburbs other district heating schemes, partly private co-operatives and partly municipal plants, were established and gradually expanded. Typically, new schemes were based on temporary boiler stations serving the areas until a scheme was big enough to afford the construction of a permanet large boiler station. In the course of the 1960s, some of the schemes were connected and a new, bigger incinerator handling the large amounts of waste from the fast growing community was built. Also, waste heat from a local carpet factory is now being fed into the system.

The extensions throughout the years have always been based on the main principle that no pipelines in a street have been laid until a sufficient number of consumers have agreed to join the system, thus assuring the economic basis of the investment.

By 1977 the total area had a DH-coverage of approx. 90%, serving some 60,000 inhabitants. The fuel was mainly oil, and it was therefore decided to build a coal- fired combined heat and power station with the primary purpose of producing heat while the electricity so to speak would be the by-product.

The back pressure plant of 90 MW_e + 175 MW_t was commissioned in 1982, and meanwhile 9 of the area's 14 district heating stations had been connected to the CHP- station through a 35 km transmission network. (Fig 5)

In 1986 the total heat production amounted to 653 GWh, being supplied from:

Incinerator and Industry	7%	base load
CHP station	87%	base load
3 n-gas fired boilers	5%	peak load
35 oil fired boilers	1%	reserve

Electricity and heat consumption generally peak at different times of the day. A heat accumulator ensures that electricity and heat supply remain independent of each other. It consists of a 33,000 m^3 insulated water tank which can store surplus heat production for periods with low electrical power requirements, e.g. at night.

The capacity of the accumulator (1030 MWh) makes it possible to utilize the plant intermittently at full load irrespective of the heat requirement. In this way, the power capacity of the plant can be fully utilized during morning and evening peak loads. In addition, operating personnel can be minimized as one- or two-shift operation is sufficient during summer and spring/autumn, respectively. As the accumulator can cover the heat supply for 5 hours at full load, short stoppages and repairs of the combined heat and power station can occur without loss of the advantage of co-generted heat and power. (Fig 6)

The accumulator has made it possible to increase the capacity of the plant by approx. 10 MW_e compared to a station without an accumulator.

The plant was built in close co-operation between Elsam (The power pool of Jutland and Funen), Vestkraft power company and the municipal heat supply companies of Herning and Ikast. It is managed by Vestkraft, which has committed to cover the heat consumption in the Herning-Ikast area as long as the demand is lower than the capacity of the plant. Within this framework the load distribution of Elsam settles the operation of the plant. At the commissioning of the plant specialists were rather sceptical to this back pressure plant, but experience has shown that the plant, equipped with the heat accumulator, is easy to integrate into the electrical system. Therefore this scepticism has been superseded by a positive attitude.

The heat demand figures of Herning Municipal Works of the heating seasons before and after the commissioning of the CHP plant illustrates the potential of energy conservation:

GWh	81/82	82/83
Oil boilers	530	70
CHP plant	0	478
Incinerator	23	29
Industry	6	6
Tot. demand	559	583

It can be derived from these figures that 50,000 tons of oil consumption was replaced by a coal consumption corresponding to 22,000 TOE. This substantial cut of energy use has benefited the environment, the supply security and the national and private economy.

The energy bill of Herning Municipal Works was reduced from 135 mill. DKK to 80 mill. DDK, and it was therefore possible to reduce the consumer price per heat unit by 1/3. Evidently that was very popular among the consumers and everybody in the area takes the district heating supply like any other public utility for granted.

The network consists of 310 km's of mains and 160 km's of service lines connecting 11,500 buildings of which 8,200 are single family houses.

Aarhus

Aarhus is the second largest town of Denmark with a population of 250,000. A DH/CHP system has been operated in the city centre since 1928, but in 1980 it was decided to extend the system to the entire urban area and to some neighbouring towns. The heat source should be two new CHP units 15 km's north of the city at an existing power plant site.

100 km's of transmission network will complete the connection of the Studstrup power plant with the city of Aarhus, a large number of suburbs and the neighbouring municipalities. (Fig 8)

The two CHP stations each have an output of 350 MW_e and 464 MW_t and the 15 km main trunk line to Aarhus has a dimension of 2 x 1000 mm.

The ultimate length of the distribution system will be 700 km plus 350 km service lines, serving 38,000 buildings.

The project has several positive effects:
The construction of the transmission network alone will lead to approx. 10,000 man-years of employment by 1995.

Serving 90% of 250,000 inhabitants' heat demand has a potential of reducing the use of 140,000 tons of gas oil and 70,000 of fuel oil per year. This consumption will be replaced by a use of 120,000 tons of coal, whereby the net energy conservation amounts to 140,000 TOE a year.

This project is the most capital demanding in Aarhus ever, and large loans have been raised, of which the major share has been financed by the European Investment Bank.

It is the pricing policy of the Municipal Works to maintain a fixed consumer price - in real terms - during the first ten years of service.

The price has been fixed at a level corresponding to approx. 60% of the price house owners have to pay to operate an individual oil burner. Even though a potential customer has to pay for a hook-up to the system and for converting his boiler room installations this is an attractive price which - according to the expectations - is adding 1000 new customers per year.

After 10 years it is the intention to reduce the consumer price due to decreasing capital cost.

Kalundborg

Kalundborg is a town with some 20,000 inhabitants situated on the west coast of Zealand about 100 km from Copenhagen. Unlike other Danish towns of the same size, Kalundborg had very little district heating until a few years ago. Therefore the development in this city is of special interest to many foreign cities.

After the first oil crisis the idea of district heating using heat derived from the turbines of the local Asnæs power plant was brought up. In 1975 a premliminary study was made and in 1977 a more detailed one was carried out. The conclusion was positive. Therefore, additional calculations were made in 1978 taking a larger supply area and a state grant of DKK 20 mill. into consideration. This analysis also showed that the project would offer competitive heat prices.

In 1979 the town council therefore decided that district heating based on combined heat and power production was to be introduced, and a detailed project was carried out.

By July 1980, heat supply commenced by means of temporary, mobile boilers which were established in areas where limited networks had been laid during the spring of 1980.

In late 1981, the pipeline to Asnæs Power Station was ready, the first unit was converted to combined heat and power production, and the idea rapidly caught on. (Fig 8)

The activity of defining the supply area was performed in close co-operation with the municipal staff. The scheme will ultimately comprise approx. 5,000 dwellings plus shops, workshops, public buildings, etc. corresponding to some 4,000 house connections. The area was divided into seven sections corresponding to the sequence of construction.

Based on aerial photographs of the town and visual investigations on the spot, the actual size and heat demand of each house was estimated, and the total building area of each subarea was calculated. It was found that the ultimate potential by year 2004 would be one million square meters of heated floor area in Kalundborg.

Originally, it was assumed that 80% of all potential consumers would be connected voluntarily by 2004. So far approx. 3,500 buildings and 700,000 m^2 have been connected. In other words: approx. 70% of the potential has so far been hooked up, whereby the expectations have been fulfilled to date.

The sequence of construction was planned to provide the largest possible heat load at the time when the power station was ready to supply heat. Therefore, area 1, which already had two minor district heating installations was extended rapidly, and part of the main line as well as connections to large consumers were established early. Later on the more time-intensive installations were made, as e.g. on the outskirts of the town. However, after about 5 years most of the ultimate network has now been installed.

In connection with the town council's decision, the utility Isefjordsværket (IFV) undertook the task to design, construct finance, and operate the plants to be established on the grounds of the Asnæs Power Station.

At the systems design, the high degree of supply security has been taken into consideration. The steam for heating the district heating water in the heat exchangers is taken from the crossover between the intermediate and the low-pressure turbines of two 268 MW turbines. The heat capacity is 58 MW per unit. (Fig 9)

Arranging the steam extraction at a lower pressure (temperature) level of the turbines would have improved the efficiency, but since the turbines were not originally designed for combined production, this was not possible.

In addition to the heat exchangers heated by steam from the turbines, there is also a heat exchanger of 40 MW which can be heated by steam from the auxiliary steam system of the plant. Thus providing a spare capacity.

The total installed heat capacity of the Asnæs Power Station is therefore 156 MW. The max. DH load of the city is calculated to be 98 MW_t, and the ultimate annual consumption to be 268 GWh.

It is agreed in a contract that IFV at any time is bound to supply the necessary quantity of heat required by the consumers, though up to the maximum load of 98 MJ/s, which is the nominal output of the power plant. It is further agreed that IFV installs, owns, and operates all district heating installations at the power plant.

Kalundborg Municipality pays IFV for the heat energy delivered, according to the two fixed and one variable charge.

First, IFV will have its capital costs in connection with the district heating installations covered by a fixed charge that is calculated as an annuity over 12 years.

Second, a monthly charge is to cover the loss of electrical capacity caused by the extraction of steam for heating purposes. By calculating this charge it is taken into account that the turbine still has a certain stand-by capacity produced by turning off the district heating supply for max. half an hour.

Third, a variable charge for the consumed energy has to be paid to the IFV. During the first 10 years of operation the district heating supply will receive the full benefit of the combined production in order to provide the lowest possible heat price that will attract a maximum of customers. This means that the consumers of district heating pay to IFV the value of the electric energy that would have been generated without extracting steam from the turbines. In other words, while consumers of electricity derive no economic gain from the combined production, they are, on the other hand, completely indemnified.

It is part of the contract, however, that a new price is to be negotiated after 10 years, and by then it is intended to share the benefit of co-generation evenly between the consumers of electricity and the consumers of district heating.

Having installed a computer-based supervision and control system the scheme is automated to a great extent. The DH Scheme is, therefore, manned only 8 hours per day. Unexpected interruption of operations arising beyond the daily working hours are registered by the supervision system connected to an alarm central at the fire brigade. In the same way telephone inquiries from customers are switched to this central. One member of the staff at the municipal works is always in charge, and will be called in for assistance.

CONCLUSION

The district heating scheme of Kalundborg, being an expression of the Danish DH/CHP tradition, has been seen as a success and model for other communities. The extension of the scheme proceeded according to the plan, the number of consumers connected is extensive and the consumers are satisfied with a competitive, stable, and reliable heat supply. The co-operation with the power station is unproblematic and the air quality of the town has been improved since the pollution from 3500 roof-level chimneys has disappeared.

DH/CHP projects often have a long and troublesome journey from the birth of an idea to reality. I hope to have shown - with these examples - that it is also possible to get such projects going efficiently and fast. We are most willing to share our experience with you.

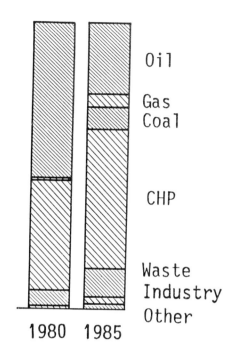

Fig. 1   Energy supply to Danish district heating
in 1980 (21.700 GWh) and in 1985 (25.600 GWh)

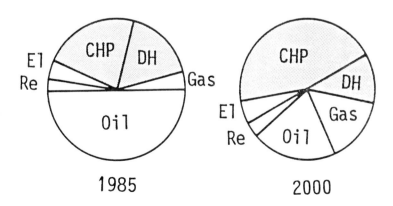

Fig. 2   Space heating by sources of energy, actual
and expected by year 2000.

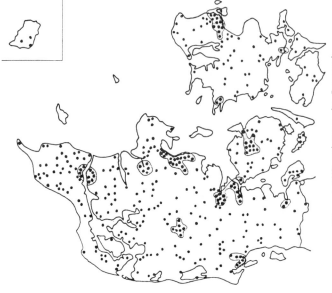

Fig. 4   DH and CHP schemes in Denmark

•   District heating scheme

◉   DH scheme based on CHP

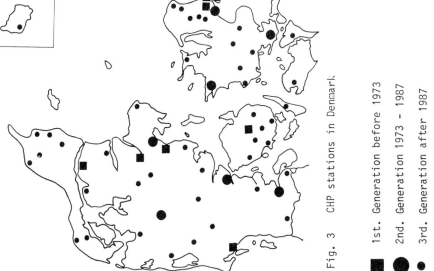

Fig. 3   CHP stations in Denmark

■   1st. Generation before 1973

●   2nd. Generation 1973 - 1987

•   3rd. Generation after 1987

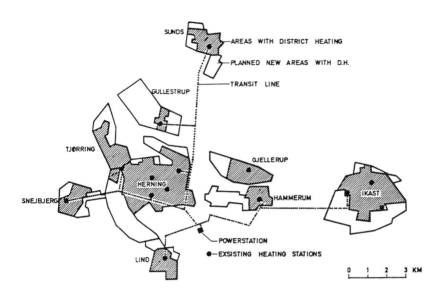

Fig. 5   The DH/CHP scheme in the Herning-Ikast region.

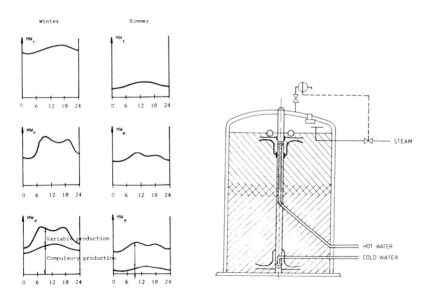

Fig. 6   Typical Danish daily load curves for heating and electricity and the heat accumulator of Herning.

362

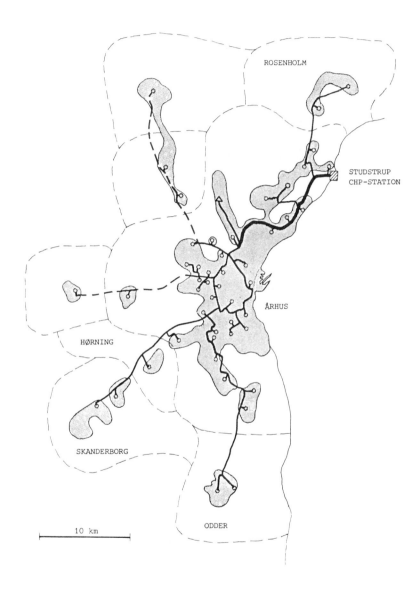

Fig. 7   The DH/CHP schemes of Aarhus, Rosenholm, Hørning,
Skanderborg and Odder municipalities

Kalundborg

Kalundborg Fjord

■	The Asnæs CHP station
–·–·–	The main supply pipe system
■	Peak load boiler stations
●	Booster stations
○	Kalundborg Municipal Works

Fig. 8   The DH/CHP scheme of Kalundborg

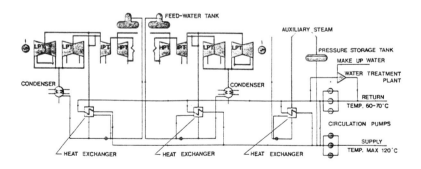

Fig. 9   The CHP installations at the Asnæs CHP station

The Potential for CHP/DH in Belfast, Edinburgh and Tyneside

R. D. Haywood* and A. Smith**

*NEI-IRD Co Ltd, Fossway, Newcastle upon Tyne
**Planning Department, Civic Centre, Newcastle upon Tyne

The technical feasibility of combined heat and power associated with district heating (CHP/DH) has been demonstrated over a number of years in cities across Europe for example Dusseldorf in the Federal Republic of germany and Malmo in Sweden, where piped hot water is as unremarkable as the supply of natural gas in this country. So far the technology has not been applied in the UK. In 1979 the Marshall Committee, having examined the likely application of CHP/DH in the UK, recommended the selection of a "lead city" for the purposes of a demonstration scheme[1]. Following the study by W.S. Atkins and Partners of the application of CHP/DH in nine cities[2] the Secretary of State for Energy announced the Government's intention to provide financial assistance to enable more detailed studies to be carried out on three schemes, which resulted in Belfast, Edinburgh and Leicester being chosen for grant aid support in January 1985. Other detailed studies, though, e.g. Sheffield and Tyneside, went ahead without Government financial assistance.

At the time of choosing the title for this paper, approximately a year ago, it had been expected that the Belfast, Edinburgh and Tyneside studies would be complete. In fact, only the Tyneside scheme has been reported on. The paper therefore will concern itself with some of the general principles that have arisen from the three studies although the detailed Tyneside results will be commented on.

The Development of Expert Systems for the Application of Heat Exchangers and Heat Pumps in Energy Conservation

T. Heppenstall and G. Halliday

NEI-International Research and Development Company Ltd

INTRODUCTION

The determination of the optimum heat exchanger design for a specific, well defined application can be regarded as a two stage process. The first stage is concerned with the selection of the most appropriate heat exchanger type, for example, the choice of a shell and tube exchanger as opposed to a plate, rotating wheel, heat pipe, etc., etc. The second stage involves the preparation of a detailed design which matches the design parameters of the chosen type to the requirements of the application. Clearly, it is not possible to completely separate the two stages of this design process but an efficient procedure would not normally involve a selection based on a detailed design of every conceivable type. Many types can be eliminated as unsuitable, for a variety of reasons, without the need for any real design effort.

The use of computer based procedures is usually restricted to the second stage of the design process. Many manufacturers have computer selection software which will identify the particular model within a range of specified requirements. General purpose design programs also exist and the best known of these, which is commercially available, is probably TASC 2, which provides a comprehensive design package for shell and tube heat exchangers. A common feature of all this kind of software is the presupposition that the particular form of exchanger under analysis is appropriate for the application. Assuming that this has been satisfactorily established, the design or selection process is amenable to conventional computer techniques because the problem is essentially the solution of well defined numerical algorithms.

The problem of type selection, i.e., the first stage of the overall design process, is different to the detailed design because it cannot easily be expressed in algorithmic terms. A satisfactory heat exchanger must accommodate the basic constraints of temperature, pressure and pressure drop imposed by the streams but, usually, many other factors must also be accounted for. These may include, for example, the propensity of the streams to foul, the availability of space, maintenance and cost considerations. In the field of heat recovery, the wider subject of this paper, this problem is exacerbated by both the very large range of possible applications and potentially suitable heat exchanger types. A view which is strongly held in some quarters of the heat recovery equipment supply industry is that the poor level of growth in UK heat recovery applications has arisen from a loss in credibility due to poor equipment choice resulting in poor performance, from what should have been viable installations.

A solution to this problem is the use of a comprehensive computer based selection procedure, but as noted earlier, the nature of the selection process does not easily correspond with the requirements of conventional computer techniques. The relationships between the various items of information cannot be expressed in similar terms to a numerical problem which is not surprising given that much of the information is not numerical. A human expert would base his analysis on deductions from a set of rules. For example, rotating regenerators are often unsuitable in Local Authority swimming pools because the necessary level of maintenance cannot be provided economically. A cost comparison between a plate heat exchanger and shell and tube will normally favour the former, unless the latter can be manufactured from mild steel. These two rules provide grounds to indicate that a plate heat exchanger would be the most satisfactory choice from plate, shell and tube and rotating regenerator for a swimming pool air conditioning system with heat recovery. The work which is described here is an extension of this form of deductive model which is implemented on a computer and incorporates a much larger set of selection rules and heat exchanger types.

The complete model is an example of what is known as an Expert System. This is a particular type of computer technique in which an expert's knowledge is encapsulated within the computer code and which contains procedures to carry out deductions in a similar manner to that which would be used by the expert. Thus, a correctly designed system should produce the same conclusions as an expert in dealing with a specific subject area. The capability of computers to act in this way has been developed over the last thirty years as part of a programme concerned with many aspects of artificial intelligence. The Expert System is one of the outcomes of this work which is receiving widespread interest outside the confines of computer research laboratories.

EXPERT SYSTEMS

Expert Systems are receiving much current publicity but it is still worthwhile to briefly review the concepts which are involved. As indicated in the Introduction, conventional computer systems are designed to deal effectively with algorithmic procedures. In the design of a heat exchanger, for example, the calculation of the necessary heat transfer area to fulfil some specific thermal requirement may be complex but it can be completely defined as a sequence of numerical procedures. In general, problems of this type can be described as:

algorithm plus data

The algorithm is a form of knowledge but it can be reasonably argued that it is not typical of the type of information which would be generally known about some particular subject area, nor does it form the basis of a widely applicable problem solving procedure. In general terms, knowledge is more likely to be of the form

All men are mortal
Socrates is a man

and the problem solving procedure is a deductive process which, in this case, provides the conclusion

Therefore Socrates is mortal

This form of problem solving can be described as

Knowledge Base plus Inferencing Procedure

The type of syllogistic model used in the example is considered to be technically deficient as a general means of representing an unambiguous logical system. However, it does illustrate the requirements of the knowledge base and the inferencing procedure. The most important feature is the symbolic nature of the relationships as opposed to numerical relationships used in conventional computing. The initial aspirations of workers in artificial intelligence was some form of electronic brain which would be capable of holding substantial quantities of human knowledge and, consequently, be capable of imitating human characteristics over a wide range of activities. This aim has long since been abandoned but techniques have been developed which allow the same concept to be applied to very narrow knowlege domains. These techniques are the basis of the so called Expert Systems, the two best known of which are concerned with medical diagnosis and geological prospecting.

Several procedures have been developed to allow the formulation of an Expert System, but the most commonly used are expert system shells and discussion will be restricted to these. A Shell is a piece of computer software which includes the facilities to incorporate a knowledge base, an inferencing procedure and a user interface. Many shells are commercially available and the most significant point of difference between them is the way in which the rules constituting the knowlege base are represented. The most common type is based on a series of statements or "rules" although the actual syntax varies from type. Other differences include the capability to perform numerical operations, the facility to interface with external programs and the form of the user interface. Some shells have the facility to accommodate uncertainty about the accuracy of the rules. Whilst this is an interesting capability from a technical point of view, its practical application is more limited. Although all the "rule-based" shells have some generic similarity there are sufficient differences to require a careful selection for a particular application.

THE NATURE OF THE HEAT EXCHANGER PROBLEM

The "best" choice of heat exchanger for a particular application should be the one which provides the most beneficial economic solution subject to the satisfactory fulfilment of relevant technical requirements. In practice, this criterion is difficult to establish since the most beneficial economic solution is difficult to define. The problem may involve cost minimisation, which may include maintenance as well as the capital and installation cost, or be concerned with the maximisation of benefit, which includes consideration of cost and economic returns. A relatively easier problem is to establish the feasible heat exchangers, i.e. those which satisfy the technical requirements, and which would be subject to the more detailed economic analysis. The approach which has been adopted in the development of the selection software has been to place emphasis on the determination of the feasible types of exchanger and provide advice, where possible, on the economic aspects.

The major potential technical constraints on the choice of heat exchangers are shown in Table 1. Table 2 shows the choice of heat exchangers which are considered to be available to meet these constraints. Certain of the rules which relate type to constraint are, to a first approximation, relatively straightforward. For example, there are maximum pressures and differential pressure which plate heat exchangers can be subjected. Fluid conditions outside the maximum values clearly eliminate plate heat exchangers from further consideration. In reality, it is difficult to establish the exact value for such parameters as maximum presssure and the uncertain area must be accommodated in any advice about suitability. Other rules are more clearly generalisations. Other types of rule present more fundamental difficulties and an example of this type would be that a plate heat exchanger is generally cheaper than a shell and tube exchanger of the same duty unless the latter can be constructed from carbon steel. A second example of this type of rule is that a single concentric tube is the most suitable form for a very low flow rate fluid passing through a large temperature difference.

The overall selection problem can be conceived as the collection of a large number of rules of the type indicated above. An input of the application requirements provides a basis for drawing references from the rules and it is this type of problem which is ideally suited to the Expert System approach.

THE MODEL DEVELOPMENT STRATEGY

Two aspects of model development must be considered. First, the selection of a suitable expert systems shell, and secondly, the determination of the selection rules. Strictly, these two requirements are interrelated since, as indicated earlier, the most suitable type of software is dependent on the nature of the rules. This would suggest that the collection of the rules should precede software selection. However, this approach would deny the use of one of the main features of the Expert System shell, which is the capability of incorporating rules in an arbitrary manner and reaching conclusions based on the current state of the knowledge base. Thus, the shell provides a convenient device for storing the rules as collected and, also, it is capable of testing the effects of the inclusion of a small number of new rules. The benefits of using the shell as part of the rule development are sufficiently high to require an initial software selection. Further, advice from experienced Expert System builders is consistent on the importance of starting on a prototype whilst reserving the option to change to a more appropriate shell at a later stage.

Software Selection

At the outset of the project a number of features concerning the requirements of the shell could be deduced. These included:

1)   A statement type of rule representation (i.e. rather than lists of attributes).

2)   An ability to perform numerical calculations.

3)   A capability to interface with external routines. (It was recognised at the outset that, for example, materials data bases would be required.)

4)   Comprehensive reporting facilities.

A brief review of the available shells was carried out and a small number were tested using a simple but relevant selection model. This exercise was of limited value because it was not possible to gain any substantial experience in the use of each of the shells and the full range of their facilities may not have been exploited. However, on the basis of the work which was possible, a shell which is marketed commercially as "Savoir" was chosen for the model development since this appeared to best fulfil all the requirements.

One of the reasons for the choice of Savoir was its comprehensive facilities to accommodate uncertain information. This allows the use of statements which can be represented as true with a specified degree of probability. For example, there is an 80% probability that certain fibres will foul a plate heat exchanger. The inferencing procedure accounts for the probabilities associated with each statement and provides a conclusion which will be true to a quantified extent. The capability to include this form of reasoning was considered to be important, particularly for the representation of fouling rules. Subsequent work has shown that the inclusion of "probabalistic" information is particularly difficult. Certain statistical factors must be included in the model and, for most applications, there is no clear procedure to derive appropriate values. The probability models which are used, for example, Bayesian probability theor may not provide acceptable representations of the inferencing processes under consideration. Finally, for the user, the knowledge that a conclusion is only probably true may be of doubtful benefit if a real decision to proceed or otherwise is required. Thus, an important reason for the choice of Savoir has proved to be considerably less useful than initially considered particularly since it has proved possible to avoid completely the use of probabalistic reasoning.

Determination of "Selection" Rules

A clearly critical requirement in the development of an Expert System is to ensure that the rules, which provide the basis for advice, are comprehensive, accurate and reflect the best current technology. Several handbooks are available which provide forms of heat exchanger selection procedures but, in general, these tend to give emphasis to equipment descriptions rather than the characteristics of application. Such guides provide a useful background but do not present the detailed knowledge possessed by practitioners engaged in the problems of exchanger selection. Thus, interviews with practising experts have provided the main source of information for the model.

Equipment manufacturers and installation/project engineers are the two main sources of expert opinion. In general, this expertise is related to very specific areas, for example, project engineers tend to specialise in one particular area such as Heating and Ventilation or Furnace Recuperation systems. Manufacturers normally offer, at most, only a few of the many types of exchanger design. Thus, a comprehensive advisory system must integrate all the various sources without creating anomalies between the various strands of Expert Opinion. Under ideal circumstances, the elicitation of knowledge from an expert would involve a series of interviews and feedback from the model to the expert as the rules were incorporated. Heat exchanger experts are widely separated geographically and the ideal procedure could not be implemented. In general, the elicitation of an expert's knowledge had to be carried out in a single meeting and it was found that the maximum attention span was about two hours.

Experts in a wide variety of heat exchanger design and application areas have been visited; the meetings with project engineers were more useful than those with the manufacturers. To some extent, this situation could be anticipated because, although the manufacturers clearly had detailed knowledge of their own products, applications were always seen in these terms and the possibility of using other types of design was not considered in detail.

The concept of a computer based selection system received widespread interest during the visits to experts and, without exception, there was full co-operation in the transfer of knowledge. The object of the meetings was clearly to encourage the expert to talk about the problems of selection. However, it was necessary to direct the interview with a view to obtaining information on specific aspects of the problem. In general, these were concerned with the limitations on the use of particular types of design, for example, pressure limitations on plate heat exchangers, and to establish the priorities in the logic of the selection procedure, for example, at what point would possible fouling considerations be included. The required form of the interview was achieved by carefully researching the area of expertise of the expert prior to the meeting. This provided a basis for determining reasonable expectations of the outcome and also allowed questions to be formulated which would contain the interview topics in the required areas. The interview procedure was refined and improved over the course of the interviews. Improvement was possible partly because the interviewer's knowledge about the selection process and the requirements of the model increased with time and this allowed the formulation of more penetrating and useful questions. Also, it became possible to demonstrate the model, which gave the expert a much clearer idea of the requirements.

The general concept of an expert system suggests that the expert's rules can be regarded as a set of heuristics in that they do not require any formal demonstration of validity.

However, it was found that, to proceed on this basis, presented certain difficulties. In particular, it was difficult to distinguish between rules which were based on technical or economic considerations. For example, certain types of furnace recuperation system are not considered at low furnace exhaust flows and, although it is possible to identify specific flow rates for selection purposes, such a rule does not indicate whether a low flow is technically unsatisfactory or whether experience has shown that it would result in an economically inviable installation. Because of the particular structure of the model, (described in the next section), this deficiency in knowledge about the basis of rules was a difficulty.

An indication of the present extent of information held on each type of heat exchanger is given in Table 2.

The Structure of the Model

The Savoir shell is capable of accommodating an arbitrary rule set arranged in an arbitrary manner. However a random organisation of rules would result in an inefficient and confusing route through the advice procedure. A structured approach presents the user with a logical series of questions and, also, it assists in the development of the model because a modular approach can be adopted.

The basic route through the selection logic is shown in Fig. 1. The initial phase of the selection procedure establishes the range of heat exchangers which would satisfy the basic environmental requirements, i.e. they would be capable of accommodating the flows, temperatures and pressures along with certain aspects of maintenance needs, space availability etc. All the mutually selected exchangers, based on these considerations, may not be suitable because, for example, the nature of the fluids may require materials of construction which are incompatable with the fabrication of certain exchangers. Similarly, certain types of fouling conditions may make some of the initial selection unsatisfactory. These aspects are involved in the second "phase" of the selection procedure.

All of these considerations may not reduce the possible selection set down to a single type. The final choice would then normally depend on cost or in some cases, efficiency. The system is able to offer some advice on these matters but formal quotations from manufacturers would normally be necessary to resolve the final choice.

The Practical Experience of Model Building.

The choice of Savoir as the prototype expert system shell has proved to be satisfactory. It has been found possible to implement all the rules and references necessary for the selection process without undue difficulty. A materials data base has been developed in Pascal which interfaces with Savoir to provide advice on materials/fluid capabilities. This works satisfactorily but, because of certain limitations of Savoir, the interface between the two languages is operationally inefficient. However, this is a programming difficulty and does not affect the operation of the system as far as the user is concerned.
The presentation of information on the computer screen is not satisfactory both in terms of the format of the questions and the final advice. A true user friendly system would require considerable improvements in this area. One final difficulty with Savoir is a limited capacity to backtrack. This prevents a particular line of enquiry being abandoned and the procedure being re-started at some intermediate stage. However, a a prototype development system, the difficulties with Savoir are of a relatively trivial nature.

The interviews with experts have been found to be satisfactory and, without exception all those contacted have been extremely co-operative and there has been a general feeling that the intention of the expert system is worthwhile. It was expected that a certain amount of contradictory advice on selection procedures would be received but there have been no serious conflicts. Definitional problems have caused more difficulty; there appears to be no standard definitions for certain heat exchanger types.

CONCLUSIONS

1. The work which has been carried out on the development of the prototype expert system has shown that the overall concept of the system is technically valid.

2. A comprehensive advisory system has been developed although further developments are possible.

3. The direct use of expert opinion to provide the information for the knowledge base has been found to be the only acceptable procedure.

4. Savoir has been found to be an acceptable shell for development purposes but would be less attractive as a delivery system.

5. General opinion amongst experts indicated that an advisory software package for heat exchanger selection would be a useful facility.

TABLE 1

THE MAJOR TECHNICAL CONSTRAINTS ON THE CHOICE OF A HEAT EXCHANGER

Operating temperature range

Air/gas volume flows

Pressure drops

Motor size and rotation speed (rotary regenerators)

Cross contamination/leakage

Differential pressure tolerated

Condensate collection

Latent/sensible heat recovery

Efficiency

Physical size and weight

Corrosion resistance

Fouling propensity/ease of cleaning

Performance control/turndown

TABLE 2

SUMMARY OF HEAT EXCHANGERS IN CURRENT RULE BASE

GAS/GAS	C	GAS/LIQUID	C	LIQUID/LIQUID	C
Type		Type		Type	
Shell and Tube	3	Economiser	1	Shell and Tube	3
Sensible Heat Wheel	4	Fuel Preheater	1	Spiral Tube	1
Enthalpy Wheel	4	Waste Heat Boilers		Close Tube	2
Parallel Plate	3	Water Tube	1	Double Pipe	3
Glass Tube	1	Fire Tube	1	Trombone	1
Recuperators		Superheater	1	Coil in Vessel	2
Finned Steel Tube	4	Fluidised Bed	1	Gasketed Plate	3
Cast Iron Tube	4	Heat Pipe	2	Spiral Plate	3
Convection	4	Shell and Tube	1	Lamella	2
Radiation	4	Direct Contact	1	Run Around Coil	2
Flue Tube	4	Heat Pump	1	Graphite Block	1
Static Regenerator	2	Air Cooled HE	1		
Run Around Coil	3	Run Around Coil	2		
Heat Pipe	2				
Recuperative Burner	2				
Heat Pump	1				

C Factor indicates extent of knowledge in current rule base.

4 = Essentially Complete
3 = Reasonably Complete
2 = Limited Information
1 = Very Limited Information

FIG. 1

SELECTION LOGIC FOR PROTOTYPE HEAT EXCHANGER SELECTION MODEL

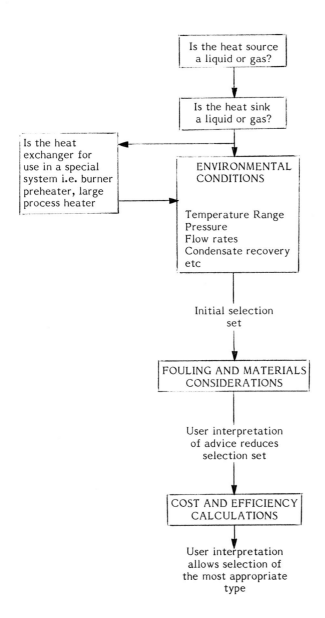

Fuel Poverty as an Outcome Measure: a Comparative Study of Energy Policies in Norway and the United Kingdom

S. Hutton*, T. Braend** and L. Warren*

*Social Policy Research Unit, University of York, York, United Kingdom
**Council for Environmental Studies, University of Oslo, Oslo, Norway

ABSTRACT

Whether low income families have difficulty in being warm at home or not can be considered as an outcome measure of the effectiveness of policies for housing, energy, and income maintenance. This study aims to compare such policies in Norway and the United Kingdom. As background, evidence of fuel poverty in a wider international context is presented. To conclude, policies from Norway which might help to relieve fuel poverty in Britain are set out.

A fuller analysis of the interview of households in Norway suggests that fuel poverty is less widespread than in Britain and is particularly associated with some extreme cases of very old housing which has not yet been renovated. We have compared the best of Britain, Newcastle, with the worst of Norway - an area of older housing in inner city Oslo. More elderly households in Britain are in fuel poverty in spite of the milder climate. The circumstances which cause this are more varied in Britain being in some cases, poor heating systems, in others the expense associated with central heating and in others lack of income. Housing seems to be less of a problem here than in Norway.

Energy Planning in Denmark—with Emphasis on Heat Supply Planning*

Lars Josephsen

Danish Energy Agency, Ministry of Energy, Denmark

Summary

The energy price development since the first socalled energy crisis in 1973 has been part of background for at general trend within the energy sector in most industrialized countries reflecting inproved efficiency in relation to conversion, transmission, distribution and consumption of energy. Industries, energy supply companies, research institutions, and individual consumers are increasingly involved in issues as energy efficiency improvements and energy saving measures.

Since the middle of the 1970's great changes in the Danish energy sector have taken place. There has been a considerable move in the fuel mix away from (imported) oil and the overall efficiency of the entire energy system has increased as the result of considerable efforts in a number of different levels. A significant contribution stems from to the expansion of district heating systems based upon combined heat and power production (CHP).

This evolution is achieved by means of a comprehensive energy planning. Based upon specified energy policy objectives this planning has been performed in a complex process including energy authorities at national and local levels, private and public energy supply companies, organizations and others.

This paper gives an outline of the objectives of the Danish energy policy and describes some results obtained through the planning and the implementation of approved plans, especially in the space heating sector planning during the decade 1976-1985.

Refuse Fired CHP Plant

J. K. Maxwell-Snape and M. J. Rowley

Kennedy & Donkin, Generation & Industrial Ltd

ABSTRACT

The traditional method for disposing of domestic refuse has been by landfill, which is relatively inexpensive when sites are plentiful and near-by. However, the reduction in the number of sites available coupled with the increase in volume of domestic refuse has made it of paramount importance that alternative means of refuse disposal be found. The very large increase in fuel costs over the last decade has changed the economics of energy recovery by refuse burning to a point where this means of disposal is competitive with most other methods.

Whilst there are various ways in which incineration of refuse can be accomplished it is proposed here to concentrate only on "mass burning of refuse" and the recovery of heat from the waste gases of combustion.

Experience to date has indicated that the performance of modern mass incineration waste-to-energy plants has been very reliable and that the recovery of heat from these can make a useful contribution to the nation's energy needs.

Designs and techniques have been constantly improving over the last few years and coupled with the lifting of legal restrictions related to the power industry, the timing would seem to be right for the development of CHP associated with refuse incineration.

A review is presented of the technical and economic aspects associated with the mass incineration of raw refuse and the recovery of the liberated energy for utilisation in combined heat and power plants. The potential and prospects for further development of the practice in the United Kingdom are assessed and reference made to initiatives which are considered necessary in order for the potential to be realised.

Boiler Innovations for Energy Saving

R. P. Ravenscroft

NEI-ICL Ltd., Newbie Works, Annan, Scotland

ABSTRACT

The fluctuations in costs of fuels have accelerated the need to widen the options available to energy users to permit the purchase of equipment which can most efficiently utilise the fuels of their choice.

The options discussed include Condensing Boilers, Boilers for Combined Heat and Power (CHP) also Cogeneration applications, Fluidised Bed and Composite Boilers which offer improved efficiencies for the conversion of energy from a variety of fuels.

A Boiler for the future is also considered which attempts to include the desirable features associated with the various options.

Constant Observation—the Keyword to Efficient Use of Energy

C. G. Rienks

*Verhaar-Uniad Technical Consultants, P.C. Hooftlaan 3, 3818 HG
Amersfoort, The Netherlands*

ABSTRACT

"A special computer that calls the central switchboards of all
buildings forming part of the system, to check and monitor the
operation of the installations."

This sums up the new management tool in the year 1987, whereby
the technical manager or a consultant can assist the user, the
building management.
In this computer age this is done, as a matter of course, through
data communication to these expert helpers.
In our organization this innovative initiative is performed by a
separate subsidiary company: 'Technical Installation Management
Consultancy'.

Electricity for Industrial Process Heating

R. Waggott

*Electricity Council Research Centre, Capenhurst,
Chester CH1 6ES, England*

ABSTRACT

The Electricity Council Research Centre, based in the North
West at Capenhurst, near Chester, aims to develop electro-
technology by researching into new and improved methods
of electricity utilisation and distribution. One of its
concerns, therefore, is the conversion of raw materials
into manufactured goods involving the application of
electrical energy at some stage of the production process.
Electrical energy in the form of heat is called electroheat
and with a wide application in manufacturing industries
it presently accounts for a substantial amount of the total
industrial electricity consumption in the U.K.[1].

Advances in manufacturing either as energy-saving
improvements to established practices or as changes in
technology demand complementary progress in the development
of new and improved techniques in electroheat. The purpose
of this short paper is to review some developments undertaken
at Capenhurst and thus point to the future role of
electroheat in process heating and its effect on improving
the competitiveness of British Industry.

R–IEE—Z

A Social Network Approach to Promote Energy Conservation in Neighbourhoods

W. H. Weenig, C. J. H. Midden and T. Schmidt

Research Center on Energy and Environment
University of Leiden, Hooigracht 15, 2312 KM Leiden, The Netherlands

Abstract

In this paper the results are described of a neighbourhood energy-saving programme which was carried out in two neighbourhoods in the Netherlands. The central research question was if and how energy conservation and home insulation can be optimized by starting from existing social networks in a neighbourhood.

For this purpose data were collected by means of a pre-test, several small surveys during the programme, and a post-test after the programme had been terminated. In addition some organizational aspects were examined.

The results clearly demonstrate the stimulating, but also sometimes inhibiting, effects of social contacts and integration within the network on the adoption of energy conserving measures.

1 AIM AND CENTRAL ISSUES OF THE PROJECT

Over the past decades many information campaigns were started to promote energy conservation by consumers. Although this has led to considerable savings, it has been widely acknowledged that with the methods employed to date a ceiling seemed to have been reached. Particularly in the rented accommodation sector and in the lower income category a stagnation in house insulation appeared. Moreover, from a Dutch national survey into attitudes, beliefs and behaviour with regard to energy conservation in the home it became evident that tenants living in rented accommodation in the lower income categories know relatively little about energy conservation and also seek little information in this field.

Apart from a few exceptions information on energy conservation and home insulation was mainly diffused through the mass media (television commercials, newspaper advertisements, leaflets, etc.). A general limitation to this approach is however that chiefly people who are already interested in the subject of the message are reached. The question therefore remained whether information methods exist which fit the situation of people in the lower income category better, methods that should not be too labour intensive and in addition could be used in future by bodies that are already active in the field of energy conservation.

A solution for this problem may possibly be found by organizing the information diffusion on energy conservation and home insulation at a more local level, e.g. at neighbourhood, district or village level. Firstly this is useful because particularly the insulation of rented accommodation is usually carried out per block. Secondly, information directed at low income groups should be as concrete and as clear as possible. The information methods that best take account of these aspects (e.g. model housing, exhibitions, practical courses) can virtually only be carried out at a local level. Finally, a local approach is interesting because a programme targeted at a neighbourhood or district can better be adapted to local characteristics and situations. On the one hand this may result in the tenants feeling more involved in the programme. On the other hand such a local approach offers possibilities to connect and to adapt directly to the existing local social network of communication and influence. Information can be distributed in a more effective way by making use of the existing interaction channels and the social relations within a housing community. How effective will depend on the type of system. It is plausible that in a network in which there are many mutual contacts, information will be diffussed faster than in a less cohesive system. Also in a tightly-knit system people will be able to encourage and stimulate each other more than in a disjointed system. On the other hand, social processes may oppose a programme, because persistent misapprehensions circulate or an organized opposition emerges. In this way it could be argued that dependent on the features of a social system several information diffusion strategies could be taken into consideration, also in the light of the financial costs.

It was therefore decided to start a project whereby in some neighbourhoods with low social economic status an information programme would be tested that would connect to the social network and other local circumstances as far as possible.
The main objective of the project was "to develop a practical guide on how information concerning energy conservation and home insulation may best be organized at a local level". This guide (which is now available from the Dutch Ministry of Housing) was aimed at the future use by municipalities and energy boards, and also by private initiators, groups of residents and landlords.
The central question of the project was "if and how diffusion of information on energy conservation and home insulation can be optimized by starting from existing social networks in a neighbourhood".

2   AIM OF THE RESEARCH PROJECT

Subsequently an information programme that could be used at neighbourhood level was developed. The programme was tuned to the social characteristics of the neighbourhood, the technical housing circumstances and the local character that the information diffusion needed to have.
The basis for the information plan was the formation of a so-called project group in the neighbourhood. The task of this project group was to look after the implementation of the information programme that had been drawn up. Parts of this programme included sending letters to residents, organizing a project on energy conservation at local primary schools, organizing an information evening on professional measures and an energy manifestation aimed at Do-It-Yourself and behaviour measures, furnishing a demonstration house, and paying home visits to residents who do not react to the proposed professional measures.
The composition of the project group was allowed to vary per neighbourhood, but it was considered desirable at any rate to certainly admit representatives of the Housing Corporation, the residents, and one or more municipal bodies (e.g. the Electricity Board, municipal information and /or the building inspectorate). This composition was advocated in order to ensure that the various interests could be adequately looked after and in order to guarantee sufficient specific expertise in both technical and also social and communicative fields.

In principle the programme was aimed at various forms of energy conservation:

1. Professional insulation: forms of insulation that are usually installed by professionals, i.e. double glazing, cavity wall insulation, and outside roof insulation;
2. Do-It-Yourself insulation: this includes caulking, insulation of pipes, double window frames, floor insulation and inside roof insulation;

3. Energy-saving behaviour in the home: this includes repeated actions such as heating, ventilating, cooking, the use of hot water, and single actions such as the purchase of energy-saving devices and lamps.

This integral approach was chosen because in this way an energy conservation package as complete as possible could be compiled and in addition disappointing results of insulation by behaviour adjustment in the "wrong" direction could be better avoided. The study was to show to what extent this combination of goals could be realized.

The information programme was carried out as an experiment in two neighbourhoods, one with a relatively cohesive social network and one with a relatively non-cohesive social network. The effects of the execution of an identical information programme were subsequently compared firstly with each other and secondly with control neighbourhoods, which in social and demographic respects were comparable to the programme neighbourhoods, but where no special energy conservation programme had taken place.

In order to be able to choose suitable experimental and control neighbourhoods for the actual experiment, first of all eight representative lower class neighbourhoods in the Netherlands were examined on a number of social aspects, the so-called "Social Diagnosis". In the Social Diagnosis aspects were examined such as the nature and the frequency of interaction between residents (both organized and informal, and also in the form of neighbour-aid), the identification with the neighbourhood and the observed solidarity and/or social control in the neighbourhood. At the same time it was examined to what extent the various neighbourhoods were directed at "the outside world" or had a more secluded character. In addition a number of demographic characteristics of the residents such as income, education and age were determined.

It became clear from the Social Diagnosis that the cohesion of a neighbourhood can be considered as a concept built up of two components, viz. "social involvement", which is expressed by the degree in which one believes residents in the neighbourhood live apart from or in solidarity with each other, and "interaction". The degree in which people feel at home in and are attached to their neighbourhood (identification) proved to be a good direct indicator of cohesion.[*]

Drachten - Midden Noord appeared to be the most cohesive, and two identical thirteen storey flats in Bergen op Zoom the least cohesive neighbourhood of the eight neighbourhoods that were examined. It was decided to carry out the information programme on energy conservation in those two neighbourhoods. In addition a neighbourhood in Breda was chosen as a cohesive control neighbourhood for Drachten and some blocks of flats in Haarlem as non-cohesive control neighbourhood for the two blocks of flats in Bergen op Zoom.

Although the research was experimental in intention, it was impossible and even undesirable to control all special circumstances and accidental factors. As a result each programme knew its own development, despite the essentially similar goals.

Due to this reason the proposed energy conserving measures were not identical in the two experimental neighbourhoods: in Drachten the options were cavity wall insulation, double glazing in the living room and double glazing in the kitchen, whereby the residents could order any combination of these three facilities. In Bergen op Zoom the proposal consisted of a 'basic package' which included wall insulation, sealing of window frames, installation of individual meters and thermo-stat radiator valves, and the renovation of the boilerhouse. Furthermore, double glazing for the living room, bedroom and/or kitchen was proposed as a separate option.

---

[*]This conclusion is based on a Multi-Dimensional Scaling analysis of a number of social network characteristics measured in the Social Diagnosis.

386

The two experimental neighbourhoods should therefore be seen as two non-arbitrary case studies instead of a purely experimental study. This is also why the two programme-neighbourhoods were not quantitatively compared with each other (i.e. with the help of statistical tests), but only qualitatively.

3   THE EXECUTION OF THE RESEARCH

Figure 1 shows the model that structured the research. The model describes the adoption process of the energy conservation measures in a number of phases. The first phase consists of *awareness* of and attention for the energy conservation programme in one's own neighbourhood, both as a whole as well as the separate parts. The second phase includes participation in the information programme, which in practice may consist of *participation* in a whole range of information activities. In this phase the resident is confronted with the actual information messages. The third phase relates to the *acceptance*. By this is meant the acceptance of the project group as a source of information as well as the acceptance of the information message itself. The fourth distinguishable phase is a *cognitive phase*, and is concerned with changes in beliefs, attitudes and knowledge with regard to the different energy conservation measures. The fifth and last phase - *the behavioural phase* - is related to the decision to give permission for the execution of professional measures or the installation of Do-It-Yourself insulation by the residents themselves. The behaviour may also refer to recurrent behaviour, such as heating or ventilation.

Beside the adoption process a number of influencing factors on the progress of the adoption process are distinguished in the research model: personal factors (e.g. existing beliefs, attitudes and knowledge concerning energy conservation measures, and demographic characteristics such as age, education, sex, etc.), social factors (e.g. type and frequency of interaction, identification with and attachment to the neighbourhood, the neighbourhood's openness to the outside world), physical and technical factors (housing characteristics, heating system and household appliances), economic factors (such as energy rates, billing modes and the availability of energy-efficient products), and finally the authorities' legislation and regulation.

Figure 1: The Research Model

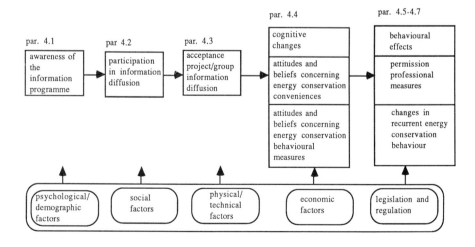

This study focuses on the analysis of the influence of social and personal factors on the adoption process.

For this purpose three surveys were taken among residents (bottom-up analysis), all by means of structured questionnaires which were taken from a representative sample of neighbourhood residents: a pre-test, several small process surveys, and a post-test, respectively held before, during and after the actual information programme.

In addition the organizational structure of the projects was analysed (top-down analysis), mainly based on written material, such as the minutes of the meetings of the project groups, the correspondence between the project group and other bodies involved, and articles on the project published in the local press, but also on data collected through systematic observation. Thus the internal functioning as well as the external functioning of the project group was evaluated. Economic and technical factors were not directly analysed, but were taken into account in the feasibility of the proposed measures and through the personal factors in the form of attitudes concerning the economic feasibility and attractiveness of the different measures proposed.

For each stage in the adoption process research questions were formulated, globally amounting to the study of the degree of impact of the different distinctive stages and what factors (demographic and social factors, as well as previously completed stages of the adoption process) have been influential. The research model indicates per stage in which paragraph one can find the results of the research concerning the stage in question.

4 RESULTS

4.1 Awareness of the projects and the information activities

On the whole the inhabitants of both neighbourhoods were well aware of the existence of the programme in general. However, various separate activities of the information programme were better known in Bergen op Zoom, mainly due to letters circulated by the project group. Whereas in Drachten the inhabitants had been informed rather more by word of mouth by neighbours; which means that, in accordance with expectation, social contacts in the neighbourhood in Drachten enhanced attention to a greater extent than in Bergen op Zoom.

We found that the degree of awareness about the project and the information activities was mainly dependent on the education level and the age of the residents, and besides on the number of contacts they had with people in the neighbourhood, in which the moderately intensive contacts played an important information-diffusing role. A more external orientation also had a positive influence.

4.2 Participation in the programme activities

The general conclusion in relation to participation in information activities is that participation was clearly better in Bergen op Zoom than in Drachten. Especially the demonstration house and the conservation proposal scored better with the inhabitants of the two blocks of flats in Bergen op Zoom. The prevailing impression is that the attention for professional insulation in Drachten was clearly at the expense of the attention for information on do-it-yourself and behaviour measures.

In both neighbourhoods the same demographic characteristics influenced participation in the information programme (i.e. education and income), but that the influence of social factors was almost opposite according to the social climate of the two neighbourhoods: in Bergen op Zoom we saw a high rate of participation by people relatively less integrated in the neighbourhood (without being socially isolated though), and who had a greater interest for more "external" matters. In Drachten, on the contrary, a more neighbourhood-orientated interest (a consequence of schoolgoing children and close contacts within the neighbourhood) led to greater participation in the programme activities.

4.3 Acceptance of the project groups and the information

In general the residents of both neighbourhoods seemed to be satisfied with the information presented to them. Nevertheless, in both neighbourhoods part of the people (one third in Drachten en one quarter in Bergen op Zoom) who attended the information evening had some complaints, which consisted mainly of obtaining no clear-cut answers to their questions, and things being presented too optimistically.

The respondents' judgements of the project groups were mostly positive in the two neighbourhoods. They had (rather) much confidence in the project groups and thought they were (fairly) expert. The latter is remarkable, considering the fact that the project group in Bergen op Zoom mainly consisted of non-professionals, in contrast with Drachten where the group included various professionals. The "lay-men dominance" in the Bergen op Zoom project group seems to have had minimal consequences for their perceived expertise.

A clear-cut conclusion concerning the factors which influenced the acceptance of the two project groups, is difficult to draw: it seems that the acceptance was influenced by quite different factors in the two neighbourhoods. With some caution however we can state that in both neighbourhoods the project group members especially gained the confidence of the younger inhabitants and of people who had shown previous interest in energy conservation. Social factors, like integration and the number of contacts in the neighbourhood also enhanced the acceptance of the two project groups.

4.4 Effects of the information: changes in attitudes and beliefs

In both neighbourhoods the information programme resulted in quite substantive mutations in attitudes with regard to energy-saving facilities. In Drachten though these changes neutralized each other on balance, in Bergen op Zoom the (presumed) influence of the information on willingness to accept wall insulation "prevailed over countercurrents". Finally, the interest in Bergen op Zoom for double glazing seemed to have been at the expense of the opinion on double window frames.

With respect to beliefs on energy-saving facilities we found that in Drachten, as far as there were any (average) changes, these were changes in opinion on the extent to which some facilities would lead to the four different consequences (i.e. energy-saving, financial consequences, consequences for living comfort and expected inconvenience of installation), while in Bergen op Zoom (at least on the average) changes were limited to alterations in the relative importance of the four consequences.

Finally, neither in Drachten nor in Bergen op Zoom could substantive changes be demonstrated in beliefs on the two consequences on which the various energy-saving behavioural measures had to be judged (i.e. financial consequences and amount of effort involved). In this respect, certainly in Bergen op Zoom no changes were to be expected because no information at all had been given on energy-saving behaviour. Also in Drachten it was no surprise that few changes were registered in beliefs on energy-saving behaviour, because information on this subject failed to reach the inhabitants (written material on this subject was very poorly read, whereas the manifestation was only attended by a very limited number of people).

Also with respect to changes in attitudes and beliefs on energy-saving facilities, the influence of demographic and social factors and earlier completed stages of the information process were examined. However, the correlations showed no consistent pattern, which made interpretation difficult. This lack of structure is probably due to the poor predictive value which the pre-programme-attitudes proved to have, on awareness and participation in the programme activities and on acceptance of the project groups. Apparently people formed their opinions during the information process, and seemed to have been scarcely influenced by their original attitudes.

4.5 Effects of the information: the adoption of professional energy-saving facilities

Percentages of adoption
There is a striking similarity in the number of adoptions of double glazing in both neighbourhoods: in Drachten 43% of the inhabitants ordered double glazing for the living room and 29% for the kitchen, in Bergen op Zoom this was 41% and 29% resp. The similarity is striking in the sense that double glazing was presented as an "extra" energy-saving option in Bergen op Zoom on top of the basic package of insulation and installation measures, whereas in Drachten it was one of two equally important possible insulation measures. Probably the more convincing way in which the financial benefits of double glazing were presented to the inhabitants can explain the relatively favourable effect in Bergen op Zoom: people there were told that the installation of double glazing would (just) result in monthly financial savings, whereas in Drachten people were given an exact amount which would have to be paid in extra rent each month, in combination with a rather complicated chart of estimated average savings in cubic metres, which made it rather difficult to figure out the final balance.

Also with respect to the other professional energy-saving measures the information in Bergen op Zoom seems to have been more successful than in Drachten: 73% of the neighbourhood residents in Bergen op Zoom said to be willing to adopt the basic package, whereas only 21% of all homes involved in Drachten placed orders for cavity wall filling. A more favourable startingpoint and a somewhat clearer information programme seem to have stimulated the higher adoption percentage of professional energy-saving measures in Bergen op Zoom.

4.6 Factors which influenced the adoption of professional insulation measures

Demographic factors
No relation between orders for double glazing and cavity wall filling and demographic characteristics could be established in Drachten. Those who did order, whether double glazing or cavity wall filling, were not different in any demographic aspect from those who did not place orders.
In Bergen op Zoom those who gave their consent had a slightly higher education level than those who did not agree. There was also a tendency that older people were more reluctant to give permission than younger people. Broken down per item (basic package, double glazing in the living room and double glazing in the bedroom and/or kitchen) both relations appeared to be valid for the adoption of the basic package and of double glazing in the living room (but not in the kitchen or bedroom). A possible explanation for the noted connection between age and adoption lies probably in the nature of one of the measures of the proposed basic package: the installation of individual meters could on balance be financially disadvantageous for the older people.

Social factors
In Drachten a negative relation existed between the number of contacts one had with inhabitants of the neighbourhood and the adoption of cavity wall filling. This relation was not found in Bergen op Zoom. These findings demonstrate the negative influence of social contacts in Drachten, as far as orders for cavity wall filling were concerned. The geographical diffusion pattern in Drachten however shows - statistically significant - that, both for double glazing and for cavity wall filling in some rows of houses a relatively high number of orders were placed, while in other rows relatively few orders were placed. The former were in general concentrated in small groups, whereby it was remarkable to see that they were grouped around local members of the project group, or constituted by people living in a row of houses sharing a common back path. Thus we may conclude that orders (for double glazing and for cavity wall filling) have at least partially been influenced by geographically determined social factors. While on the whole the adoption of cavity wall filling in the neighbourhood was negatively affected by social contacts, it seems that on a very local level there has been a positive influence (or at least: no negative influence).

In Drachten both the positive and the negative advices have influenced orders for cavity wall filling. With respect to the adoption of double glazing only positive advices seem to have played a part. The latter was also found in Bergen op Zoom, but here advice concerning the basic package did not seem to have influenced the extent of consent. Apparently people in Bergen op Zoom made up their mind about the basic package independent from only advice. On the other hand, in Drachten, where controversial information on cavity wall filling circulated in the neighbourhood, social comparisons may have played a more important role in decision making, in the sense that people listened more to what other residents said about the idea. This was indicated by the fact that in Drachten advice - both on double glazing and on cavity wall filling - was to an important degree obtained from local relatives and other residents.

Furthermore, the expressed importance attached to the opinion of neighbours and other residents was positively related to the adoption of professional insulation measures: in Drachten of double glazing, in Bergen op Zoom of the basic package. Also in Drachten more orders for double glazing were placed by people who had relatives in the neighbourhood.

Finally, in Drachten the degree of integration into the neighbourhood also seems to have influenced orders for professional insulation: the "less integrated" residents placed on average more orders for double glazing, probably because they had been given less advice by others in the neighbourhood, as being less integrated, and/or, if they were given any advice, thought the information less important. In other words: in Drachten, a somewhat independent attitude towards the neighbourhood seems to have increased orders for cavity wall filling, but not for double glazing.

Attitudes
In general, both in Drachten and in Bergen op Zoom, pre-programme-attitudes had very little predictive value for ordering any kind of professional insulation measures. Apparently the pre-attitudes were rather instable, which suggests that final decision on taking conservation actions were mainly induced by the programme.

Opinions on the various professional measures after the information programme showed a clear distinction between those who had placed orders and those who had not, both in Bergen op Zoom and in Drachten. In both neighbourhoods the differences in attitude between the two groups were greatest with regard to double glazing. The expected inconvenience caused by the installation appeared to be a major factor in the decision to adopt any insulation measures in both neighbourhoods.

Participation in information activities
Both in Drachten and in Bergen op Zoom visitors of the information evening adopted more energy-saving measures; this holds for double as well as for cavity wall filling.and the basic package, respectively.In Bergen op Zoom all information activities seemed to have had at least some positive influence whereby the demonstration house and the savings proposition only affected the basic package. In Drachten no influence could be demonstrated from any of the other information activities.

Acceptance
In Drachten no relation was found between acceptance of the project group and the adoption of cavity wall filling and/or double glazing. In Bergen op Zoom the perceived expertise of the project group and the confidence one had in it positively affected the adoption of double glazing but not of the basic package. The credibility of the project group seems to have been of importance for extra orders, beyond the "minimum package" as outlined in the basic package. In Drachten there were no such "extra options".

Conclusion
In conclusion we can say that, next to the persuasiveness of the information and the members of the project groups, the different social characters of both neighbourhoods have influenced the adoption process in various ways.

4.7 Effects of the information: installed do-it-yourself measures and changes in energy-saving behaviour

Both in Drachten and in Bergen op Zoom very few residents installed do-it-yourself insulation during the project. There were also few changes in recurring energy-saving behaviour (i.e. cooking, washing, heating, lighting, ventilation). Changes were statistically insignificant in both neighbourhoods, compared with the pre-test and with the control-neighbourhoods.
Intentions to install do-it-yourself insulation and to save energy by means of behavioural measures - as measured in the process surveys - dropped as well during the information programme.
In paragraph 4.4 we already indicated that few changes in energy-saving behaviour had to be expected in the two neighbourhoods: in Bergen op Zoom because no information at all was given about this subject, and in Drachten because especially those activities failed to reach the residents' attention. This goes for information on do-it-yourself measures as well.

4.8 The influence of social and demographic characteristics during the information process

Social influence as demonstrated by the effect of integration into the neighbourhood, received advice, and the importance attached to the opinion of other neighbourhood residents, was primarily important in the final decision to give permission for insulation, although some social influence could also be registered in earlier stages of the information process. In the cohesive Drachten neighbourhood the influence of the social network on adoption was so strong that the effects of the other factors, as observed in earlier stages, like education and age, but also of previous interaction on energy conservation, were no longer discernible in this (last) stage.

Thus in Drachten the influence of demographic characteristics was only perceptible during the information process, whereas social influence seems to have been the deciding factor in the eventual decision-making. In Bergen op Zoom there was not such a dominant influence from the social network on decisions, and the influence of individual demographic characteristics persisted until the last stage (i.e. permission).

4.9 Organization Analysis

In both neighbourhoods, but mostly in Drachten, implementation of all three objectives simultaneously (the stimulation of professional, do-it-yourself and behavioural measures) proved difficult to achieve: the greatest emphasis was unintentionally put on professional insulation, while much less attention was paid to do-it-yourself and behavioural measures. However, this does not imply that in one programme the three objectives would be fundamentally incompatible. But in our opinion the professional and the do-it-yourself measures need to be presented as different options in one single package.

It proved necessary to have enough representatives of the local residents in the project-group in order to tune the project as much as possible to the wishes and needs living in the neighbourhood, and to take into account possible existing resistance. Adequate representation of all social sections of the neighbourhood also proved to be important. In Bergen op Zoom, but especially in Drachten, some parts of the neighbourhood did not have a representative in the project group, and consequently in these parts less professional insulation was ordered.
We will stress the importance of the availability of a certain amount of technical and information diffusion skills and knowledge. In Drachten the lack of information expertise was strongly felt, while in Bergen op Zoom technical knowledge was insufficient. In the latter neighbourhood an

392

attempt was made to solve the problem by calling in outside expertise. It proved necessary to be cautious in the choice of external advisers.

Although it is wise to include professional 'experts' in the project group it must be avoided that a "rift" develops between the "professionals" on the one hand and the "lay" residents on the other hand. Especially in Drachten a (presumed) pattern of division of tasks proved discouraging for the representatives of the inhabitants (decisions taken by professionals to be carried out by inhabitants).

5 CONCLUSIONS AND RECOMMENDATIONS

The principal conclusions and recommendations derived from the research are the following:

1. The neighbourhood as direct social environment proved to be an important factor in the residents' decision to take energy-saving measures. In the cohesive neighbourhood effects of social influence appeared to be stronger than in the non-cohesive neighbourhood.
2. Social cohesion should be seen as a neighbourhood characteristic with several aspects, of which the most important ones can be described as social involvement and interaction (in nature and frequency). In a number of ways the social network can be used in the programme, in particular in drawing the residents' attention to the programme, and in encouraging them to take part in it, but also, e.g., to recruit project-group members. It is of the utmost importance that the programme takes into account existing social relations, like the tenant-landlord relation, the presence of neighbourhood segments, the influence on opinion of prominent residents, existing formal or informal organizations of residents, e.g. in the form of neighbourhood committees or tenants' associations.
3. Categories of inhabitants which are usually difficult to reach like the elderly and those with a low education level, can be encouraged to action through a neighbourhood-orientated approach. In the cohesive neighbourhood social integration and social contacts even neutralized the afore-mentioned demographic differences.
4. Neighbourhood-orientated energy conservation projects like the ones described here have strong effects, so much so that pre-attitudes prevailing before the project have little or no predictive value for the final decisions. The effects will not necessarily go in the desired direction though. Neighbourhood-orientated information affects the social agenda and creates a process of opinion-making in which counter-currents can easily evolve, and in which counter-arguments can become popular, irrespective of their accuracy. This kind of counter-processes can develop particularly easily in a cohesive neighbourhood. This leads us to the recommendation that information messages should not have a once-only character. An alert attitude is indispensable: a project group should constantly check whether misunderstandings are not disturbing the process of opinion forming, threatening to develop into insurmountable obstacles.
5. The number of contacts in a neighbourhood increases awareness of the project. If the interaction level is low, special attention should be given to personal information.
6. Participation of residents in the project group is a great help in bringing the project as much as possible in tune with the needs and wishes of the residents, and also to optimize their involvement, and to remove or prevent resistance.
7. Knowing project group members stimulates participation in the information programme. It also seems to stimulate the adoption of professional insulation measures if project group members live in the immediate vicinity. Representatives of the residents in the project group increase awareness and legitimacy of the project group, without affecting its perceived expertise.
   The representation should comprise as much as possible all social neighbourhood segments. These could, but do not necessarily have to, coincide with a geographical division. In this respect a social diagnosis of the neighbourhood at the outset of the project is indispensable
8. The selection and demarcation of the target group and of the objectives of the project in its size and scope should be brought in accordance with the composition and manpower of the

project group and the available resources. Too large or too heterogeneous a target group could lead to problems during the project. Experiences in both projects suggest that if do-it-yourself measures are included in the objectives next to professional measures, these need to be given a lot of attention in the programme. These aspects become too easily of a secondary importance. If do-it-yourself and professional measures are combined in one programme, the two sets of measures must be integrated in one option package, whereby the different measures have a complementary character.

9. Information about behavioural measures concerning heating and ventilation is also threatened by underexposure. Information on this subject is best provided after the insulation has been installed. It is only then that the resident is confronted with the problems inherent to an insulated house. Moreover, all the trouble involved in the adoption of a particular insulation package and its application are over at that stage. Behavioural information should be directed at the prevention of disappointment over the effects of certain insulation measures. Other research has shown it would be sensible to pay sufficient attention to feedback in the information activities, e.g. in the form of a "fuel-cost club". Individual meters appear to be indispensable to make behavioural information sensible.

10. The various tasks of the project group require both technical and information knowledge and skills. While advice and support can be sought on certain points, some degree of knowledge of these aspects is absolutely indispensable.

Author Index

Subject Index